Ethnomathematics and Culturally Responsive Pedagogy

Lutfi Incikabi
https://orcid.org/0000-0002-7912-780X
Kastamonu University, Turkey

Hartono Tjoe
https://orcid.org/0000-0002-2728-8484
The Pennsylvania State University, USA

Fadime Ulusoy
https://orcid.org/0000-0003-3393-8778
Kastamonu University, Turkey

IGI Global
Scientific Publishing
Publishing Tomorrow's Research Today

Vice President of Editorial Melissa Wagner
Director of Acquisitions Mikaela Felty
Director of Book Development Jocelynn Hessler
Production Manager Mike Brehm
Cover Design Jose Rosado

Published in the United States of America by
IGI Global Scientific Publishing
701 East Chocolate Avenue
Hershey, PA, 17033, USA
Tel: 717-533-8845 | Fax: 717-533-7115
Website: https://www.igi-global.com E-mail: cust@igi-global.com

Library of Congress Cataloging-in-Publication Data

LCCN: 2026036877 (CIP Data Pending)
ISBN13: 979-8-2600-3485-9
Isbn13Softcover: 979-8-2600-3486-6
EISBN13: 979-8-2600-3487-3
British Cataloguing in Publication Data
A Cataloguing in Publication record for this book is available from the British Library.

Table of Contents

Detailed Table of Contents

The socio-educational theory, critical field, and practice approaches presented in this chapter make up the conceptual framework for culturally responsive mathematics teacher education. This reevaluates the teaching of maths as something socially mediated, culturally situated, shaped by the use of language, identity and interaction. The chapter has four main parts that detail the socio-cultural foundations of mathematics education and teacher preparation for culturally relevant pedagogical practices synthesis of conceptual framework. Ways to improve urban education are designing culturally relevant tasks, teaching that encourages discourse and equitable participation. To illustrate theory, chapter provides examples from the classroom, and classroom resources like reflective prompts, the task evaluation checklist and participation mapping. The need for teacher training to deal with divergent classrooms in mathematics is further brought about in a research framework, where mention was made of knowledge and understanding of mathematics as an essential quality of the teacher.

This chapter proposes a conceptual and pedagogical framework for integrating socio-scientific issues into mathematical modelling from socio-critical and socio-cultural perspectives in mathematics education. It argues that issues such as cloning, climate change, energy use, health, and technology should not be treated only as ethical or scientific discussion topics, but as complex public problems that can be structured, examined, and justified through mathematical reasoning. The chapter first establishes theoretical connections among socio-scientific issues, mathematical modelling, critical citizenship, students' lifeworlds, and classroom participation. It then shows how modelling processes such as identifying variables, making assumptions, mathematizing, interpreting, validating, and revising models can support

academic concepts. It demonstrates that cultural context facilitates deeper conceptual understanding of transformational geometry.

This study aims to introduce the theoretical frameworks of ethnomathematics and out-of-school learning and to present examples of mathematical activities that can be applied in museums and cultural heritage sites related to middle school mathematics, grounded in these frameworks. First, the theoretical foundations of ethnomathematics and out-of-school learning approaches are discussed in detail. Then, implicit mathematical elements embedded in architectural decorations, handicrafts, and cultural motifs found in museums and heritage sites in Kastamonu are identified. These elements are linked to middle school mathematics learning areas, numbers and operations, algebra, geometry and measurement, data processing, and probability, incorporating them into activity designs. The study provides a theoretical framework for integrating ethnomathematics into out-of-school learning and offers teachers examples of applications related to cultural heritage. It is believed that the proposed activities will contribute to making the cultural and out-of-school aspects of mathematics more visible.

This chapter aims to investigate the use of mathematical tasks related to the calculation of zakat, fasting duration, food distribution (iftar), ketupat, bedug, and others to teach the basic concepts of mathematics, such as fractions, percentages, time management, and geometry. This social context-based learning aims not only to enhance students' mathematical competences but also to introduce them to social values such as justice, responsibility, and sharing. This study identifies the main challenges in designing context-based tasks, such as task complexity, student acceptance of the social context, and teaching time constraints. This chapter also provides solutions and recommendations to improve task design, as well as proposing future research directions involving learning technology and long-term evaluation in context-based education. Thus, context-based social mathematics learning is expected to introduce social values that can enrich students' lives in society.

This chapter uses Rumah Lontiok, a traditional elevated timber house of the Kampar Malay community in Riau Province, Indonesia, as the cultural basis for an ethnomathematical pedagogical framework. The authors conducted content analysis at three heritage sites with laser-assisted field documentation, and then validated the resulting framework through a two-round expert panel of five specialists (overall validity score 3.76 out of 4.00). The study identified five solid geometry domains: the tiang pillar network, the lantai panggung floor platform, the cosmologically organized interior rooms, the curved bumbungan melentik ridge, and the integrated composite solid of the whole building. Domains 1 to 3 align with Kurikulum Merdeka of Phase C core outcomes, whereas domains 4 and 5 serve as enrichment. Van Hiele's (1986) theory provides the structural backbone for the four-stage framework. Bishop's (1988) and Gerdes' (1988, 1994) contributions extend it further. The framework was then translated into six classroom learning activities for grades 5 to 6.

This chapter examines how the indigenous game Wera can enhance the teaching and learning of number sense in primary mathematics in Namibia. Guided by Lev Vygotsky's social constructivist theory, the study views learning as a socially and culturally mediated process. A qualitative approach was employed using document analysis, lesson observations, semi-structured interviews, and stimulated recall with teachers, learners, and indigenous knowledge holders. Data were analysed thematically. The findings show that Wera strengthens counting skills and supports learners in connecting indigenous counting systems with the Hindu-Arabic numeral system. The intervention also resulted in a mean gain of 23.68 points across number-

Preface

INTRODUCTION

Mathematics education is inevitably linked to the cultural, linguistic and social contexts in which learning occurs. Mathematics has long been regarded as a universal and context-independent discipline. However, over the past few decades, research has increasingly shown that mathematical knowledge is influenced by human experiences and cultural practices. People in communities throughout the world have invented sophisticated mathematical practices through architecture, art, crafts, games, trade, agriculture, navigation and everyday problem solving. As mathematics education seeks to meet the needs of culturally and linguistically diverse classrooms, it has become more important to recognize these different forms of mathematical knowledge. This handbook, *Ethnomathematics and Culturally Responsive Pedagogy*, was developed in response to this growing need by bringing together research and practice that show culture can serve not only as a context for learning mathematics, but also as a source of mathematical understanding itself.

The chapters are set in different educational contexts and draw on different research traditions but share a common goal to investigate how the teaching of mathematics can be more culturally relevant, equitable and meaningful through the use of ethnomathematics and culturally responsive pedagogy. The authors acknowledge cultural knowledge, local practices, indigenous customs and community experiences as useful intellectual resources for mathematical investigation, rather than as just another example or decorative aspect to mathematics teaching. Therefore, the chapters show how culturally-based pedagogies can enhance mathematical thinking, while at the same time enhancing students' participation, identity and sense of belonging.

The handbook reflects the diversity of contemporary ethnomathematics research by presenting studies and instructional designs from a range of geographical and cultural contexts. Readers will encounter examples drawn from traditional Anatolian motifs, Turkish architectural heritage, indigenous games in Namibia, Rumah Lontiok architecture in Indonesia, museums and cultural heritage sites, Islamic cultural practices during Ramadan, socio-scientific mathematical modelling, and technology-supported collaborative learning environments. Although these environments differ, they all exemplify a common principle: formal mathematical concepts are intertwined with the cultural experiences and communal knowledge of the learners, which facilitates meaningful mathematics education.

A distinguishing feature of this volume is its emphasis on translating theory into educational practice. Beyond discussing conceptual perspectives on ethnomathematics and culturally responsive pedagogy, the chapters provide concrete instructional frameworks, curriculum analyses, design principles, classroom tasks, teacher education models, and practical tools that facilitate application across diverse educational environments. From preparing prospective teachers to designing culturally responsive lessons, to analyzing textbooks, developing mathematical modelling activities, and creating learning experiences grounded in indigenous knowledge, the contributors provide multiple pathways through which culturally responsive mathematics education can be enacted in classrooms.

The chapters also show that culturally responsive mathematics education can take many forms. It compasses a wide range of instructional possibilities, including design-based research, ethnomodeling, mathematical modelling, digital learning environments, out-of-school mathematics education, curriculum analysis, teacher professional learning, and culturally embedded task design. These contributions illustrate how a variety of pedagogical approaches can have a common goal: to help students learn mathematics through contexts that are authentic, culturally relevant, and intellectually demanding.

This handbook is designed for mathematics education researchers, teacher educators, classroom teachers, curriculum developers, graduate students, and policy makers interested in creating more inclusive mathematics learning environments. The book offers researchers current theoretical perspectives and empirical studies that advance the discourse of ethnomathematics and culturally responsive pedagogy. Teacher educators and practitioners will benefit from classroom examples, instructional models, and practical design principles that can readily be adapted to their own contexts. Curriculum developers and educational leaders will find frameworks for infusing cultural knowledge into mathematics curricula without compromising mathematical integrity and academic rigor.

Ethnomathematics and Culturally Responsive Pedagogy seeks to cultivate a more inclusive vision of mathematics education through the collaborative work of scholars and practitioners across diverse educational contexts. The handbook acknowledges that students enter mathematics classrooms with rich cultural knowledge, experiences, and perspectives that can serve as powerful tools for learning. When we take an ethnomathematical view and a culturally responsive approach, we can help develop a space in which all students—regardless of language background, cultural identity, gender, or personal experience—can develop confidence, agency, and meaningful connections with mathematics.

As editors, we hope this handbook has the potential to contribute to the international dialogue about the role of culture in mathematics education. The chapters show that acknowledging cultural diversity does not weaken the universal nature of mathematics. Rather it enriches the teaching of mathematics by recognizing the many ways in which human societies create, communicate and use mathematical ideas. This volume brings together scholars from different countries, cultures and educational traditions, offering both theoretical insight and practical inspiration for educators who wish to develop mathematics classrooms where learners see their own cultures, identities and experiences reflected as legitimate and valuable resources for mathematical thinking. We hope that *Ethnomathematics and Culturally Responsive Pedagogy* will inspire ongoing research, dialogue, and collaboration among scholars and practitioners committed to developing mathematics education that is culturally sustainable, socially responsive, and educationally meaningful to all learners.

ORGANIZATION OF THE BOOK

Chapter 1: Culturally Responsive Mathematics Teacher Education via Design Principles and Classroom Vignettes: Preparing Teachers for Responsive Practice

This chapter establishes a conceptual foundation for culturally responsive mathematics teacher education by examining mathematics teaching as a socially mediated and culturally situated practice. The chapter explores how language, identity, interaction, and cultural experiences influence mathematical learning and teacher preparation. Through a synthesis of socio-cultural, critical, and practice-based perspectives, it presents strategies for preparing teachers to design culturally relevant tasks, facilitate equitable classroom discourse, and support diverse learners. Drawing upon classroom vignettes and practical resources, including reflective prompts, task evaluation tools, and participation mapping strategies, the chapter provides actionable guidance for teacher educators seeking to develop responsive mathematics teaching practices. By emphasizing the importance of teachers' cultural awareness and pedagogical knowledge, this chapter highlights the essential role of teacher education in creating inclusive mathematics learning environments.

Chapter 2: Integration of Socioscientific Issues into Mathematical Modelling Problems: Theoretical Connections, Classroom Dialogues, and Implications for Mathematics Teaching

This chapter examines the integration of socioscientific issues into mathematical modeling as a means of connecting mathematical reasoning with real-world social challenges. Moving beyond traditional applications of mathematics, the chapter presents a socio-critical and socio-cultural framework in which issues such as climate change, health, technology, energy use, and biotechnology become opportunities for students to engage in mathematical inquiry and critical reflection. By exploring processes such as identifying variables, constructing assumptions, mathematizing situations, interpreting results, and validating models, the chapter demonstrates how mathematical modeling can support students in analyzing uncertainty, evidence, equity, and societal consequences. Through classroom dialogue examples, the chapter illustrates how socioscientific modeling encourages meaningful participation, critical citizenship, and deeper connections between mathematics and students' lived experiences.

Chapter 3: Escape Rooms as Socio-Cultural Learning Environments in Primary School Mathematics: Designing Collaborative and Digitally Mediated Practices

This chapter explores the potential of digital escape rooms as innovative socio-cultural learning environments that promote collaboration and mathematical engagement among elementary learners. Through the implementation of digitally mediated tasks focused on angles, geometric shapes, spatial reasoning, and directional thinking, the chapter investigates how students collaboratively construct mathematical understanding. The findings highlight four key dimensions of collaborative mathematical activity: participation structures, iterative experimentation, artifact-supported regulation, and planning-oriented reasoning. By demonstrating how students negotiate ideas, coordinate actions, and engage in shared problem-solving, this chapter provides insights into designing technology-enhanced mathematics experiences that support dialogue, participation, and collective mathematical agency.

Chapter 4: Bridging Ethnomathematics and School Mathematics—Concept Mapping Across Program–Textbook–Task Levels: C–T–M–E Depth Analysis of TYMM Textbooks

This chapter examines the relationship between ethnomathematics and culturally responsive mathematics education through an analysis of curriculum documents and mathematics textbooks developed under Türkiye's educational reform initiatives. Utilizing the C–T–M–E model and a systematic framework for evaluating cultural integration, the chapter investigates how cultural elements are represented and incorporated across curriculum, textbooks, and mathematical tasks. The analysis reveals varying levels of integration, ranging from surface-level cultural references to transformative approaches that connect mathematical learning with cultural identity and meaning-making. By providing concept maps, analytical tools, and task transformation models, the chapter offers valuable resources for curriculum designers and educators seeking to move beyond tokenistic inclusion toward more meaningful culturally responsive mathematics practices.

Chapter 5: Ethnomathematics for Teaching Geometry and Measurement: A Design-Based Research Approach

This chapter presents a design-based research approach for integrating ethnomathematics into geometry and measurement instruction through culturally meaningful learning experiences. Challenging the perception of mathematics as culturally neutral, the chapter demonstrates how traditional Anatolian motifs can serve as authentic contexts for exploring mathematical concepts such as polygons, symmetry, and transformations. Through a structured process of analysis, design, implementation, and refinement, cultural artifacts are transformed into curriculum-aligned mathematical activities. The chapter introduces four ethnomathematical design principles—cultural authenticity, mathematical transparency, exploratory construction, and structural decomposition—which provide guidance for educators seeking to incorporate cultural contexts into mathematics teaching. By positioning cultural practices as sources of mathematical knowledge rather than decorative additions, the chapter advances approaches for creating more relevant and engaging mathematics instruction.

Chapter 6: Ethnomodeling as a Bridge: Connecting Local Cultural Knowledge to Academic Geometric Transformations

This chapter explores ethnomodeling as an approach for connecting students' cultural knowledge with formal mathematical concepts through the study of traditional Turkish artistic patterns. Focusing on geometric transformations within rug and ceramic motifs, the chapter demonstrates how students apply mathematical reasoning to analyze translation, reflection, and rotation while engaging with culturally meaningful designs. Through qualitative classroom investigation, the chapter reveals how cultural contexts support deeper conceptual understanding and encourage students to view mathematics as connected to their communities and experiences. By positioning ethnomodeling as a bridge between local knowledge and academic mathematics, this chapter highlights its potential to foster inclusive learning environments where students' cultural resources contribute to mathematical exploration and creativity.

Chapter 7: Ethnomathematical Approach to Out-of-School Mathematics Education: Museums and Heritage Sites

This chapter examines the role of museums and cultural heritage sites as meaningful environments for ethnomathematical learning. By connecting out-of-school education with ethnomathematical perspectives, the chapter identifies mathematical ideas embedded within architectural designs, traditional crafts, and cultural motifs found in heritage settings. These cultural elements are linked to middle school mathematics concepts, including numbers and operations, algebra, geometry and measurement, data analysis, and probability. The chapter provides examples of activity designs that enable educators to incorporate community knowledge and cultural heritage into mathematics instruction. Through these approaches, the chapter demonstrates how mathematics can become more visible and meaningful when learners explore the mathematical structures embedded within their surrounding environments.

Chapter 8: Designing Collaborative Mathematics Tasks as a Professional Learning Space for Teachers: A Case Study from the Perspective of Islamic Culture during Ramadan

This chapter investigates the design and implementation of culturally grounded mathematics tasks that incorporate Islamic cultural contexts, particularly experiences associated with Ramadan. By examining mathematical activities related to concepts such as zakat, fasting duration, food distribution, geometric patterns, and resource sharing, the chapter demonstrates how contextualized tasks can connect mathematical learning with social values, including responsibility, fairness, and community engagement. The chapter also explores the challenges teachers encounter when designing socially meaningful mathematical tasks, including complexity, student acceptance, and instructional time constraints. Through practical recommendations and future directions, the chapter emphasizes the potential of culturally responsive task design to enrich mathematics learning while supporting broader social and ethical development.

Chapter 9: Solid Geometry in Rumah Lontiok Architecture of Kampar: An Ethnomathematical Analysis and Pedagogical Framework for Elementary Education

This chapter presents an ethnomathematical exploration of Rumah Lontiok, a traditional Kampar Malay architectural structure, as a foundation for developing culturally responsive geometry instruction. Through detailed analysis of architectural features, the chapter identifies multiple connections between traditional construction practices and solid geometry concepts, including spatial structures, geometric forms, and composite solids. Building upon ethnomathematics, Van Hiele's theory of geometric thinking, and culturally responsive pedagogical perspectives, the chapter develops a structured framework for integrating cultural heritage into elementary mathematics classrooms. By translating architectural knowledge into classroom activities, the chapter demonstrates how indigenous cultural practices can serve as meaningful resources for strengthening students' geometric understanding and appreciation of local knowledge systems.

Chapter 10: The Use of the Indigenous Game of Wera to Support the Acquisition of Number Sense in Namibian Primary Schools

This chapter explores how indigenous games can support culturally sustaining mathematics education by examining the use of Wera as a tool for developing number sense among primary students in Namibia. Grounded in social constructivist perspectives, the chapter positions learning as a culturally mediated process in which indigenous knowledge and formal mathematics can be meaningfully connected. Through classroom observations, interviews, and analysis of learning experiences, the chapter demonstrates how Wera supports counting skills, strengthens connections between indigenous and formal numerical systems, and enhances students' mathematical understanding. The findings highlight the potential of indigenous games to transform mathematics curricula by recognizing community knowledge as a valuable resource for teaching and learning. Through this contribution, the chapter advances discussions surrounding culturally responsive pedagogy and the integration of local practices into mathematics education.

CONCLUSION

Ethnomathematics and Culturally Responsive Pedagogy brings together a diverse collection of perspectives that highlight the transformative potential of recognizing mathematics as a cultural, social, and human endeavor. Across the chapters, the contributors challenge the assumption that mathematics is independent of cultural contexts by demonstrating how mathematical knowledge is developed, communicated, and experienced through language, traditions, identities, communities, and lived experiences. Through theoretical explorations, empirical investigations, and practical instructional frameworks, this handbook advances a broader understanding of mathematics education as a space where equity, inclusion, and meaningful participation can be actively cultivated.

The chapters presented in this volume illustrate the ways in which ethnomathematics and culturally responsive pedagogy provide valuable pathways for reimagining mathematics teaching and learning. By connecting community knowledge, cultural practices, and formal mathematical concepts, the contributions demonstrate how educators can create learning environments that affirm students' identities while strengthening mathematical understanding. From culturally grounded mathematical modeling and digital learning environments to indigenous games, heritage-based activities, and classroom task design, the chapters reveal the diverse ways in which cultural contexts can serve as powerful foundations for mathematical exploration.

A central theme throughout this handbook is the importance of moving beyond superficial representations of culture toward deeper forms of culturally responsive practice. Meaningful integration of cultural knowledge requires educators to examine how mathematics is presented, whose knowledge is valued, and how students are positioned as participants in mathematical communities. The contributions emphasize that culturally responsive mathematics education is not simply about incorporating cultural examples into existing curricula; rather, it involves transforming instructional approaches, assessment practices, classroom interactions, and curriculum designs to recognize and sustain diverse ways of knowing.

The chapters also highlight the critical role of teachers and teacher education programs in advancing culturally responsive mathematics education. Preparing educators to engage with diverse classrooms requires opportunities to develop cultural awareness, design meaningful mathematical tasks, facilitate inclusive discourse, and critically reflect on issues of identity and equity. The frameworks, design principles, classroom examples, and practical tools presented throughout this volume provide important resources for supporting educators as they navigate the complexities of teaching mathematics in culturally diverse environments.

Through its exploration of ethnomathematical perspectives across multiple geographical and cultural contexts, this handbook demonstrates that mathematics is not only a subject of study but also a reflection of human creativity, problem-solving, and collective knowledge-building. Traditional practices, indigenous knowledge systems, community experiences, and cultural artifacts all represent valuable sources of mathematical understanding. Recognizing these connections allows mathematics education to become more relevant, engaging, and accessible for learners while promoting respect for diverse intellectual traditions.

As Editors, we hope that *Ethnomathematics and Culturally Responsive Pedagogy* serves as a meaningful resource for researchers, educators, policymakers, and teacher educators committed to advancing equitable mathematics education. By bringing together scholarship and practice, this volume contributes to ongoing efforts to develop mathematics classrooms where students' identities, cultures, and experiences are recognized as essential components of learning. Through continued engagement with ethnomathematics and culturally responsive approaches, mathematics education can move toward a more inclusive future—one in which all learners are empowered to see themselves as capable mathematical thinkers and active contributors to the creation of mathematical knowledge.

Chapter 1
Culturally Responsive Mathematics Teacher Education via Design Principles and Classroom Vignettes:
Preparing Teachers for Responsive Practice

Arzu Aydoğan Yenmez
http://orcid.org/0000-0001-8595-3262
Niğde Ömer Halisdemir University, Turkey

ABSTRACT

The socio-educational theory, critical field, and practice approaches presented in this chapter make up the conceptual framework for culturally responsive mathematics teacher education. This reevaluates the teaching of maths as something socially mediated, culturally situated, shaped by the use of language, identity and interaction. The chapter has four main parts that detail the socio-cultural foundations of mathematics education and teacher preparation for culturally relevant pedagogical practices synthesis of conceptual framework. Ways to improve urban education are designing culturally relevant tasks, teaching that encourages discourse and equitable participation. To illustrate theory, chapter provides examples from the classroom, and classroom resources like reflective prompts, the task evaluation checklist and participation mapping. The need for teacher training to deal with divergent class-

DOI: 10.4018/979-8-2600-3485-9.ch001

rooms in mathematics is further brought about in a research framework, where mention was made of knowledge and understanding of mathematics as an essential quality of the teacher.

INTRODUCTION

Over the years, maths education has gone through a significant theoretical and pedagogical turn from largely individual and cognitive models of learning to more holistic socio-cultural, critical and practice-based perspectives. There has been recognition in the minds of mathematicians and educationists that mathematical knowledge is not merely the body of truth that happens to exist somewhere. Mathematics is a human activity that is historically developed, culturally situated and socially mediated (D'Ambrosio, 2006; Lerman, 1998; Radford, 1997). Learning maths these days is often no more viewed simply as a matter of mastering abstract ideas or procedural skills; instead, a growing number see learning as participation in socially organized practices through which it is believed meaning is constructed (Lave & Wenger, 1991; Vygotsky, 1978). As described by Atweh et al. (2004), the next theoretical orientation was shaped by wider socio-cultural and social justice orientations. The orientation emphasizes how mathematics procedures are embedded in our society and power relations.

The impact of this rethinking is great on classrooms and what teaching can do in them. From the socio-cultural perspective, Mathematics classrooms are not seen as neutral spaces. They arise out of cultural norms, institutional structures and power relations (Abreu, 2000; Hall, 2007). The manner in which students are attracted to mathematical ideas is affected by their various backgrounds. However, their teaching strategies have generally not acknowledged such resources, instead favouring dominant forms of knowledge and participation. Many students, particularly those on the margins, see the disconnection between math and their identity and their life (Gutiérrez, 2013; Ladson-Billings, 1995). Research on identity in mathematics education provides further evidence linking students' opportunities to learn to how they are positioned in classroom practices and whose knowledge gets legitimized (Aguirre et al., 2013).

Culturally responsive teaching (CRT) has emerged as a key framework for rethinking the mathematics curriculum through these challenges. According to Gay (2018), CRT stresses the importance of using the unique cultural background of learners to promote everyone's participatory and learning equity. This viewpoint disrupts the view of mathematics was seen as universally applicable. Through this idea mathematical knowledge will be culturally disembedded and value-neutral (Gutiérrez, 2013; Sleeter & Delgado Bernal, 2004). Thus D'Ambrosio's (2006) ethnomathematics perspective shows that mathematics is not limited to formal practices but is found in all cultures and daily life giving legitimacy to this knowledge. These viewpoints also align with principles for equitable and meaningful mathematics teaching practices to diverse learners (National Council of Teachers of Mathematics, 2014).

At the same time, studies on language and discourse have emphasized that communication is central to mathematical learning. According to Moschkovich (2010), mathematical knowledge is built through interaction, negotiation, and the use of various semiotic resources. In multilingual classrooms that have a diverse population, the language practices of students should not be regarded as barriers, but as resources for meaning making (Song & Coppersmith, 2020). This importance of communication echoes the socio-cultural and human-centered design perspectives which highlight the role of context, interaction and experience in the production of meaning and engagement (Moalosi et al., 2004; Moalosi, 2007; Rosenman & Gero, 1998).

There is a need for a complete rethink in teacher preparation due to these theoretical advancements. To prepare teachers for culturally responsive mathematics instruction requires more than content knowledge and general pedagogical knowledge. Such preparation must develop teachers who draw on students' cultural resources, who critically reflect on their own beliefs and practices, and design learning environments that are equitable and inclusive (Nolan & Xenofontos, 2023; Siliunas et al., 2024). The cognitive and emotional nature of this process can make it a complex transformation. Teachers navigate tensions, challenges and opportunities across educational contexts. Furthermore, studies indicate that there are practical challenges to implementing socio-cultural curricula that require careful planning in classroom design and support (Kinuthia, 2012).

Besides, teacher education shouldn't only focus on creating individuals with competencies but also on the capabilities of conditions and systems. Curriculum design, assessment policies, and institutional norms correspondingly reflect socio-cultural artifacts that exhibit particular values and power relations (Arnesen et al., 2010; Binns, 2015; Rezat, 2024). If there is no critical analysis of the impact of these structures, implementing culturally responsive teaching will be limited. Furthermore, recent studies suggest grounding theoretical knowledge in authentic classroom experiences through practice-based approaches that enable teachers to develop practices in context (Chou et al., 2018; Godino et al., 2024). Moreover, transitions across educational levels – from school to university, say – highlight the need for coherent and context-sensitive approaches (Di Martino et al., 2023). In light of this, a demand is being called forth for frameworks of culturally resonant mathematics education that gives insight into both theory and practice. Even if the literature provides rich conceptual ideas, executing them in practical classroom instructions is a daunting task. To bridge this gap, it is necessary to create design principles, pedagogical tools and examples that connect theory to practice in a meaningful way. Similarly, the broader socio-cultural design research focus on sustainability, context awareness and user-centered in complex systems (Fatani et al., 2017) can be beneficial for such activities.

This chapter answers this need by proposing a model for practising culturally responsive mathematics teacher education that is built on socio-cultural theory, but also draws from critical and practice-based perspectives. The chapter has four interconnected components. Initially it treated the socio-cultural precepts for mathematics education. It pointed out how interaction, culture, language, and power creates a learning environment. Following this, it will further explore the preparation of teachers for culturally responsive instruction with a focus on their knowledge, beliefs, and competencies for equitable teaching. It is also accurate that socially-minded pedagogy must be used in combination with cross-curricular and contextual links to encourage deeper knowledge. Fourth, it offer socio-cultural design principle to translate the theories into an actionable guidance for practice. The chapter also guides the use of educational vignettes, and implementation tools on how to teach in the classroom while creating and enhancing shared semantics between participants. The vignettes show situated examples of culturally responsive teaching. The tools e.g. reflective prompts, task evaluation frameworks, questions of discourse analysis, and participation mapping are structured ways for analysing teaching practice or improving it. All of these aim to help teacher educators and teacher candidates develop the knowledge, skills, and dispositions to create mathematics learning environments that are inclusive, meaningful, and socially just.

In sum, this chapter establishes culturally responsive mathematics teacher education as a transformative process beyond technical competence. There is need for a redefinition of teaching as a socially, culturally and politically situated practice and teachers' role as agents of change who can challenge inequities and promote more inclusive forms of mathematical participation (Gay, 2018; Gutiérrez, 2013; Müller, 2025). Through its grounded nature, which incorporates practice in the understanding of theory, the chapter adds its weight to valuable efforts to reimagine mathematics education, especially in light of contemporary classrooms and their uses.

BACKGROUND

Socio-Cultural Foundations in Mathematics Education

In the last few decades, the theory and explanation of learning in mathematics education have shifted from purely cognitive and individualistic descriptions to more socio-cultural explanations. A shift is taking place from understanding knowledge as something abstract and decontextualized, a set of truths that people can make deductions from. Instead, knowledge has become viewed as a human activity, a cultural practice. That is to say, learning here is not simply an internal mental process but a participation in shared and socially organized activity where meaning is generated, negotiated, shared and transformed. In essence, socio-cultural theory posits that all cognition is mediated by social and cultural tools. The view that higher psychological processes development first on social plane and then individual was presented by Vygotsky (1978) in one of the first and most influential formulations. The zone of proximal development demonstrates the usefulness of guided participation, scaffolding, and collaboration for learning processes. In other words, students learn mathematics through conversations, shared problem-solving and interactions with their more knowledgeable peers or teacher – rather than through individual reasoning. Lave and Wenger (1991) built on Vygotsky's work to create the concept of situated learning and described learning to be legitimate peripheral participation in communities of practice. Based on this framework, as learners become more competent, their roles become more central and less peripheral. For mathematics education, this lens focuses attention on classrooms as communities that instantiate, negotiate, and legitimize mathematical practices. To learn mathematics is, then, to engage in the discourse, norms and practices of the mathematics community rather than just the learning of procedures and concepts.

The classroom as a socio-cultural system in mathematics is being increasingly conceived because of these theoretical developments. According to socio-cultural perspectives, classrooms are not neutral places for the transmission of knowledge. Instead, they are sites that are shaped by culture and powered through institutions (Abreu, 2000; Hall, 2007). Mathematical meaning arises from discourse, gestures and representations within a shared activity. Thus, the nature of learning must be examined not only at the level of classroom and curriculum but also at that of society and culture. Linking of these macro and micro dimensions is seen important in socio-cultural research in mathematics education. Abreu (2000) maintains that students' mathematics will not be comprehended without a consideration of the broader cultural context that they are embedded in. For example, students use everyday forms of mathematical knowledge learned in the home or community, which is often quite different from the formal practices of school mathematics that are valued at school. This dissonance shapes our participation, identity and what we are able to achieve, creating tensions. The implications of this perspective require an expansion of focus and analysis that incorporates cognitive, social, cultural, and institutional forces.

The socio-cultural theory has impacted maths education because it disputes the transculturality of maths. D'Ambrosio's (2006) elaboration of the ethnomathematics perspective involves mathematical knowledge is culturally situated and different societies develop distinct mathematical practices according to their needs, values and experiences. Through this stance, the conception of what is termed 'mathematics' is expanded and the dominance of Western mathematics in formal education is undermined. Similarly, Radford (1997) emphasizes how the meanings associated with mathematical ideas are not universal, but rather social, historical, and culturally constructed. Research into language and discourse in mathematics education supports this point of view. As stated by Moschkovich (2010), students contribute to the construction of mathematical meaning through language, especially in multilingual contexts. Mathematical discourse is not merely an instrument for the communication of ideas, but rather a means of creating meaning. Students can only participate in mathematical discussions if they can draw on everyday and mathematical language practices, which are all culturally mediated. As a result, socio-cultural approaches highlight the need to view linguistic diversity as part of mathematics learning. Socio-cultural perspectives have examined identity and power in mathematics education; increasingly foregrounding culture and language has contributed significantly to this work. Gutiérrez (2013)'s description of the "sociopolitical turn" extends socio-cultural theory with an explicit focus on equity and access and social justice. This perspective articulates that mathematics education is not neutral; rather, the opposite is true. The practices of classrooms can either reproduce existing hierarchies or provide opportunities for equal participation. For example, students from marginalized communities may be excluded when their cultural knowledge and way of knowing are not recognized in the dominant norms of school mathematics (Gay, 2018; Ladson-Billings, 1995).

Teaching and teacher education are impacted positively by these findings. Teachers are important actors in mediating the classroom socio-cultural environment that shapes the norms, expectations and interactional patterns for learning. As revealed by the research, the beliefs, knowledge and pedagogical practices of the teachers can either support or hinder the participation of students in the mathematical practices (Agbozo et al., 2024; Siliunas et al., 2024). Culturally responsive teaching seeks to connect students' cultural backgrounds to what is being taught in ways that promote conceptual understanding and belonging (Gay, 2018). These approaches acknowledge and validate varying forms of knowledge; hence they go against the deficit views which hold that learners are at a disadvantage due to their backgrounds. In addition, socio-cultural theory has effects on curriculum design and educational policy. The curriculum is not a mere content that needs to be delivered, but a cultural message that carries with it certain values, assumptions and power relations. According to Binns (2015), the design of a curriculum is based on social context which is built in institutional context. Likewise, Arnesen et al. (2010) policy frameworks highlight underlying teacher competencies in that cultural diversity and inclusion. The curriculum and pedagogy need to rethought in order to make them responsive to the diverse cultural contexts of the learner. The scope of socio-cultural perspectives has recently widened through the engagement of additional developments from critical theory and sustainability. Müller (2025) proposes a critical pragmatic approach that combines socio-cultural analysis with ethical and practical considerations, with mathematics education playing a role in solving global problems. As a result, mathematics teaching must go beyond handing down knowledge and strive to develop the critical awareness of students as well as their agency and responsibility

In conclusion, socio-cultural foundations in mathematics education offer a framework that allows us to understand learning as socially situated, culturally mediated, and historically based. Incorporating viewpoints from Vygotsky, communities of practice, ethnomathematics, sociopolitical, and more, this view illuminates the construction and use of mathematical knowledge. Significantly, it highlights that creating equitable and inclusive mathematics classrooms demands not only pedagogical innovation, but also critical consideration of the cultural and social assumptions underlying education (Gutiérrez, 2013; Müller, 2025).

Teacher Preparation for Culturally Responsive Instruction

Due to the increase in cultural, linguistic and socio-economic diversity in classes, teachers' expectations have changed radically. Culturally responsive instruction (CRI) has become a central pillar for effective and equitable teaching. In this regard, teacher preparation is not just the provision of knowledge of the subject matter and general pedagogical skills anymore. Educators must also develop the capacity to notice, create, and cultivate students' diverse cultural resources as part of the learning process (Gay, 2018; Siliunas et al., 2024). As a result, preparing teachers for culturally responsive instruction necessitates a substantial redesign of teacher education informed by socio-cultural theory and focused on equity, identity and social justice. Culturally responsive teaching is grounded on socio-cultural perspectives that frame learning as a socially mediated and culturally situated phenomenon (Lave & Wenger, 1991; Vygotsky, 1978). In this light, students come to the classroom not as blank slates but as persons whose ways of knowing, communicating, and acting are shaped by culture. Students' interpretations of mathematical meanings, engagement with talk in classrooms, and meaning-making are shaped by these resources (D'Ambrosio, 2006; Moschkovich, 2010). As such, teacher preparation must enable teachers to perceive these resources not as hindrances but as assets that can benefit the context.

Ladson-Billings (1995) culturally relevant pedagogy framework can serve as the basis on which to understand the aims of culturally responsive instruction. In this framework, teaching effectively requires supporting students' success, competence, and consciousness all at the same time. The expanded Triadic Model shows that teachers have a role to play more than that of delivering the content. They also aid in helping the learners in formation of identity and critical awareness. The problematizing view disavows the traditional idea that mathematics is neutral and universal in mathematics education. This view ultimately shows how knowledge gets produced historically and has power and ideology (Gutiérrez, 2013). In this conceptualization, teacher preparation is a deeply transformational act. Professional development for teachers is more than getting new knowledge and skills. Specifically, it involves changing their beliefs, values, and identities. Studies have consistently illustrated the great influence that teachers' beliefs about their students, especially those from disadvantaged backgrounds, have on teaching practices and learning opportunities (Agbozo et al., 2024). Limited perspectives that view students' languages and culture as deficits set low expectations and discriminate against those students. On the other hand, culturally responsive approaches value what every student brings to the classroom as an asset (Gay, 2018).

Efforts in teacher education programs should be intentional to transform such deeply held beliefs. According to Nolan and Xenofontos (2023), this is conceptualized through the COFRI (Challenges, Opportunities, Fears, Resistance, and Insights) framework that highlights the emotional and cognitive aspects of teachers' learning. The dynamics encapsulated in COFRI include challenges, possibilities, anxiety, resistance and insights. According to their work, becoming a culturally responsive teacher is not a straightforward journey but one filled with tensions and contradictions. Consequently, it is essential for teacher education programs to provide opportunities for reflection, dialogue and critique. This will surely allow upcoming teachers to challenge their assumptions and build more inclusive outlooks. Teachers should develop pedagogical competencies for culturally and contextually responsive teaching in addition to changing beliefs. According to Hernandez et al. (2013), content integration, knowledge construction, prejudice reduction, social justice and academic development can make up a multidimensional model. The measurements provide the science of practice for culturally responsive teaching. For example, content integration uses examples and contexts from different cultures in Mathematics classes, while knowledge construction relates new ideas to life experiences. The elements of prejudice reduction and social justice seek to create a more inclusive and equitable learning environment.

Experiencing and understanding these concepts first-hand will lead to a greater understanding. To link theory to practice, teacher education must be practice based. According to Chou et al. (2018), transformative teacher preparation programs that combine coursework with field-based experiences can build teachers' capacity to implement culturally responsive practices. Researchers found that a teacher's developmental progression begins with awareness of inclusion and progresses towards developing a positive attitude, deepening understanding, and enacting culturally responsive teaching practices. The changing nature of education means that it is important to provide real opportunities to learn in context for teachers to experiment and reflect on their practice. A very important aspect of teacher preparation is the development of teacher self-efficacy. Teachers confident in their capacity to meet culturally diverse need are likely to employ innovative and inclusive instructional strategies (Siliunas et al., 2024). On the other hand, traditional, teacher-centered approaches may be adopted because of low self-efficacy to majority of learners. Hsiao's (2015) view is that culturally responsive teaching competencies are not merely instructional competencies but also a rich array of relational and affective competencies that involve building trust; establishing high expectation and creating a sense of belonging. These elements are particularly crucial in mathematics classrooms since students' identities and attitudes can heavily impact their participation and success in the subject.

Language is at the center of culturally responsive instruction, especially mathematics education, where discourse is a means of making meaning. As Moschkovich (2010) points out, maths understanding is rooted in the use of language and not as apart from it as one might think. In multilingual classrooms, teachers should help students with everyday and academic language so that language diversity doesn't hamper learning in any way. Song and Coppersmith (2020) note that many teachers cannot meet the linguistic needs of students, especially those who are English learners because they are not adequately prepared. This finding shows that student teachers need education to prepare them to teach mathematics taking into consideration the language and culture. Although there is growing focus on culturally responsive instruction, a large gap remains in preparing teachers to deploy them. There's a gap between the theory taught in medical schools and what is practiced in clinics. Although they may take courses relating to culture and equity, transferring that knowledge to classroom practice is not so easy (Nolan & Xenofontos, 2023). Teacher's constricted curriculum, set tests as well as institutional weightage may prevent them from adapting their lessons to the students' needs. Individual and system level resistance to change can form a barrier to the operation of culturally responsive pedagogy.

Complex structural and institutional factors influence teacher preparation. Teacher education programs may be under-resourced, undermanned, and under-skilled to train all teachers to be culturally responsive. Arnesen et al. (2010), suggests that defining better-defined competencies to meet cultural diversity and a better linkage of policy-curriculum-teacher education is essential for promoting intercultural understanding. The policies and strategies of organizations and agencies dealing with equalization of learning opportunities must be aligned with the pedagogy. If not, efforts may be fragmented and inconsistent. An expanding body of mathematics education research indicates that we need to accommodate socio-cultural, critical or ethical perspectives to intelligent challenges. Müller (2025) advocates a critical pragmatic approach with theories, practices, and ethos that focuses on agency and responsibility for teachers and students; both have a role.

This view aligns with culturally responsive teaching that aims at enhancing academic performances and empowering students as active participants in the society. In conclusion, teacher preparation for culturally responsive instruction is a complex process; it requires teachers to deeply transform not just technically skilled only. Teachers will ultimately need knowledge of cultures that includes critical awareness, reflective dispositions and practical competencies to enable inclusive and equitable learning environments. If teacher education programmes are integrated with socio-cultural theory, critical pedagogy and practice-based approaches, they could respond better to the needs of a diversely situated student. This kind of preparation is essential to the ongoing struggle for equity & social justice in mathematics education, which is to provide equitable access & opportunities in mathematics for all learners (Gay, 2018; Gutiérrez, 2013).

Conceptual Foundations of Culturally Responsive Mathematics Teacher Education

Culturally responsive mathematics teacher education (CR-MTE) will be a socially just response to globalisation that broadens the perspective of mathematics education. This framework exemplifies a major movement away from decontextualized models of teacher education toward a view of teaching as culturally situated, socially mediated and politically informed. The preparation of mathematics teachers is not only concerned with their utility but also their identity, equity, power, and meaning-making that may be invoked during the teaching and learning process in various contexts (Gay, 2018; Gutiérrez, 2013). In this description, it's stated that human knowledge acquisition is a socially constructed process, drawing upon Vygotsky socio-cultural theory of learning. Learning, as opposed to acquisition, is depicted as participation within usage sets associated with specific practices, which is made possible through mediation. Mathematical knowledge cannot be conceived, from this perspective, as a static body of truths. Rather, it is a dynamic fluid construct; it emerges through social interaction and social activity collectively. Radford (1997) has also discussed this view stating that the mathematical idea is historically and culturally constituted, and thus it is not taken as universal and context independent.

Teacher knowledge must be reconceptualised due to this socio- cultural orientation. Teachers need to be equipped to not only know mathematics content but also to view and act upon the culturally and socially embedded nature of learning environments. Abreu (2000) stresses the importance of relating macro socio-cultural contexts (community practices, social norms) and micro classroom interactions. In terms of teacher education, this means that future teachers need to learn to notice how the out-of-school experience, cultural practice, and identity of students inform their engagement with mathematics. A key aspect of CR-MTE is the acknowledgment of mathematics as culturally relevant knowledge. Ethnomathematics, according to D'Ambrosio (2006), is a significant important notion that explains the relationship between mathematics and culture. It challenges Western mathematical epistemologies and opens room for different kinds of mathematical reasoning and mathematical representation. In other words, teacher education requires a shift in focus from seeing diversity as a problem to being managed towards diversity as a resource to be used for teaching. This means teachers must periodically ask themselves what kind of knowledge gets prioritized in the curriculum and what gets sidelined.

This framework provides a viable basis for the implementation of culturally responsive teaching (CRT), enabling us to give these ideas practical form. According to Gay (2018), CRT is used in learning where student's cultural knowledge, the experiences they bring and their communication styles are made use of. An analogous argument was made by Ladson-Billings (1995). Pedagogy must help with education achievement, cultural intelligence and social awareness. Teaching is defined through these frameworks as relational and dependent on context. In this way, a pedagogy responsive to the students identities and experiences is essential for meaningful learning to occur. A combination of CRT and socio-cultural theory leads to a richer understanding of teaching that contains cognitive and sociopolitical aspects. Gutiérrez (2013) calls this a "sociopolitical turn," emphasizing the importance of the power, identity and access dimensions of mathematical learning. Math classes can be viewed as "neutral turf". Either they reproduce social inequalities, or they disrupt them. Therefore, teacher education must equip teachers to notice unequal practices and develop learning environments that achieve inclusivity and justice.

A very important addition to this conceptual framework is the role of language and discourse in mathematising. Centering language within mathematics, socio-cultural research shows the importance of discourse in the construction of mathematical understanding (Moschkovich, 2010). In different forms, human language serves as a mediation tool for the development of mathematical ideas. Teachers need to be able to hold conversations and give equal importance to all forms of expression. Whether in one of the teacher education programs, designing the curriculum is focusing not just on what, but how it is communicated and negotiated in a range of contexts and languages. The understanding of teacher identity as a dynamic process is also embraced by the CR-MTE framework. According to Nolan and Xenofontos (2023), teacher learning is not a process of acquiring but be-coming as the process of teacher learning occurs through experiences, reflections, and interactions within a context. The COFRI (Challenges, Opportunities, Fears, Resistance, and Insights)

Framework illustrates that this process may encounter challenges and resistance and provide insights, highlighting the emotional and cognitive complexities of teacher development. Teacher education should provide opportunities for critical reflection of students which will enable them to analyse their own cultural identities, biases and positionalities. CR-MTE has to be set within broader institutional and systemic contexts along with individual transformation. The significance of policies, curricula and assessment in education is not just an issue for schools, but for the entire education system (Arnesen et al., 2010; Binns, 2015). The system structures mirror the dominant culture and can either facilitate or impede the implementation of culturally responsive practices. In conclusion, conceptual framework has to include the interaction between teacher agency and systemic conditions. Also, it would be meaningful only if there is alignment at multiple levels. Theoretical developments have provided additional impetus to the framework in various ways. Müller (2025) makes the case for a pragmatic critique that connects the theoretical and practical in mathematics education. Teachers are equipped with the critical capacity to question and the practical capacity to innovate. Such integration is particularly relevant to CR-MTE, where the challenge is not limited to being aware of cultural diversity but also to using this awareness in classroom practice.

In addition, this activity theory expanded the conceptual foundations of CR-MTE, Learning and Teaching is conceived as part of a complex system involving many actors, tools, and context in activity theory (Godino et al., 2024). Teacher education may be thought of as an activity system shaped by interplay among teacher candidates, educators, curricula and institutional structures. Taking a systems-oriented view therefore helps understand some of the tensions and contradictions involved in implementing culturally responsive practices. Drawing upon the comprehensive body of work we have reviewed, we can conceptualise culturally responsive mathematics teacher education as involving the following interrelated elements: (1) socio-cultural understandings of learning as participation in culturally mediated practices; (2) mathematics knowledge as historically and culturally situated; (3) equity, identity and social justice; (4) language and discourse as central to meaning-making; (5) teacher identity through reflection and transformative processes; and (6) institutional and systemic contexts that shape our practices. These components dynamically interact, influencing teacher understanding and enactment of culturally responsive instruction. The theoretical framework will aid in converting socio-cultural and culturally responsive ideas into design principles for teachers training. It guides curricula, teaching methods, and assessments that meet culturally responsive objectives. Ensuring that teacher preparation is intellectually rigorous and practically meaningful creates a clear relationship between theory and practice.

To make the proposed framework more explicit and practically accessible, Table 1 summarizes the model of culturally responsive mathematics teacher education developed in this chapter. The table presents the major dimensions of the model, their core components, and their functions within teacher education. In this way, it clarifies how the conceptual foundations, socio-cultural design principles, and practical tools work together to support the preparation of reflective, equity-oriented, and culturally responsive mathematics teachers.

Table 1. Proposed model of culturally responsive mathematics teacher education

Model Dimension	Core Components	Key Features / Requirements	Function within the Model
1. Conceptual Foundations	Socio-cultural learning; culturally situated mathematical knowledge; equity, identity, and social justice; language and discourse; teacher identity and reflection; institutional and systemic contexts	Mathematics learning is understood as socially mediated, culturally situated, and shaped by power relations. Teacher education should acknowledge students' cultural resources, language practices, identities, and the broader structures that influence teaching.	Provides the theoretical and philosophical basis of culturally responsive mathematics teacher education.
2. Socio-Cultural Design Principles for Practice	Culturally meaningful task design; discourse as a resource; equitable participation structures; identity affirmation and mathematical agency; equity-oriented assessment; practice-based and contextualized teacher learning; critical reflection and sociopolitical awareness	Instruction should connect mathematics with students' lived experiences, support multiple forms of participation and expression, affirm identities, promote agency, and encourage teachers to critically examine inequities in practice.	Translates the theoretical foundations into actionable pedagogical principles for teacher preparation and mathematics instruction.
3. Practical Tools for Teacher Education	Reflective prompts; equity-oriented task evaluation checklist; discourse analysis questions; participation mapping tool	Teacher candidates should be provided with structured tools to examine their instructional choices, analyze classroom interaction, evaluate tasks, monitor participation, and refine culturally responsive practices.	Operationalizes the model by supporting reflection, planning, implementation, and evaluation in teacher education contexts.
4. Intended Outcomes	Reflective, equity-oriented, culturally responsive mathematics teachers	Teachers are expected to design, implement, and critically evaluate mathematics instruction that is inclusive, context-sensitive, participatory, and socially just.	Represents the overarching aim of the model and the expected contribution of culturally responsive mathematics teacher education.

Note. The proposed model brings together the conceptual foundations, practice-oriented design principles, and implementation tools discussed throughout the chapter. These dimensions are presented as interconnected rather than sequential, collectively supporting the preparation of culturally responsive mathematics teachers.

Among the dimensions summarized in Table 1, three priorities are particularly central to the proposed model. First, culturally responsive mathematics teacher education should move beyond a general recognition of diversity and provide teacher candidates with concrete ways of translating cultural, linguistic, and experiential resources into mathematically meaningful instruction. Second, the model emphasizes that equity-oriented teaching must balance cultural responsiveness with sustained attention to conceptual depth, participation, and mathematical agency, rather than reducing responsiveness to the superficial use of familiar contexts. Third, the model highlights that culturally responsive practice cannot be understood solely as an individual teacher competency; it is also shaped by institutional structures, curriculum expectations, assessment practices, and broader systems of power. These priorities clarify the most critical issues addressed by the model and provide a sharper foundation for the practice-oriented design principles discussed in the following section.

SOCIO-CULTURAL DESIGN PRINCIPLES FOR PRACTICE

This chapter contributes to the articulation of socio-cultural design principles that can be interpreted as idea(ology) guiding teacher educators and teacher candidates. The socio-cultural theory, culturally responsive pedagogy and sociopolitical perspectives provide a sound conceptual foundation but can transform situations only if they are operationalized. The following principles build on earlier discussions by introducing socio-cultural theory, ethnomathematics, critical pedagogy, language research, and recent research on teacher education to provide a comprehensive framework for culturally responsive mathematics teacher education.

1. Culturally Meaningful Task Design

Creating mathematical tasks that are culturally meaningful is a key feature of the socio-cultural approach to mathematics education. In short, ethno-mathematics refers to any mathematical ideas of people that is culturally based. This presents a challenge to the traditional presentation of mathematics as abstract and context-free, presenting mathematics as human activity-based. Culturally meaningful tasks enhance student engagement and understanding by connecting formal mathematical ideas to students' intuitive understandings. For instance, tasks can create links with practices such as trading, architecture, agriculture or local measurement systems, allowing students to perceive mathematics as relevant. Research (Abreu, 2000) shows that students' interpretation and activities in mathematics are heavily influenced by their previous knowledge and cultural experiences. This shows in approaches. In addition, task design that is culturally meaningful promotes equity by affirming that different knowledge systems can be mathematically sound and contest Western superiority (D'Ambrosio, 2006; Gutiérrez, 2013). This method additionally adheres to the guidelines of culturally responsive teaching by making the link between the curriculum and students' identity (Gay, 2018). For teacher education, this principle calls for preparing teachers to critically examine curricular materials and redesign activities in culturally relevant ways. Prospective teachers must develop the ability to identify culturally relevant contexts, integrate them into mathematical instruction, and reflect on how these tasks support both conceptual understanding and identity development.

2. Discourse as a Resource

According to socio-cultural theory, language and discourse are key mediators of learning. According to Moschkovich (2010), students' engagement in explaining and negotiating mathematical ideas with each other plays a major role in making meaning. As such, the linguistic practices of students must be used as resources, rather than viewed as obstacles. In the classroom, students speak in everyday language and are multilingual. Students also speak in culturally specific ways. Studies show that using these resources can enhance understanding of mathematics and participation, particularly in settings with diverse languages (Moschkovich, 2010; Song & Coppersmith, 2020). As such, Teachers must create spaces where expression is valued, whether verbal, symbolic, visual, gestural and more. Moreover, robust discourse practices should be designed to foster deeper reasoning and co-construction of knowledge. Strategies that teachers may use include open-ended questioning, revoicing, and scaffolding to support students in articulating their thinking. The practices are consistent with Vygotsky (1978) on an emphasis on guided interaction and co-construction. For teacher education programs, using discourse facilitation skills is paramount. This entails assisting aspiring teachers in grasping how to decipher contributions from students, backing multilingual learners, and developing classroom settings that stimulate meaningful dialogue.

3. Equitable Participation Structures

Socio-cultural and sociopolitical perspectives of education emphasizes the notion of equitable participation. Access to participation in social practices is understood as a prerequisite for learning. However, participation is unevenly distributed and significantly influenced by race, language, gender, and socio-economic status (Gutiérrez, 2013; Lave & Wenger, 1991). Creating classroom norms and routines that allow students to imagine, engage, contribute and be viewed as competent participants in a lesson is designing equitable participation structures. This could include group work, structured discussions or any strategies that divest authority from the teacher's centre and redistribute it elsewhere. It is not only access, but also recognition and power, which a critical perspective on participation entails. Students need to not only be incorporated into the math activity but for their contributions to be legitimated (Gutiérrez, 2013). Teacher must actively to disrupt participation patterns through which a few voices dominate. Investigation of social practice theory further states that participation is shaped by institutional and cultural contexts (Binns, 2015). As such, participation structures at both the classroom and system level are required. Teacher education programs ought to prepare educators who will able to judge participation patterns, inequalities, and equitable participation techniques.

4. Identity Affirmation and Mathematical Agency

Learning can be interested based on identity, who sees themselves in identity who relate with in maths and who sees them as identity relevant and accessible (Ladson-Billings, 1995). Culturally responsive teaching (CRT) encourages teachers and students to affirm each other's different cultural identities (Gay, 2018). This allows students to feel that they belong. Identity affirmation means recognizing and valuing the background, experience and ways of knowing of the student. This can be done through culturally relevant content, inclusive classroom practices, and supportive relationships. When students' identities are reflected in the curriculum, they are more likely to engage with mathematics learning. Mathematical agency lets students act confidently, successfully and independently in the practices of mathematicians. In order to develop this agency we must broaden the meaning of competence to include various kinds of thinking and solving diverse problems (Gutiérrez, 2013). Teachers need to provide students with the chance to own their learning, voice their opinions, and solve meaningful problems. Essential role of teacher education in supporting identity affirmation and agency. Prospective teachers must be encouraged to reflect on their own identities and biases, as well as to develop strategies for fostering inclusive and empowering learning environments (Nolan & Xenofontos, 2023). This process engages cognitive and emotional processes, emphasizing the transformative quality of teacher learning.

5. Equity-Oriented Assessment

The assessment practices are quite powerful as they determine what is valued in mathematics education and how students are positioned as learners. According to Ladson-Billings (1995), assessing math concept understanding is a form of assessment that often follows protocols and standardized. Assessment oriented toward equity seeks to curtail these setbacks by recognising various types of understanding and focusing on growth, reasoning and participation. The different strategies of formative assessment, which include continuous feedback, peer assessment, and reflection opportunities, offer students multiple ways of demonstrating understanding. In view of socio-culture, assessment should be part of the learning process and not be allowed to segregate. This supports the notion that learning is a socially mediated process that takes place all the time (Vygotsky, 1978). Consequently, teachers must plan assessment practices that support learning and offer productive feedback. Furthermore, evaluating one's own ideas is something designed to encourage critical reflection. Evaluation criteria should be questioned by teachers as a means through which assessment practices undermine and reinforce inequities (Gutiérrez, 2013). Teacher preparation ought to take over designing assessments that are inclusive, culturally responsive, and in-line with the curriculum.

6. Practice-Based and Contextualized Teacher Learning

Another principle that comes out of the literature is the need for practice-based and contextualized teacher learning. Research evidence reveals that the acquisition of culturally responsive teaching cannot be taught; rather, it must be experienced in real classroom contexts (Chou et al., 2018). A practice-based approach involves connecting coursework and field experiences so teacher candidates relate theory to practice. In this way, the teachers acquire practical skills and develop reflective disposition and responsiveness towards students. Furthermore, teacher learning is a socio-cultural process that occurs through the interaction of the mentor, peers, and institution (Godino et al., 2024). Teacher education programs must create spaces for collaboration to promote continuous reflection and growth.

7. Critical Reflection and Sociopolitical Awareness

Ultimately, culturally relevant mathematics instruction demands a dedication to critical thought and sociopolitical consciousness. According to Gutiérrez (2013), a new perspective in mathematics education recognizes that power, equity and justice are fundamental issues for teaching. It is important that teachers re-evaluate their practice, interrogate the dominant narratives and address inequities in the classroom. Become aware of cultural, social, institutional, and other factors that influence learning opportunities and outcomes. Critical reflection plays a significant role in ethical and responsible pedagogical development. Müller (2025) states that mathematics education must develop a critical and pragmatic approach to education to solve the socio-political problems of society. Teacher education programs should aim at developing not only teacher competency but also critical consciousness and social responsibility.

Interpretation of Design Principles

As a group, these social design principles are constructively aligned with and provide actionable guidance for culturally responsive mathematics teacher education. The authors underscore the significance of anchoring mathematics in students' cultural contexts, drawing on discourse as a resource, ensuring equitable participation, affirming identities, implementing inclusive assessment, supporting practice-based learning and fostering critical awareness. Most importantly, they are all connected and must be enacted together to create meaningful and equitable learning spaces. Through a blend of theory and practical suggestions, the paper shows how mathematics teaching and learning can be transformed in ways that are culturally relevant and socially just (Gay, 2018; Gutiérrez, 2013).

CLASSROOM VIGNETTES

The classroom vignettes presented in this section are idealized, author-constructed pedagogical scenarios rather than empirical cases drawn from original classroom data. They were developed to illustrate how the socio-cultural design principles proposed in this chapter might appear in mathematics teaching and teacher education contexts. While the vignettes are informed by the literature on culturally responsive pedagogy, socio-cultural theory, multilingual mathematical discourse, participation, and identity, they should be read as analytical and instructional tools rather than as reports of observed classroom episodes. Their purpose is to support reflection, discussion, and pedagogical interpretation among teacher educators and teacher candidates. Classroom vignettes can be powerful analytical tools informed by relevant theoretical and empirical literature, facilitating understanding of culturally responsive mathematics teaching and its dynamic, situated, and often complex nature. The stories do not only show the 'successful' practices but the negotiations, tensions and teaching decisions through which teachers try to align instruction with the cultural, linguistic and experiential resources of students (Gay, 2018; Gutiérrez, 2013). In sociology and culture perspective, vignettes function as mediational artefacts that reveal meaning-making, participation and identity construction processes in a mathematics classroom (Lave & Wenger, 1991; Vygotsky, 1978). These tools enable teacher candidates and educators to examine not only what happens in classrooms but why it happens, thereby promoting a deeper understanding of pedagogy.

Vignette 1: Reframing Proportional Reasoning through Community Contexts

The teacher candidate could have used more complicated situations where children are more likely to be reached at the same time as they are at the supermarket. It signifies an intentional effort to connect abstract mathematical frameworks with informal, culturally contextualized activities (D'Ambrosio, 2006).

The first task: " Is 3 kilograms at 70 lira a better deal than 1 kilogram at 25 lira?". It works as an entry point for culture and cognition. Instead of using symbols to define ratios, the teacher exploits the students' already present economic reasoning which is proportional. This supports Abreu's (2000) idea that students possess rich mathematical knowledge contingent on the situation, which is underused in schools.

Most significantly, the responses by the students indicate spontaneous mathematization:

- Changing price per unit to total price
- Comparing values in different contexts
- To derive a strategy: "compare prices per unit every time".

This sequence shows how tasks that are meaningful in a student's culture develop from informal reasoning to formal mathematical concepts or ideas within the framework of socio-cultural learning theory.

On further analysis there may be potential teaching issues. For example:

- What methods does the teacher employ to help students understand proportionality as a concept rather than a specific context?
- How do we address misunderstandings when informal reasoning is in conflict with formal definitions?

As this vignette shows, it's not just a matter of "making it relatable." It's about thinking carefully about the relationship between context and abstraction. According to the vignette, linking mathematical objects to familiar cultural contexts has a way of increasing engagement and deepening conceptual understanding. As such, choosing a context in task design should not be thought of as a mere pedagogical decision but rather an epistemological necessity in the teaching of mathematics (Abreu, 2000; D'Ambrosio, 2006).

Vignette 2: Multilingual Discourse in Geometric Reasoning

According to Vygotsky, language is not a prerequisite for learning; rather, it is a resource through which learning takes place (Moschkovich, 2010). The learner's utilization of:

- Mother tongue
- salute

- Picturesque description.

exhibits a multimodal approach to meaning-making where meaning is spread across semiotic systems. The teacher's reply: "Could someone repeat this in math language?" holds considerable importance. It shows a teaching change that:

1. Authenticates the original contribution.
2. Fosters group interpretation.
3. Connects informal and formal conversations.

This connects to Moschkovich's (2010) claim that mathematical discourse should not only be formal but also else a rich diversity of language and modal forms. The phrasing strategy. We are discussing the rotation phase. Helps the students connect their human intuitive understanding with the technical mathematical terminology. From a teacher education lens, this vignette demonstrates key skills:

- Recognizing meaning from unusual expressions.
- Enabling translation between representations.
- Ensuring access while keeping up cognitive demand concurrently.

At the same time, it poses significant queries:

- When does scaffolding become over-simplification?
- How can educators balance language assistance with conceptual difficulty?

Thus, it shows us that discourse happens not only in a classroom but also a site of power, access and understanding (Gutiérrez, 2013). As we can see from the vignette, it is not necessary for learners of mathematics to speak the same language. Multilingual and multimodal communication environments facilitate the making of meaning. They re-position the teacher from being a transmitter of knowledge to being a vehicle for establishing new collective meanings (Moschkovich, 2010; Vygotsky, 1978).

Vignette 3: Negotiating Participation and Authority

Mathematics classrooms exhibit uneven participation by learners, which is shaped by the implicit norms in the classroom, power dynamics, and social positioning (Lave & Wenger, 1991). This small case presents a clear instance of how teachers can intervene to change these processes.

The teacher's action: "Share with us your idea first…"

It is not neutral encouragement – it is a relocation of power. The teacher's choice to prioritise a hesitant student's voice disrupts dominant patterns of participation and opens up space for other contributions. It shows that afterwards:

- The construction of knowledge in collaboration.
- Verification of incomplete reasoning.
- Elaboration of concepts through the interaction of equals.

This involves the person's family life, wider community or society and also the wider participation in the social life and processes of humanity. Yet, from a socio-political stance, it also works toward redressing disparities that seem embedded in classroom discourse (Gutiérrez, 2013). Significantly, the last statement of the teacher: " It reflects strong maths." Overhauls proficiency through worth:

- Connections throughout representations
- Team Thinking
- Limited but worthwhile contributions.

The extension of what counts as "mathematical ability" is at the heart of Culturally Relevant Teaching. This vignette shines a light on teacher candidates:

- Participation must follow a well-thought-out design.
- Equity requires measures to be put in place.
- The ability to lead in a classroom may change.

The vignette shows that equitable participation cannot happen spontaneously but needs to be structured intentionally by purposeful pedagogical interventions; thus, the redistribution of mathematical authority makes visible the social and political nature of classroom learning environments (Gutiérrez, 2013; Lave & Wenger, 1991).

Vignette 4: Identity and Mathematical Agency

This short piece demonstrates one of the most significant yet often neglected aspects of mathematics education, namely identity and learning. By associating the patterns of weaving with sequences, the student is engaging in mathematical reasoning:

- Inculcating cultural values in our students.
- Establishing a credible identity as an expert.
- Connecting one's personal identity with academic content.

The reaction of the teacher. "Such a lovely bond." is crucial in confirming one's identity. It indicates that the student's culture is not marginal but central for the learning of mathematics. This is consistent with the students' need to be seen in their cultural competence (Ladson-Billings, 1995) and validate Gay (2018). The student also explains this: "It repeats all four steps… " This shows the evolution of mathematical agency-the ability to comprehend, generalise, and relay mathematical ideas. From a socio-cultural standpoint, this moment symbolizes:

- The combination of tradition and knowledge is important in any society to achieve its developmental goal.
- Changing individual identity through the participation process
- The growth of the definition of mathematics.

This illustrates the importance of teacher education:

- Identity crucial to learning; not an add-on.
- Open up to new epistemologies to learn cultures
- Agency occurs when students see themselves in the discipline.

Through this vignette, students' cultural identities are shown to be central, rather than peripheral, to learning mathematics. Thus, the need to reconceptualize who is seen as mathematical and what counts as mathematical competence to include cultural expressions is important. This is one way to see students as legitimate mathematical agents (Gay, 2018; Ladson-Billings, 1995).

PRACTICAL TOOLS FOR TEACHER EDUCATION

For socio-cultural, culturally responsive principles to go further than theory and into practice, teacher education programs must offer you structured, research-informed supports for reflection, planning, implementing, and evaluating. These tools are mediational means (Vygotsky, 1978) that help teacher learning as the teacher candidates systematically analyse their own practices, inform their pedagogical choices, and continuously adjust their practices in culturally diverse classrooms. In other words, these should not be seen as rigid tools but as dynamic sets of resources that are capable of evolving in use, reflection and context (Godino et al., 2024).

1. Reflective Prompts

Reflective prompts can form the basis of the critical consciousness and professional growth of teacher candidates. Teachers use reflection, as supported by socio-cultural and critical pedagogical theory. By reflecting, teachers examine their beliefs, values, and assumptions with reference to larger cultural and sociopolitical formations (Gutiérrez, 2013; Nolan & Xenofontos, 2023). Although, on the surface, reflection might involve considering whether a lesson was "successful," culturally responsive teacher education asks for helping educators learn to engage in deep, critical reflection as they begin to explore power, representation, and equity. With that in mind, reflective prompts should be developed by teacher educators to encourage teacher candidates to move beyond technical considerations. Instead, the prompts should encourage them to examine how their instructional choices shape students' opportunities to learn and participate.

For example, the prompt:

How does this task connect to students' cultural and everyday experiences?

invites teachers to consider the extent to which mathematical tasks are situated within meaningful contexts. Teachers are urged to reflect on the extent. This is linked to ethnomathematical views, which highlight the significance of connecting formal mathematics to practical experiences (D'Ambrosio, 2006). A teacher candidate could reflect: "I became aware that my task was using examples unknown to my students. In future, I could include contexts like local markets or transport networks."

Similarly, the prompt:

"Who's knowledge is represented in this lesson and who is missing?"

encourages teachers to critically examine curriculum content and uncover implicit biases. This sort of reflection can help to counter the dominance of particular cultures in mathematics education, allowing for greater inclusivity (Gay, 2018). Additional prompts, such as: " How do you acknowledge and support students' identities? What are the ways students can participate meaningfully?" influence educators to view identity and participation as important aspects of learning. The questions respond to socio-cultural and sociopolitical perspectives that identify engagement as a function of identity, recognition and belonging (Gutiérrez, 2013; Ladson-Billings, 1995). In practice, reflective prompts can be used in lesson plans, post-lesson reflections, peer talks, and mentorship. As teacher candidates engage in reflective practices, they learn to adjust their teaching over time in response to their students and the context.

2. Task Evaluation Checklist (Equity-Oriented)

The task evaluation checklist offers a tool for evaluating instructional materials through the lenses of equity and socio-culture. Both reflective prompts and checklists facilitate internal reflection, but the latter provides more structured and observable insights about the design and implementation of a task. At the most basic level, this checklist operationalizes key principles of culturally responsive teaching by converting those principles into criteria. The question, for example:

" Does the task relate to the real-world or community contexts?"

> ensures that tasks are not abstract and decontextualized but grounded in students' lived experiences. Research suggests that contextualisation promotes engagement and understanding of concepts (Abreu, 2000).

The criterion:

"Does it permit the use of various solutions?"

It highlights the value of different perspectives in problem-solving. Mathematics teaching has tended to endorse one 'right' answer and the methods that lead to it and therefore marginalise those few students whose methods do not fit the mould. When students are given opportunities to attack a problem using different methods their understanding of mathematics deepens, and they see themselves as mathematicians (Gutiérrez, 2013).

Likewise:

"Are diverse representations encouraged?"

recognizes that mathematical understanding is constructed through multiple modes, including visual, symbolic, verbal, and gestural representations (Moschkovich, 2010). Encouraging varied representations enhances inclusivity for multilingual learners and students with diverse learning styles.

The item on the checklist:

"Does it facilitate conversations and working together?"

aligns with socio-cultural views of learning as a social process (Vygotsky, 1978). When students participate in interactive task, they co-construct knowledge, share opinion and construct reasoning.

Finally:

"Does it confirm students' cultural knowledge?"

ensures that tasks recognize and view students' backgrounds as authentic knowledge sources. In culturally responsive teaching, it is crucial that students feel as if they belong (Gay, 2018). This checklist can be used during lesson planning, curriculum analysis, and peer review activities in teacher education. Most importantly, it should not be applied rigidly but used as a framework, which should help create critical thought and iterative improvements in instructional task.

3. Discourse Analysis Questions

The use of discourse analysis tools enables teacher candidates to analyse classroom discourse and how it shapes mathematical learning. For the socio-cultural perspective, it is through discourse that knowledge is constructed (Moschkovich, 2010; Vygotsky, 1978).

The query: " Who speaks and who stays quiet?"

highlights the issue of inequities across participation patterns Students' participation within the discussions may not be equal. Some groups of students dominate the conversation while others remain marginalized according to research (Gutiérrez, 2013). Unless teachers analyze who is speaking in the class and who is not, they won't know.

The prompt: "How are the contributions of students recognized or extended?"

emphasizes the teacher's role in shaping discourse. Practices like revoicing, probing, and linking students' ideas can provide validation and deepen understanding.

To illustrate, an educator may remark: " You're saying the slope shows how fast the amount of increases and decreases. Please try the above paraphrasing. ". Not only do these moves give extra credence to the conceptual development of students but they also position them as legitimate contributors to mathematical knowledge.

The question: "How Is Language and Representation Used in the Text?" Teachers are encouraged to consider the diversity of classroom communication. Students can communicate their ideas using everyday language, gestures, diagrams, and home languages. Recognizing these forms as valid helps to create settings for inclusive learning (Moschkovich, 2010).

Finally: " In what ways does the teacher help students articulate their ideas effectively?"

emphasizes instructional support to facilitate discourse. When students practice scaffolding, they tend to articulate their ideas, while revoicing ensures others hear them. In teacher education, one can include the use of video analysis, observation of classroom, and discussion of video as discourse analysis. By focusing on the role of discourse in learning, these practices can help teacher candidates understand how they can embed this knowledge in their teaching.

4. Participation Mapping Tool

The participation mapping tool offers both a visual and an analytical framework for examining patterns of student engagement during classroom interactions. Whereas discourse analysis concentrates on the content and quality of interaction, participation mapping highlights who participates, how frequently they participate, and in what forms. A typical participation map may include categories such as frequency of participation, type of contribution such as explanation, questioning, or agreement, and the nature of the teacher's response. An example of a participation map is presented in Table 2.

Table 2. Example of a participation map

Student	Frequency	Type of Contribution	Teacher Response
A	High	Explains reasoning	Extended
B	Low	Short answers	Minimal
C	Medium	Asks questions	Encouraged

By analyzing such data, teachers can identify patterns of inclusion/exclusion. Every student should receive feedback. For example observed more silence or least feedback to a set of students. According to Lave and Wenger (1991), from a sociocultural perspective, participation is not just about talking, but being recognized as a legitimate participant.

Consequently, it is essential to examine qualitative elements, including whether students feel their contributions matter and their role within the classroom. This tool can be employed during practicum experiences, video analysis, and peer observations in teacher education. This helps teacher candidates develop a more sophisticated understanding of the classroom's dynamics and help them create strategies to advance equitable participation.

Synthesis of Practical Tools

By using these practical tools together, the reflective prompts, task evaluation checklists, discourse analysis questions and participation mapping can provide a larger set of tools for operationalizing culturally responsive mathematics teaching. Help teacher candidates transcend instinctual or personalized behavior and enact more intentional, reflective, and equity-oriented practices. Significantly, these tools are linked. The reflective phase influences the task's design, which in turn affects the discourse, participation, and deeper questions of identity and equity what the participations reveal. Using these tools in a mutually reinforcing manner will help teacher candidates acquire the knowledge, skills and dispositions required to develop inclusive and meaningful mathematics learning environments (Gay, 2018; Gutiérrez, 2013). Ultimately, such resources can generate not only competent practitioners but also reflective and critical professionals who can respond to the complexities of a multicultural and diverse classroom. As a result of that, these classroom snippets, coupled with practical tools, illuminate what culturally responsive mathematics teaching is not. In addition, they together highlight what either individual (isolated) strategies for culturally responsive teaching and learning may overlook my added emphasis. The scenarios illustrated articulate the socio-cultural principles concerning how learning is happening, which is situated, discursive and relational. At the same time, the practical tools offer systematic ways in which the principles can be assessed, applied and refined. Crucially, in addition to serving as technical supports, the tools are mediational means that mediate teachers' interpretation of classroom activity, pedagogical choices and the positioning of students in mathematical practices (Vygotsky, 1978). Reflective questionnaires lead to the cultivating of critical consciousness. Task evaluation frameworks are designed to be equitable and fair. Discourse analysis tools help map patterns of meaning-making. Participation mapping sheds light on the often-invisible dynamics of inclusion and exclusion. Thus, the tools work together to help teachers move from instinctive practice to a much more analytical, intentional and equity-oriented approach in the instruction. Furthermore, these tools show how everything is interconnected throughout teaching. The following sequence is important for effective collaborative learning: reflection informs task design; task design shapes the discourse; discourse influences participation; and participation, in turn, shapes identity and access to learning. The reciprocal nature of this relationship shows that culturally responsive teaching requires pedagogical decisions that are not set in stone but rather set in motion within social and cultural contexts (Godino et al., 2024). From a teacher education perspective, the use of these tools appear to be critical in bridging the ever-widening gap between theory and practice. The socio-cultural and culturally responsive frameworks are highly useful concepts within education, but what impact they have depends upon how they are translated into classroom action. Teacher education programs can support the development of pedagogically-skilled teachers who are critically reflective and socially responsive practitioners by embedding these tools within coursework, practicum experiences, and professional development. (Nolan & Xenofontos, 2023; Siliunas et al., 2024). In conclusion, this conceptual framework would position educators as agents who can identify and change inequities in mathematical education. It highlights that equitable teaching is not a standardized process but involves contextual, critical and responsive practice. As such, it reaffirms the central assertion of this chapter: that culturally responsive mathematics teacher education must be envisaged as a transformative process, namely, one that reshapes meanings of teaching, learning, and mathematical participation in ways that are inclusive, meaningful and socially just (Gay, 2018; Gutiérrez, 2013).

CONCLUSION

This chapter seeks to provide a comprehensive and integrative reconceptualization of culturally responsive mathematics teacher education through the lenses of socio-cultural theory, critical pedagogy, and practice. Research has shown that teaching-and-learning of mathematics are not neutral or merely cognitive, therefore, they are situated within cultural, social, linguistic and political contexts which shape the construction, communication and legitimation of knowledge (Gutiérrez, 2013; Lave & Wenger, 1991; Vygotsky, 1978). In this connection, the chapter resembles bigger socio-cultural and social justice perspectives that hold learning cannot be separated from issues of equity, participation, and power (Atweh et al., 2004).

The chapter has signified the significance of socio-cultural perspectives in reinterpreting the nature of mathematics education at a theoretical level. Based on ethnomathematics, discourse studies, and historical epistemology, it has viewed mathematical knowledge as a culturally situated and dynamic practice generated through interaction and participation (D'Ambrosio, 2006; Moschkovich, 2010; Radford, 1997). This perspective contradicts main universalist ideas and is in line with work that highlights identity and equity of students used differently to make sense of maths (Aguirre et al., 2013). It also includes transitional contexts within mathematics education, where continuity across learning settings requires attention to socio-cultural and identity-related issues (Di Martino et al., 2023). Informed by these pillars the chapter conceptualized teacher preparation as a transformative process that transcends technical competency. Teachers can be prepared for culturally responsive teaching through fostering critical awareness, reflective dispositions and engagement with identity, equity and power (Gay, 2018; Nolan & Xenofontos, 2023; Siliunas et al., 2024). Also, it takes awareness of the systemic and institutional constraints (Arnesen et al., 2010; Binns, 2015) that influence possibilities in education. Furthermore, it takes awareness of practical constraints in implementing socio-cultural curricula and designs (Kinuthia, 2012). The conceptual framework that is discussed in this chapter brings together socio-cultural theory, culturally responsive pedagogy and sociopolitical perspectives to form a coherent foundation for mathematics teacher education. This framework sees teaching practices as relational, contextual, and ethical practices that are discursive, identity-mediated and ultimately shaped by wider systems of power and inequality, e.g. Ladson-Billings (1995), Sleeter & Delgado Bernal (2004), Gutiérrez (2013). Equitable mathematics learning environments are built not only by being responsive to students' cultural backgrounds but also by focusing on cultivating agency, inclusion, and social justice consistent with principles for effective mathematics teaching (National Council of Teachers of Mathematics, 2014).

This chapter concretizes the above-mentioned theoretical insights into practical empirical insights such as socio-cultural design principles to operationalize theories. This means that the principles of task design that these principles promote, the use of discourse as a resource, equitable participation structures, affirmation of identity, assessment for equity and critical reflection reflect a tradition of socio-cultural and human-centered design thinking (Moalosi et al., 2004; Moalosi, 2007; Rosenman & Gero, 1998). They emphasize that instructional design must be responsive to cultural contexts and user needs, a perspective also supported by research on sustainable socio-cultural design frameworks (Fatani et al., 2017). According to Gay (2018), D'Ambrosio (2006), and Moschkovich (2010), culturally responsive teaching is not a set of practices but rather a coherent and systemic approach requiring alignment across multiple dimensions of practice. The illustrative classroom vignettes in this chapter have shown how these principles may be pedagogically interpreted and enacted in plausible educational situations, rather than serving as reports of original empirical classroom data. The teachers in the examples mediated discourse, redistributed participation, and affirmed students' identities, making visible the relational and sociopolitical dimensions of classroom practice (Gutiérrez, 2013; Lave & Wenger, 1991). It is also important to recognize students as active participants whose cultural and experiential resources play a role in making meaning of mathematics. Along with these examples, the chapter has put forward some tools like reflective prompts, checklist for evaluating tasks and activities, framework for analysing classroom discourse, mapping participation tool etc that can help us implement culturally responsive practices in a systematic way. These tools are used to mediate the movement of teachers from their instinctive decision making towards more reasoned, analytical and equity oriented practice (Godino et al., 2024; Vygotsky, 1978). They also contribute to the formation of reflective and critical professional identities that are essential for the maintenance of cultural responsiveness over time (Nolan & Xenofontos, 2023).

All factors looked at in the chapter reveal a unified and systemic process of mathematics education. A greater understanding of the socio-cultural, institutional, design context in general and theories practices, curriculum and policy in specific, cannot liberate from implementation (Chou et al., 2018; Rezat, 2024). This perspective shows research in mathematics education that is increasingly sensitive to applying theory to practical issues and is resolving it. In summary, the chapter positions culturally responsive mathematics teacher education as a transformative project that can re-engender the purposes and practice of mathematics. The centre outlines an approach to teaching anti-racism in ways that foreground equity, identity and agentic power (Gutiérrez, 2013; Müller, 2025). It asks that educators challenge monocultural models of thought and action. It envisions teachers as reflective practitioners and agents of change capable of contributing to more inclusive, meaningful and socially just learning environments.

RECOMMENDATIONS

This chapter calls for culturally responsive mathematics teacher education to be positioned centrally, not peripherally in teacher education policy and practice. Cultural responsiveness should not be seen as an added pedagogical consideration; rather, it may be regarded as an essential principle to prepare teachers to work effectively and equitably in diverse mathematics classrooms. A major recommendation is for teacher training programs to develop a more explicit and sustained link between socio-cultural theory and classroom practice. It is necessary to provide teacher candidates with systematic opportunities to design and evaluate culturally significant mathematical tasks, analyze classroom discourse and participation, while further reflecting on the ways through which knowledge, identity and power play out in mathematics teaching. With this in mind, the tools offered in this chapter present a useful means of translating understanding into teaching practice. This chapter highlights the necessity of preparing teachers to see language, culture and student experience as intellectual resources for learning mathematics. As such, mathematics teacher education should prepare future teachers to create classrooms that expand access to discourse, affirm multiple identities, and legitimize multiple forms of reasoning and participation. An orientation such as this serves to enable purposeful engagement and mathematical agency. In addition, more attention should be paid to equity-oriented assessments. In order for culturally responsive teaching to be put into practice coherently, assessment practices must go beyond narrow and decontextualized measures of achievement to one that values reasoning, representation, growth, and participation. Teacher education programs should therefore train teachers to make assessment practices that are academically sound but also inclusive and socially responsive. In conclusion, in order to take culturally responsive mathematics teacher education to the next level; an insitutional and policy-level intervention is required. A stronger alignment of school structures, curriculum, assessment and teacher education is necessary for sustainable change. Only through such coherencess culturally responsive mathematics education can become a lasting systemic dimension of education instead of being a localized personal involvement.

REFERENCES

Agbozo, K. K., Agyei, E., & Arthur, Y. D. (2024). Socio-cultural foundations in teacher education: Fostering egalitarianism among preservice teachers in the mathematics classroom. *Golden Ratio of Social Science and Education, 4*(2), 136–146. DOI: 10.52970/grsse.v4i2.339

Aguirre, J. M., Mayfield-Ingram, K., & Martin, D. B. (2013). *The impact of identity in K–8 mathematics: Rethinking equity-based practices*. National Council of Teachers of Mathematics.

Arnesen, A.-L., Hadjitheodoulou-Loizidou, P., Allan, J., Trasberg, K., Chavdarova-Kostova, S., Furch, E., Vallianatos, A., Dumont, B., & Qiriazi, V. (2010). *Policies and practices for teaching socio-cultural diversity: A framework of teacher competences for engaging with diversity*. Council of Europe Publishing.

Atweh, B., Meaney, T., McMurchy-Pilkington, C., Neyland, J., & Trinick, T. (2004). Social justice and socio-cultural perspectives in mathematics education. In Perry, B., Anthony, G., & Diezmann, C. (Eds.), *Research in mathematics education in Australasia 2000–2003* (pp. 29–52). Post Pressed.

Binns, C. (2015). What can 'social practice' theory and 'socio-cultural' theory contribute to our understanding of the processes of module design? *Journal of Further and Higher Education, 39*(5), 758–775. DOI: 10.1080/0309877X.2014.953462

Chou, P.-I., Su, M.-H., & Wang, Y.-T. (2018). Transforming teacher preparation for culturally responsive teaching in Taiwan. *Teaching and Teacher Education, 75*, 116–127. DOI: 10.1016/j.tate.2018.06.013

D'Ambrosio, U. (2006). *Ethnomathematics: Link between traditions and modernity*. Sense Publishers. DOI: 10.1163/9789460911415

de Abreu, G. (2000). Relationships between macro and micro socio-cultural contexts: Implications for the study of interactions in the mathematics classroom. *Educational Studies in Mathematics, 41*(1), 1–29. DOI: 10.1023/A:1003875728720

Di Martino, P., Gregorio, F., & Iannone, P. (2023). The transition from school to university in mathematics education research: New trends and ideas from a systematic literature review. *Educational Studies in Mathematics, 113*(1), 7–34. DOI: 10.1007/s10649-022-10194-w PMID: 36408299

Fatani, K., Mohamed, M., & Al-Khateeb, S. (2017). Sustainable socio-cultural guidelines for neighborhood design in Jeddah. *Procedia Environmental Sciences, 37*, 584–593. DOI: 10.1016/j.proenv.2017.03.045

Gay, G. (2018). *Culturally responsive teaching: Theory, research, and practice* (3rd ed.). Teachers College Press.

Godino, J. D., Batanero, C., Burgos, M., & Wilhelmi, M. R. (2024). Understanding the onto-semiotic approach in mathematics education through the lens of the cultural historical activity theory. *ZDM Mathematics Education, 56*(6), 1331–1344. DOI: 10.1007/s11858-024-01590-y

Gutiérrez, R. (2013). The sociopolitical turn in mathematics education. *Journal for Research in Mathematics Education, 44*(1), 37–68. DOI: 10.5951/jresematheduc.44.1.0037

Hall, A. (2007). Vygotsky goes online: Learning design from a socio-cultural perspective. *Learning and Socio-cultural Theory: Exploring Modern Vygotskian Perspectives International Workshop, 1*(1). http://ro.uow.edu.au/llrg/vol1/iss1/6(open in a new window)

Hernandez, C. M., Morales, A. R., & Shroyer, M. G. (2013). The development of a model of culturally responsive science and mathematics teaching. *Cultural Studies of Science Education, 8*(4), 803–820. DOI: 10.1007/s11422-013-9544-1

Hsiao, Y. J. (2015). The culturally responsive teacher preparedness scale: An exploratory study. *Contemporary Issues in Education Research, 8*(4), 241–250. DOI: 10.19030/cier.v8i4.9432

Kinuthia, W. (2012). Reported challenges for socio-cultural curricula in instructional design. *Multicultural Education & Technology Journal, 6*(2), 87–99. DOI: 10.1108/17504971211236281

Ladson-Billings, G. (1995). Toward a theory of culturally relevant pedagogy. *American Educational Research Journal, 32*(3), 465–491. DOI: 10.3102/00028312032003465

Lave, J., & Wenger, E. (1991). *Situated learning: Legitimate peripheral participation.* Cambridge University Press. DOI: 10.1017/CBO9780511815355

Lerman, S. (1998). Research on socio-cultural perspectives of mathematics teaching and learning. In Kilpatrick, J., & Sierpinska, A. (Eds.), *Mathematics education as a research domain: A search for identity* (pp. 333–350). Kluwer Academic Publisher., DOI: 10.1007/978-94-011-5196-2_6

Moalosi, R. (2007). *The impact of socio-cultural factors upon human-centred design in Botswana* (Doctoral dissertation, Queensland University of Technology).

Moalosi, R., Popovic, V., & Hudson, A. (2004) Socio-cultural factors that impact upon human-centred design in Botswana. In *Proceedings Design Research Society International Conference* 1–11. https://eprints.qut.edu.au/2740/

Moschkovich, J. N. (2010). *Language and mathematics education: Multiple perspectives and directions for research.* Information Age Publishing. DOI: 10.1108/978-1-61735-161-7

Müller, D. (2025). Towards a critical pragmatic philosophy of sustainable mathematics education. arXiv preprint arXiv:2504.17149.

National Council of Teachers of Mathematics. (2014). *Principles to actions: Ensuring mathematical success for all.* Author.

Nolan, K. T., & Xenofontos, C. (2023). The role of insights in becoming a culturally responsive mathematics teacher. *Education Sciences, 13*(10), 1028. DOI: 10.3390/educsci13101028

Radford, L. (1997). On psychology, historical epistemology, and the teaching of mathematics: Towards a socio-cultural history of mathematics. *For the Learning of Mathematics, 17*(1), 26–33.

Rezat, S. (2024). Research on curriculum resources in mathematics education: A survey of the field. *ZDM Mathematics Education, 56*(2), 223–237. DOI: 10.1007/s11858-024-01559-x

Rosenman, M. A., & Gero, J. S. (1998). Purpose and function in design: From the socio-cultural to the techno-physical. *Design Studies, 19*(2), 161–186. DOI: 10.1016/S0142-694X(97)00033-1

Siliunas, V., Wan, G., & Edejer, E. (2024). Teacher preparation for culturally responsive teaching: Implications for teacher education. *Action in Teacher Education, 46*(4), 313–331. DOI: 10.1080/01626620.2024.2358912

Sleeter, C. E., & Delgado Bernal, D. (2004). Critical pedagogy, critical race theory, and antiracist education: Implications for multicultural education. In Banks, J. A., & Banks, C. A. M. (Eds.), *Handbook of research on multicultural education* (2nd ed., pp. 240–258). Jossey-Bass.

Song, K. H., & Coppersmith, S. A. (2020). Working toward linguistically and culturally responsive math teaching through a year-long urban teacher training program for English learners. *Journal of Urban Mathematics Education, 13*(2), 60–86. DOI: 10.21423/jume-v13i2a409

Vygotsky, L. S. (1978). *Mind in society: The development of higher psychological processes*. Harvard University Press.

ADDITIONAL READING

Bartell, T. G. (Ed.). (2018). *Toward equity and social justice in mathematics education*. Springer., DOI: 10.1007/978-3-319-92907-1

Bishop, A. J. (1991). *Mathematical enculturation: A cultural perspective on mathematics education*. Kluwer Academic Publishers.

Bonner, E. P. (2009). Achieving success with African American learners: A framework for culturally responsive mathematics teaching. *Childhood Education, 86*(1), 2–6. DOI: 10.1080/00094056.2009.10523100

Gutstein, E. (2012). *Reading and writing the world with mathematics: Toward a pedagogy for social justice*. Routledge. DOI: 10.4324/9780203112946

Gutstein, E., & Peterson, B. (Eds.). (2005). *Rethinking mathematics: Teaching social justice by the numbers* (2nd ed.). Rethinking Schools.

Hodge, L. L., & Cobb, P. (2019). Two views of culture and their implications for mathematics teaching and learning. *Urban Education, 54*(6), 860–884. DOI: 10.1177/0042085916641173

Hunter, R., Hunter, J., Anthony, G., & McChesney, K. (2018). Developing mathematical inquiry communities: Enacting culturally responsive, culturally sustaining, ambitious mathematics teaching. *Set: Research Information for Teachers*, (2), 25–32. DOI: 10.18296/set.0106

Nasir, N. S., Hand, V., & Taylor, E. V. (2008). Culture and mathematics in school: Boundaries between "cultural" and "domain" knowledge in the mathematics classroom and beyond. *Review of Research in Education, 32*(1), 187–240. DOI: 10.3102/0091732X07308962

White, D. Y. (2003). Promoting productive mathematical classroom discourse with diverse students. *The Journal of Mathematical Behavior, 22*(1), 37–53. DOI: 10.1016/S0732-3123(03)00003-8

KEY TERMS AND DEFINITIONS

Culturally Responsive Mathematics Teacher Education: A transformative approach to teacher education that creates math teachers who can design, enact, and critically reflect on instruction that is responsive to students' cultural, linguistic, and social resources that enables equity and inclusion.

Culturally Responsive Teaching: An instruction that pays attention to the students' backgrounds, experiences, communication modes, and brings that connection during the teaching and learning process for better learning, achievement, and participation.

Equitable Participation: A condition within the classroom where all learners are continuously supported to engage with mathematical activity, take part in discussions and are accepted as legitimate and capable participants.

Mathematical Discourse: The communicative process through which mathematical meanings are constructed, negotiated and refined through the use of spoken language, written representation, interaction and multiple semiotic representations.

Socio-Cultural Theory: A theory of learning that conceptualizes learning as socially mediated, culturally situated, and historically grounded process that focuses on making meaning through the operation involvement, through language, through tools and through people.

Mathematical Agency: Learners' ability to work with confidence and authority on mathematical activities through the expression of ideas, justification of reasoning, decision making and positioning themselves as competent doers of mathematics.

Ethnomathematics: A perspective on mathematics as culturally situated knowledge. Diverse communities develop mathematical practices in relation to their social contexts, everyday life, and culture.

Identity Affirmation: A method used to teach that students' cultural experiences, identities, and know-how can be utilized to produce meaning in mathematics teaching and learning.

Chapter 2
Integration of Socioscientific Issues Into Mathematical Modelling Problems:
Theoretical Connections, Classroom Dialogues, and Implications for Mathematics Teaching

Semahat Incikabi
http://orcid.org/0000-0002-7686-1996
Sinop University, Turkey

ABSTRACT

This chapter proposes a conceptual and pedagogical framework for integrating socio-scientific issues into mathematical modelling from socio-critical and socio-cultural perspectives in mathematics education. It argues that issues such as cloning, climate change, energy use, health, and technology should not be treated only as ethical or scientific discussion topics, but as complex public problems that can be structured, examined, and justified through mathematical reasoning. The chapter first establishes theoretical connections among socio-scientific issues, mathematical modelling, critical citizenship, students' lifeworlds, and classroom participation. It then shows how modelling processes such as identifying variables, making assumptions, mathematizing, interpreting, validating, and revising models can support students in examining uncertainty, risk, evidence, equity, and social consequences.

DOI: 10.4018/979-8-2600-3485-9.ch002

Through classroom dialogues, the chapter illustrates how SSI-based modelling can promote meaningful mathematical participation, reflective discussion, and critical decision-making.

INTRODUCTION

Socioscientific issues (SSI) are open-ended and often controversial issues in which scientific knowledge is intertwined with ethical, cultural, and societal considerations and that require public decision-making (Zeidler et al., 2005; Simonneaux, 2007; Zeidler & Sadler, 2008). In contrast, mathematical modelling (MM) provides opportunities to structure such complex issues through variables, relationships, assumptions, and quantitative comparisons. Well-structured socioscientific issues can bring together different domains such as reading skills, science content, social studies, mathematics, and art, while also providing students with real-life experiences related to moral reasoning, epistemological development, and peer discussion (Zeidler & Nichols, 2009). From this perspective, the integration of socioscientific issues into mathematical modelling processes may enable students to use mathematics as a tool to establish relationships among different types of knowledge, value systems, and social consequences, and to participate in multidimensional decision-making processes.

There is an important distinction between merely incorporating real-life contexts into mathematical modelling and integrating socioscientific issues into mathematical modelling together with multidimensional decision-making processes. Zeidler et al. (2005), for example, conceptualize SSI within a multidimensional pedagogical framework involving cognitive–moral reasoning, socio-cultural interaction, understandings of the nature of science, classroom discourse processes, and students' developmental characteristics. In addition, addressing real problems connected to students' own life worlds, cultural experiences, and social contexts plays a significant role in bringing socioscientific issues into mathematical modelling. Such problems may involve global issues such as climate change or nuclear energy, but they may also include local social issues that students directly encounter in their daily lives or issues such as cloning that can easily be associated with everyday experiences. When the world itself is considered as a shared cultural environment in the broadest sense, global socioscientific problems can become common spaces for discussion and meaning making for students. Realistic problem situations that connect with students' shared experiences can therefore provide more meaningful and powerful learning environments for mathematical modelling in terms of discussion, model construction, negotiation, and relating multiple perspectives.

On the other hand, both formal and informal reasoning structures are used in science education literature when discussing socioscientific issues. Researchers draw on these structures depending on the aims of the study and the nature of the discussion. While premises in formal argumentation are generally treated within more stable and linear structures, in informal argumentation individuals may revise their premises based on personal experiences, beliefs, and information gathered from different knowledge sources (Eş & Varol, 2019). At this point, mathematical modelling—which inherently has the potential to support both formal and informal reasoning processes—provides an appropriate context for discussing socioscientific issues. The cyclical structure of modelling, which allows iterative forward and backward transitions (Ferri, 2006, 2007; Doerr et al., 2017), enables individuals to reconsider their assumptions throughout the process, move between different forms of knowledge, and continuously reconstruct their models through intra- and inter-group discussions. Within this process, the stages of understanding, simplifying, model construction, mathematical work, interpretation, validation, and presentation simultaneously support more systematic and structured forms of formal reasoning while also creating opportunities for flexible and multidimensional informal discussions. Accordingly, mathematical modelling may provide a more suitable environment for reasoning and decision-making in line with the multidimensional and dynamic nature of socioscientific issues.

Although there are studies within the mathematical modelling literature that address socially contextualized problem situations such as environmental issues, sustainability, energy use, or oil spills (e.g., Solares-Rojas et al., 2022; Steffensen & Kasari, 2023), in many of these studies the problem situations are not directly conceptualized within a socioscientific issues framework; socio-critical and socio-cultural perspectives are not addressed together within a holistic theoretical framework; and ethical justification, social decision-making processes, and multidimensional structures such as SEE-STEP (Social, Economic, Environmental, Scientific, Technological, Ethical, and Political) do not systematically occupy a central place in the discussions. On the other hand, within the science education literature, SSI is conceptualized as a multidimensional theoretical field associated with argumentation and discussion, the nature of science, informal reasoning, socioscientific reasoning, scientific literacy, moral reasoning, democratic citizenship, and multidimensional decision-making structures (Eş & Öztürk, 2021; Sadler, 2004, 2009; Zeidler, 2003, 2014; Zeidler et al., 2005). In addition, within the mathematics education literature, mathematical modelling has been discussed through socio-critical mathematical modelling (e.g., Barbosa, 2006; Rosa & Orey, 2015) as well as theoretical and practical studies related to socio-cultural mathematical modelling (e.g., Rosa & Orey, 2025). However, studies that address socioscientific issues together with mathematical modelling processes through socio-critical and socio-cultural perspectives remain quite limited.

Considering this theoretical background and the limitations in the literature, this chapter opens a discussion on how socioscientific issues may be related to mathematical modelling processes. The central argument of the chapter is that socioscientific issues should not only remain within the ethical and social discussions of science education, but can also be addressed together with mathematical modelling through socio-critical and socio-cultural mathematics education perspectives. In this regard, the first original contribution of the chapter is that it theorizes the integration of socioscientific issues into mathematical modelling through these two mathematics education perspectives together.The second original contribution is the proposal of four discussion dimensions aimed at making classroom discussions emerging during mathematical modelling processes more visible and systematic, based on the common features arising from the intersection of the distinctive nature of socioscientific issues and the cyclical, discussion-oriented, and multidimensional structure of mathematical modelling.In this context, examining socioscientific issues in mathematical modelling processes through discussion dimensions shaped around multidimensional model development and decision-making processes involving SEE-STEP components, assumptions and model construction, social interaction and negotiation processes, and justice, equity, and social consequences constitutes one of the original aspects of this chapter. Thus, the chapter establishes an integrated theoretical relationship among socioscientific issues, mathematical modelling, and socio-critical/socio-cultural mathematics education literature, and presents a holistic framework that reveals the theoretical foundations, pedagogical implications, and emerging discussion dimensions of this integration.

The chapter first examines the social, cultural, and pedagogical dimensions of mathematics education within socio-critical and socio-cultural mathematics education perspectives. It then discusses the main characteristics and distinctive features of socioscientific issues and their relationship with mathematical modelling. Subsequently, socio-critical and socio-cultural perspectives on mathematical modelling are explored, followed by the rationale for integrating socioscientific issues into mathematical modelling and the conceptual and pedagogical framework underlying this integration. In the second part, classroom dialogues and instructional vignettes are used to discuss how this approach appears in educational settings, with particular attention to assumptions, multidimensional decision-making processes, social interactions, justice, equity, and social consequences. Finally, the implications of the chapter for mathematics education, teacher education, and classroom practices are evaluated.

THEORETICAL CONNECTIONS

Mathematics Education from Socio-Critical and Sociocultural Perspectives

An increasingly influential line of research in mathematics education challenges the assumption that mathematics teaching can be considered independent of context, values, and social relations. This line of inquiry in mathematics education argues that mathematics should be examined not only through its conceptual structure and procedural power, but also in relation to school practices, classroom discourses, students' identities and forms of participation, cultural and social positioning, citizenship, inequality, and power relations. The dominant perception of school mathematics often presents mathematics as a neutral discipline positioned outside social, cultural, and political relations. However, as Apple (2011) emphasizes, teaching is a political act that shapes how students read the world and position themselves within society. Therefore, rather than viewing mathematics education as a field limited to the transmission of technical knowledge and skills, it should be understood as an educational practice that gains meaning within historical, cultural, and political contexts (Ernest et al., 2016; Gibbs, 2019; Gibbs & Park, 2022; Makramalla, 2022; Skovsmose, 2005, 2022).

This perspective raises questions not only about how mathematical knowledge is represented in schools, but also about how it is experienced and interpreted by students. Accordingly, the nature of mathematics education is closely related not only to the purposes for which mathematics is used, but also to the ways in which it is constructed as a space for thinking, valuing, and participation for students. Defining mathematics solely in terms of utility and instrumentality may narrow students' cognitive and affective relationships with mathematics; in contrast, addressing mathematics together with its creative, aesthetic, and intellectual dimensions makes both its universal power and its applicability across diverse contexts more visible (Whitcombe, 1988).From this perspective, mathematics education should be viewed as a space for meaning-making in which students can understand, question, and discuss the role, function, and consequences of mathematics within both local cultural contexts and globalized societies, while also relating mathematical knowledge to social life (de Castro, 2024; Greer & Mukhopadhyay, 2015; Orey & Rosa, 2010; Rosa & Orey, 2021).

One of the most prominent theoretical strands of this broader line of inquiry is the socio-critical approach to mathematics education. The socio-critical approach positions mathematics education not as the transmission of a pure, objective, and value-free body of knowledge, but as a political educational field shaped by democracy, inequality, participation, and power relations (Ernest et al., 2016; Valero,

2004, 2020; Vithal, 2004). From this perspective, mathematical knowledge in school is not merely a set of problems to be solved or rules to be learned; it is also a school practice related to whose voices are heard, which forms of knowledge are legitimized, which students are recognized as successful, and which life opportunities are supported. Therefore, one of the central questions of critical mathematics education concerns the opportunities mathematics education opens for students, as well as the limitations it produces. Skovsmose's framework similarly shows that mathematics has the potential to empower students, while also carrying risks of exclusion, labeling, and the reproduction of social inequalities (Skovsmose, 2005, 2013, 2022). Thus, the central concern of socio-critical mathematics education is not only whether students can perform mathematical procedures, but also whether they can question how mathematics produces authority in schools and society, which claims it strengthens, and which experiences it renders invisible (Skovsmose, 2005; Valero, 2004).

Within this approach, critique is understood not as a closed method that produces definitive answers, but as a responsibility to think in favor of justice, equity, and democracy under conditions of uncertainty. Informed by Freirean pedagogy, feminist scholarship, anti-racist research, and critical educational theories, this orientation opens up discussion on how mathematical knowledge functions in the social world (Skovsmose, 2014; Stinson & Bullock, 2012). In this context, reflective competence refers to a capacity that goes beyond knowing mathematics: recognizing the power relations constructed through numbers, measures, graphs, achievement indicators, and model results; critically evaluating these relations; and transforming mathematical knowledge into an interrogative citizenship practice for a more just public life (Gibbs, 2019; Gibbs & Park, 2022; Skovsmose, 2020a). Accordingly, the socio-critical perspective approaches mathematics learning not merely as a matter of cognitive development, but in relation to students' possibilities for participating in and transforming the social world.

The sociocultural approach, on the other hand, explains mathematics learning not as a narrow cognitive process occurring solely within the individual mind, but as a process of meaning-making mediated through language, culture, classroom interactions, social norms, and students' life worlds (Gates, 2020; Valero, 2004). This framework is associated with the reflections of Vygotskian sociocultural psychology and Lave and Wenger's participation- and community-based understanding of learning in mathematics education (Atweh et al., 2001; Gutiérrez, 2013; Kieran et al., 2002; Lerman, 2006; Morgan, 2014). The sociocultural turn that became increasingly visible in mathematics education research from the 1980s onward showed that explaining learning solely in terms of individual performance, conceptual development, or the production of correct answers was insufficient. Students' relationships with mathematics are shaped by classroom discourses, forms of participation, teacher expectations, family and community experiences, linguistic resources, and school culture (Atweh et al., 2001; Gates, 2020; Lerman, 2006).

From this perspective, the mathematics classroom is not an empty space in which knowledge is transmitted in a ready-made form; rather, it is a cultural space in which norms are constructed regarding what counts as a valid solution, a reasonable explanation, acceptable participation, or a "successful student" (Cotton & Hardy, 2004; Morgan, 2014; Valero, 2004). The ways in which these norms are established are closely related to how the linguistic, cultural, and social resources that students bring into the classroom are recognized, utilized, or constrained within the teaching process. Therefore, the resources derived from students' families, language, social class, ethnic backgrounds, local cultures, and everyday experiences should not be viewed as secondary elements to be excluded from mathematics learning; rather, they should be considered direct components of mathematical meaning-making (de Abreu, 2020; Orey & Rosa, 2010). From this standpoint, sociocultural mathematics education approaches students not as passive subjects lacking mathematical knowledge, but as individuals who participate in mathematical discourse through the cultural resources, experiences, and meaning systems they bring into the classroom. Such participation does not merely imply adapting to an existing discourse; it also involves transforming that discourse and contributing to the reconstruction of the classroom's shared learning culture (Atweh et al., 2001; Darragh, 2016; Kieran et al., 2002; Valero, 2004).

On the other hand, the strengthening of the sociocultural perspective in mathematics education has gradually led to a more explicit sociopolitical orientation. Gutiérrez describes this shift as a "sociopolitical turn" that emerged after the establishment of the sociocultural perspective in the field, placing greater emphasis on issues of identity, representation, power, and inequality (Gutiérrez, 2013). This turn, influenced by critical theory, critical race theory, and post-structuralist approaches, has focused on examining how inequalities related to race, class, language, citizenship status, and social positioning are produced within mathematics education (Atweh et al., 2001; Gutiérrez, 2013; Louie & Zhan, 2022; Martin, 2019). The growing emphasis on mathematics teaching for social justice, identity-oriented mathematics education, and sociopolitical transformation in a globalized world reflects this orientation (Atweh et al., 2001; Gutiérrez, 2013; Gutstein, 2003; Morgan, 2014; Xenofontos et al., 2021). Thus, sociocultural mathematics education extends beyond being merely a framework for explaining students' meaning-making processes and instead opens discussion on the social conditions under which these processes become possible or constrained.

At this point, socio-critical and sociocultural approaches may be considered complementary yet distinct theoretical orientations. The sociocultural approach makes visible how mathematical meaning is constructed through language, discourse, interaction, and participation practices (Atweh et al., 2001; Lerman, 2001, 2006). The socio-critical approach, however, problematizes whose interests these practices serve, which students gain mathematical authority, which forms of knowledge are excluded, and how these processes relate to social justice (Jorgensen et al., 2014; le Roux & Swanson, 2021; Morgan, 2014). In other words, while the sociocultural perspective explains classroom meaning-making and participation processes, the socio-critical perspective examines how these processes are related to power, inequality, and democracy. When considered together, mathematics education can be understood as a social practice in which identities are constructed, participation is regulated, categories of success and failure are produced, and students' cultural and social positions, life worlds, and encounters with social problems are incorporated into mathematical meaning-making processes alongside issues of power, inequality, and justice.

This combined perspective demonstrates the need for educational contexts in mathematics education that can simultaneously engage with students' life worlds, cultural resources, and encounters with social problems. The integration of socio-scientific issues into mathematics education provides a particularly strong starting point in this regard. SSI are issues that can connect with students' everyday lives, values, cultural experiences, and social positions, while also involving dimensions of ethics, justice, risk, uncertainty, evidence, and public decision-making (Simonneaux, 2007; Zeidler et al., 2005; Zeidler & Sadler, 2008). For this reason, SSI bring together within the same context both the meaning-making, identity, and participation processes emphasized by the sociocultural approach and the questions of power, inequality, democracy, and social responsibility highlighted by the socio-critical approach.

What Are Socioscientific Issues? Their Characteristics and Distinctive Features

Socioscientific issues (SSI) are controversial issues directly related to scientific knowledge while simultaneously involving ethical, political, economic, environmental, and social dimensions. Cloning, stem cell research, genetically modified foods, global climate change, land-use decisions, the release of pollutants into the environment, nuclear/radiation debates, alternative energy, chemical risks, and health and consumption choices are among the prominent examples of SSI discussed in the literature (Högström et al., 2025; Sadler & Zeidler, 2005). These issues are addressed as public dilemmas in which competing perspectives are developed by different stakeholders and which produce consequences in biology, the environment, health, economics, technology, or other areas of social life (Simonneaux, 2007). What fundamentally distinguishes SSI is not merely that they involve scientific content; rather, they are also open-ended, ill-structured, multidimensional, and uncertainty-laden problems that cannot be reduced to a single correct solution (Simonneaux, 2007; Zeidler & Sadler, 2008). For this reason, SSI differ from purely technical problems that can be solved solely through factual knowledge; instead, they simultaneously involve ethical tensions, moral reasoning, social values, and often conflicting interests (Zeidler et al., 2005; Zeidler & Sadler, 2008).

This multidimensional structure of SSI requires systematic examination of which dimensions students consider during decision-making processes. One framework developed for this purpose is the SEE-STEP model (Eş & Öztürk, 2021). Developed based on Chang Rundgren and Rundgren's (2010) SEE-SEP framework, this model treats technology as a separate category and enables the analysis of SSI-related decisions through dimensions of sociology/culture, environment, economy, science, technology, ethics/morality, and policy. In this respect, SEE-STEP serves as an analytical framework that makes visible the dimensions through which students think about issues such as nuclear energy, genetically modified organisms (GMOs), and global warming (Eş & Öztürk, 2021).

At this point, it should be noted that not every controversial or ill-defined problem can be considered an SSI. What specifically distinguishes SSI is that they bring together scientific knowledge, ethical evaluation, social responsibility, and public decision-making processes within the same problem context (Zeidler et al., 2005; Zeidler & Sadler, 2008). Such issues do not merely require individuals to understand scientific claims; they also require them to weigh evidence, evaluate different perspectives, manage uncertainty, and construct justified judgments at both personal and societal levels (Zeidler et al., 2005). Since the quality of citizens' decisions regarding public issues is of great importance in democratic societies, individuals' ability to justify their views on socioscientific issues and articulate these views with supporting reasons is considered a central competency (Zeidler & Keefer, 2003). In this regard, SSI can be viewed not only as a thematic component of contemporary science education but also as a distinctive pedagogical domain in which scientific literacy, ethical sensitivity, reflective judgment, and public responsibility are simultaneously examined (Sadler & Zeidler, 2004; Zeidler et al., 2005).

However, the multidimensional structure of SSI makes them pedagogically valuable not only for science education but also for mathematics education. Socioscientific issues are often intertwined with data, ratios, risk, probability, graphs, models, and comparative decision-making processes. Therefore, integrating SSI into mathematics education may contribute to students' ability not only to discuss social and ethical issues, but also to structure, justify, and critically examine these issues mathematically.

In this context, the relationship between SSI and mathematics education becomes particularly visible in the potential of SSI to move mathematical thinking beyond procedural operations, abstract reasoning, or arriving at correct answers. From a sociocultural perspective, SSI allow students to construct mathematical meaning through their own life worlds, experiences, values, and social contexts. Such issues are not distant or abstract topics for students; rather, they are closely connected to everyday life, public debates, and personal positioning (Zeidler et al., 2005; Zeidler & Sadler, 2008). Zeidler's theoretical framework for SSI also supports this interpretation. Drawing upon areas such as cognitive and moral development, emotional reasoning, character education, socio-moral discourse, and the nature of science, this framework situates SSI within a sociocultural perspective (Zeidler, 2014). The entry points proposed in the same study—namely the nature of science, cultural issues, discourse, and issue-based approaches—are also regarded as pedagogical opportunities that contribute to the development of scientific literacy from a functional perspective (Zeidler, 2014).

Within this framework, SSI contexts should not be viewed merely as learning environments in which students read data, compare ratios, or produce quantitative results. In these contexts, students interpret data through their own cultural and social experiences, compare their ideas with the perspectives of others, collectively construct rational thinking through classroom discourse, and engage in the development of justified decisions (Simonneaux, 2007; Zeidler & Sadler, 2008). In this way, within the context of mathematics education, SSI become an interactive learning space in which students construct meaning, engage in justification, encounter different perspectives, and learn to discuss public issues.

At this point, it is particularly important to emphasize that incorporating mathematics into SSI contexts does not mean marginalizing ethical, emotional, or experiential dimensions. On the contrary, when SSI are addressed solely within science education, discussions may sometimes remain at the level of ethical sensitivity, personal experience, intuitive judgment, or emotional reaction. Indeed, studies show that students' socioscientific decision-making processes are influenced by different reasoning patterns, particularly context-sensitive intuitive and emotional evaluations (Sadler & Zeidler, 2005; Zeidler et al., 2005). The aim of the approach proposed here is not to exclude these dimensions, but rather to connect them with a more explicit, examinable, and justifiable decision-making framework. From this perspective, mathematics emerges not merely as a tool for calculation, but as a form of reasoning that enables the comparison of claims, the weighing of evidence, the visualization of relationships among variables, the clarification of assumptions, and the discussion of decisions in more well-grounded ways.

In this regard, SSI can be considered a powerful pedagogical opportunity capable of bringing together sociocultural meaning-making processes and socio-critical inquiry and citizenship dimensions within mathematics education. However, how this opportunity can be realized in mathematics classrooms should be discussed not only at a theoretical level, but also through classroom interactions and learning processes. The central argument emphasized in this chapter is that integrating socioscientific issues into mathematics education does not simply mean selecting interesting real-life topics. Rather, the main issue is making visible the data, quantitative representations, comparisons, assumptions, and justification processes through which social decisions are constructed. In this respect, addressing SSI together with mathematical tools and quantitative reasoning processes may contribute not only to students' discussion of ethical and social issues, but also to their ability to structure, question, and justify these issues mathematically. The classroom dialogues and implementation vignettes presented in the following sections of the chapter are intended to illustrate how this theoretical potential may become concretized in instructional settings. Therefore, the integration of SSI into mathematical modelling should aim not merely to repeat ethical and social discussions already present in science education, but to make visible the mathematical foundations through which these discussions are constructed and critically examined.

Addressing socioscientific issues together with mathematical tools and quantitative reasoning processes may also provide science educators with a more functional interdisciplinary approach that connects ethical and social discussions with more well-grounded decision-making frameworks. At the same time, the mathematics education literature emphasizes that mathematical models produce consequences for social life and that citizens' critical participation in these processes is essential for democracy (Gibbs & Park, 2022). This emphasis opens a productive area of discussion for mathematics educators regarding how social decisions are constructed through quantitative representations, indicators, calculations, and modelling processes, and how these processes can be critically examined. In this respect, the proposed framework moves the relationship between science and mathematics education beyond mere content integration and instead reconceptualizes it as a shared space of reasoning in which social issues can be discussed in more comparable, justifiable, and public ways. The relationships discussed in this chapter are summarized in the following figure as a theoretical and pedagogical framework for the integration of SSI into mathematics education.

Figure 1. Theoretical and pedagogical framework for the integration of socioscientific issues into mathematics education

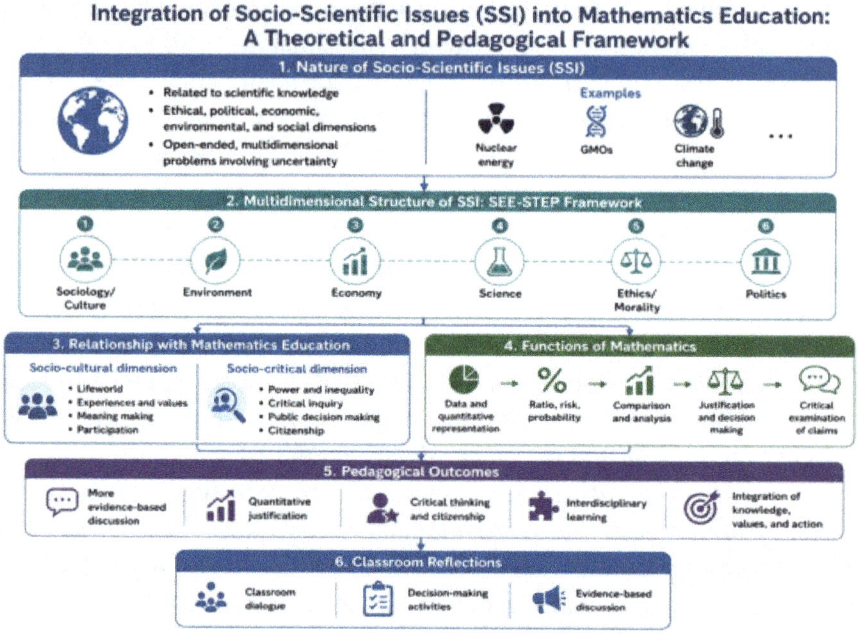

The framework illustrates that the integration of socioscientific issues into mathematics education cannot be reduced to a single dimension; rather, it should be understood through the interrelated relationships among sociocultural meaning-making, socio-critical inquiry, mathematical reasoning, and classroom practices.

Mathematical Modelling from Socio-Critical and Socio-Cultural Perspectives

Mathematical modelling is generally defined as a bidirectional translation process established between the real world and the mathematical world. In this process, a real-life situation serves as the starting point; the situation is simplified and structured into a mathematical model, mathematical operations are carried out on the model, and the resulting outcomes are then transferred back to the real context for interpretation and validation (Blum & Ferri, 2009; Ferri, 2006). However, modelling should not be regarded merely as an activity of applying mathematical knowledge to real-life problems. It is also considered a learning domain that develops in close interaction with teaching and learning processes, establishes its own didactical framework, and enables students to recognize the individual, social, and cultural functions of mathematics (Kaiser et al., 2006; Kaiser & Sriraman, 2006).

The modelling literature includes various approaches characterized by different aims, emphases, and epistemological orientations. Kaiser and Sriraman (2006) classify approaches to mathematical modelling according to their central purposes and theoretical backgrounds, distinguishing among realistic/applied, contextual, educational, socio-critical, epistemological, and cognitive modelling perspectives. In more recent studies, Kaiser (2020) argues that more process-oriented and multidimensional approaches to modelling have gained prominence through the inclusion of metacognitive dimensions as well as socio-critical and sociocultural issues within modelling processes. Within this framework, the development of modelling competencies is associated not only with mathematical achievement, but also with students' capacities to interpret, question, and justify real-life situations, particularly when mathematics education aims to support responsible citizenship (Kaiser, 2020).

Socio-critical modelling considers a critical examination of mathematical models and their consequences, the modelling of social, political, and economic problems, the empowerment of students, and the development of reflective discourse as central concerns (Barbosa, 2006; Blomhøj, 2009; Kaiser & Sriraman, 2006). Barbosa's conceptualization of "modelling as critique" also demonstrates that modelling can be used in school mathematics to promote conscious critique and reflection (Barbosa, 2006). In this context, reflective discussions are indispensable components of socio-critical modelling processes (Barbosa, 2006; Blomhøj, 2009; Kaiser, 2006; Kaiser & Sriraman, 2006).

According to Rosa and Orey, mathematical modelling based on a sociocultural perspective brings local practices and academic mathematics into dialogue and acknowledges that students' identities and cultural positions shape their ways of mathematical thinking (Rosa & Orey, 2025). In this context, mathematical modelling is not merely a process of carrying out mathematical procedures; rather, it is a learning space in which students construct meaning within cultural contexts and relate their own experiences to mathematical thinking. Therefore, SSI contexts are important in mathematical modelling not only in terms of identifying variables, using data, and producing model outcomes, but also in terms of enabling students to discuss different perspectives, negotiate uncertainties, and justify their decisions through social interaction. From the perspective of mathematics education, this is also consistent with sociocultural studies showing that students' participation in school mathematics is related to the mathematical practices they encounter across home, school, and community contexts (de Abreu, 2020; Herzig & Steinthorsdottir, 2020). Taken together, these two approaches reposition mathematical modelling as a socially responsible and critical learning process that is also connected to students' experiences. The following schematic representation summarizes the position of

SSI contexts at the intersection of socio-critical and sociocultural perspectives on mathematical modelling:

Figure 2. The position of SSI contexts at the intersection of socio-critical and sociocultural mathematical modelling perspectives

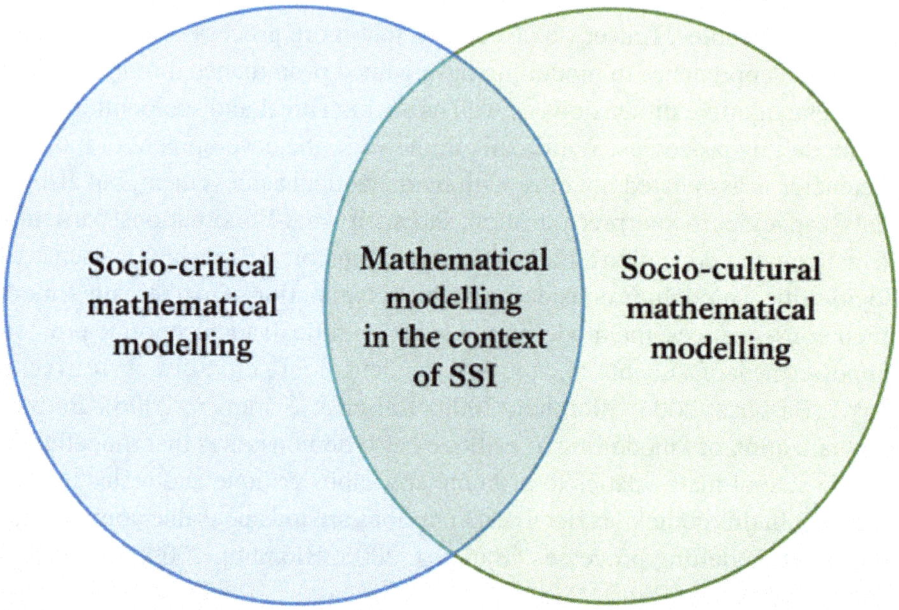

The modelling cycle follows flexible pathways that include understanding the problem, simplifying, mathematizing, carrying out mathematical work, interpreting, validating, and presenting the model, as well as revising the model when necessary (Blum & Leiss, 2007). Therefore, modelling should be understood not as the production of a single model, but as a dynamic activity that involves meaning-making, assumption development, the structuring of quantitative relationships, the evaluation of results, and, when necessary, reconsideration. The mathematical modelling cycle provides an important pedagogical structure that helps make this complex process more observable and discussable for students. In this respect, mathematical modelling makes the uncertain, multidimensional, and value-laden nature of SSI more manageable in classroom settings, while also allowing students to move among different solution pathways and re-justify their decisions.

This is because the nature of mathematical modelling involves working with the complexity embedded in real-life problems (Doerr et al., 2017). Such complexity requires the use of modelling-specific stages, such as understanding the situation, identifying variables, establishing relationships, comparing quantitative data, interpreting results, and revising the model when necessary (Anhalt & Cortez, 2016; Blomhøj, 2009; Kaiser, 2020). However, these stages do not automatically resolve the fragmented and multidimensional structure of SSI; rather, they make it possible to address this structure in ways that are observable, discussable, and justifiable in the classroom. For this reason, the modelling cycle should not be viewed as a closed and fixed scheme, but as a flexible process that allows returns, shifts, and movement among alternative pathways (Ferri, 2006, 2007; Doerr et al., 2017). This flexible structure has the potential to transform the uncertainty and multidimensionality inherent in SSI into a pedagogically manageable, discussable, and justifiable learning process.

In conclusion, the integration of socioscientific issues into mathematical modelling offers an interdisciplinary and bidirectional contribution. On the one hand, it enables problems directly related to students' own life contexts, cultural experiences, and social realities to be brought into mathematical modelling processes, thereby opening a new and more authentic domain of realistic contexts for mathematical modelling and more strongly reflecting the connection between modelling and real life. On the other hand, it contributes to supporting complex socioscientific issues discussed in science education with mathematical reasoning, quantitative justification, and data-based inquiry processes, thereby enabling the development of more rational, comparable, and justifiable decision-making processes. Addressing SSI in classroom settings should not be limited to students' intuitive, emotional, or ethical responses; rather, it should also enable these responses to be connected with evidence, data, variables, assumptions, and justification processes.

Rationales and Importance of Integrating SSI into Mathematical Modelling

One of the primary reasons necessitating the integration of socioscientific issues into mathematical modelling is the complexity inherent in these issues. Socioscientific issues are open-ended, ill-defined, debatable, interdisciplinary, and value-laden problems; therefore, they are difficult to reduce to a single correct solution and often require judgment-based decisions among alternatives that involve limited advantages and disadvantages (Abd-El-Khalick, 2003; Sadler & Zeidler, 2005; Simonneaux, 2007). Negotiating such problems activates both the cognitive and affective dimensions of everyday reasoning; personal experiences, emotional and social evaluations, moral priorities, and perceptions of the complexity of the issue may all influence decision-making processes (Sadler & Zeidler, 2005; Zeidler et al., 2005). At this point, the pedagogical structure provided by the mathematical modelling cycle becomes particularly important.

A second important rationale for integrating SSI into mathematical modelling is the possibility of bringing together rational data use and critical thinking within the same learning process. The SSI literature demonstrates that students' socioscientific decision-making processes are influenced not only by analytical reasoning but also by intuitive and emotional forms of reasoning (Sadler & Zeidler, 2005; Zeidler et al., 2005). At the same time, the development of functional scientific literacy requires the support of intellectual habits such as skepticism, open-mindedness, evaluation of evidence, data-based inquiry, and acceptance of uncertainty (Sadler & Zeidler, 2004; Zeidler et al., 2005). Addressing SSI through mathematical modelling does not position these dimensions in opposition to one another; rather, it can connect ethical and emotional reasoning with a more explicit, examinable, and publicly discussable framework of justification. In this process, mathematics becomes prominent through its capacity to compare data, examine alternative scenarios, weigh claims, make assumptions visible, and justify decisions (de Oliveira Vidal et al., 2025; Gibbs & Park, 2022).

Third, the integration of SSI into mathematical modelling is important because it makes visible the role of mathematics in social life. Mathematical models, indicators, calculations, and algorithmic systems are increasingly used in shaping decisions in areas such as health, economics, environment, technology, and public policy (Blomhøj, 2009; de Oliveira Vidal et al., 2025; Skovsmose, 2020b). This situation requires students not only to construct models, but also to discuss which variables the model prioritizes, which dimensions it leaves in the background, which criteria it legitimizes, and what kinds of social consequences it may produce. Such inquiry transforms modelling from being merely an activity of producing mathematical representations into a process that also opens the social circulation of models and their influence on decision-making processes to critical examination.

The fourth rationale concerns democratic participation and classroom dialogue. As Barbosa argues, reflecting on the role of mathematics in society should be considered a distinct opportunity for modelling; this opportunity should be informed by studies focusing on the sociocultural dimensions of mathematics and the critical nature of mathematical models (Barbosa, 2006). Similarly, increasing teachers' awareness of social issues strengthens the potential for connecting mathematics teaching with critical participation and democratic dialogue (Caldeira, 2009). Therefore, the integration of SSI into modelling is not only related to topic selection; it also has the potential to transform the nature of classroom interaction. In this process, students are positioned not as individuals who passively accept ready-made conclusions, but as subjects who ask questions, weigh alternatives, evaluate public consequences, and justify their decisions. In this way, modelling can become a learning space that makes visible the relationship among thinking, discussion, and action in relation to social issues.

Fifth, this integration has the potential to critically reframe students' relationships with mathematics, culture, and technology. This discussion reminds us that culture is not limited solely to local or traditional practices; rather, it is a multilayered structure shaped together with technology, emotions, social life, and ideological orientations. Indeed, drawing on White's cultural perspective, Bishop (2020) argues that cultures require technological, emotional, social, and ideological components in order to develop. This emphasis demonstrates that culture in mathematics education should be addressed not only through students' past experiences, but also through their contemporary relationships with technology, social issues, and values. This necessity becomes even more evident in today's world, where scientific knowledge and technological developments increasingly shape everyday life, communities, and global decision-making processes. As scientific knowledge rapidly advances and new technologies emerge, it becomes increasingly important for individuals to understand the effects of these developments on themselves, their communities, and global life (Berkowitz & Simmons, 2003).

Taken together, these rationales suggest that the integration of SSI into mathematical modelling should not be viewed merely as an interdisciplinary expansion of content, but rather as a theoretical opening that allows the pedagogical purposes of mathematical modelling to be reconsidered through socio-critical and sociocultural perspectives. For this reason, the integration of SSI into mathematical modelling can be regarded as a powerful pedagogical opportunity for both socio-critical and sociocultural mathematics education.

Discussion Dimensions for SSI-Based Mathematical Modelling

The following section proposes four discussion dimensions in order to examine more systematically the classroom discussions emerging during SSI-based mathematical modelling processes. These dimensions were identified based on data obtained from classroom observations, intra- and inter-group interactions throughout the SSI-based mathematical modelling process, group reports submitted at the end of the process, the researcher's field notes, and analyses of these data sources. Throughout the process, attention was given to which variables pre-service teachers prioritized, how they constructed justifications, which social and ethical priorities they emphasized, and how they conducted negotiation processes. In addition, the dimensions were developed by considering the multidimensional nature of socioscientific issues, the cyclical and discussion-oriented nature of mathematical modelling, and the common intersections of socio-critical and socio-cultural perspectives in mathematics education. These dimensions are intended to provide researchers with an analytical framework for examining classroom discussions, teacher educators with a structure for organizing instructional processes, and teachers with a theoretical and pedagogical framework for planning, implementing, and evaluating SSI-based modelling activities. Figure 3 presents the discussion dimensions integrated into the SSI-based mathematical modelling process.

Figure 3. Discussion dimensions and their theoretical relationships in ssi-based mathematical modelling

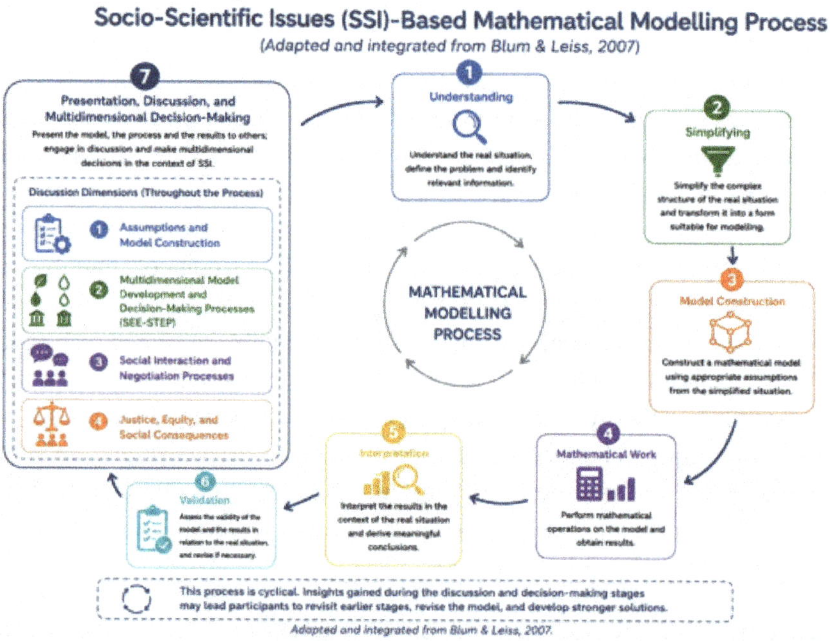

The framework was developed based on the mathematical modelling cycle proposed by Blum and Leiss (2007) and expanded in line with the discussions that emerged during classroom observations and implementation processes. Accordingly, the dimensions of "Assumptions and Model Construction," "Multidimensional Model Development and Decision-Making Processes (SEE-STEP)," "Social Interaction and Negotiation Processes," and "Justice, Equity, and Social Consequences" appear to structure classroom discussions arising at different stages of the modelling process. Although the first two dimensions are closely related, they focus on different aspects of the modelling process. The "Assumptions and Model Construction" dimension primarily focuses on how the model is initially structured. In this dimension, attention is given to how pre-service teachers define the problem, which variables they include in or exclude from the model, which forms of information they consider reliable, and which ethical or social priorities they place at the center of the modelling process. Accordingly, this dimension makes visible the simplification, framing, and structuring decisions that emerge during the early stages of modelling. In contrast, the "Multidimensional Model Development and Decision-Making Processes (SEE-STEP)" dimension focuses on how the constructed model is developed through different forms of justification and multidimensional evaluations. The primary concern here is how pre-service teachers establish relationships among scientific, ethical, economic, environmental, political, technological, and socio-cultural dimensions and how they incorporate these dimensions into their decision-making processes. Therefore, while the first dimension focuses more on the foundational logic and assumptive framework of the model, the second dimension examines how the model is developed within multidimensional reasoning and decision-making processes. The framework also emphasizes that the mathematical modelling process is not linear; rather, it involves a cyclical structure open to back-and-forth transitions, revisions, and multidimensional discussions.

Assumptions and Model Construction

A substantial part of the classroom discussions emerging during SSI-based mathematical modelling processes revolves around how the problem is defined, which variables are taken into consideration, which data are regarded as reliable, and upon which assumptions the model is constructed. Since socioscientific issues are often open-ended, ill-structured, controversial, and capable of generating different social consequences, there is rarely a single predetermined correct approach regarding what counts as sufficient information, which risks should be prioritized, or which ethical boundaries should be considered (Zeidler et al., 2005; Sadler & Zeidler, 2005). Therefore, within the mathematical modelling process, simplifying, constraining, and making the complexity of real life representable becomes necessary to a certain extent.

In this context, assumptions should not be viewed merely as technical simplifications; rather, they may also be considered constructive choices that reveal which dimensions pre-service teachers regard as important, which social or ethical issues they prioritize, and which relationships they seek to make visible within the problem situation. Since mathematical models can be viewed not only as technical calculation tools but also as representational systems that make visible certain assumptions, priorities, and social acceptances (Skovsmose, 2005; Gibbs & Park, 2022), the assumptions and model construction dimension was treated as an analytical dimension for examining how pre-service teachers structure a socioscientific issue through particular variables, which elements they exclude from the model, and which social, ethical, or scientific priorities they place at the center of the modelling process. From this perspective, the dimension contributes to making more visible the relationship between mathematical structuring and social and ethical priorities in SSI-based mathematical modelling processes.

Multidimensional Model Development and Decision-Making Processes (SEE-STEP)

Decision-making processes related to socioscientific issues often require the simultaneous consideration of scientific knowledge together with ethical, social, economic, environmental, and political justifications (Zeidler et al., 2005; Zeidler & Sadler, 2008; Simonneaux, 2007). Therefore, in this study, the dimension of multidimensional model-based decision-making was used to examine through which scientific, technological, ethical, economic, environmental, political, and sociocultural variables preservice teachers constructed their mathematical models. In making these justification dimensions visible, the study drew upon the SEE-STEP framework developed by Eş and Öztürk (2021), which enables socioscientific issues to be examined multidimensionally through the dimensions of Sociology/Culture, Environment, Economy, Science, Technology, Ethics/Morality, and Policy.

Social Interaction and Negotiation Processes

In SSI-based mathematical modelling processes, social interaction can be considered not merely a natural outcome of group work, but a fundamental process shaping how the problem is interpreted, which variables are regarded as important, which assumptions are accepted, and how model outcomes are interpreted. Sociocultural perspectives emphasize that mathematical meaning is produced through language, discourse, interaction, and participation practices (Lerman, 2001, 2006; Sfard, 2008). Therefore, this dimension focuses not only on the mathematical structure of the models developed by preservice teachers in the context of SSI-based mathematical modelling problems, but also on how disagreements, persuasion processes, personal experiences, dominant forms of participation, and repositioning become visible within the negotiations and discussions emerging during the modelling process.

Justice, Equity, and Social Consequences

One of the major dimensions emerging in SSI-based mathematical modelling discussions concerns the kinds of consequences developed models may produce for different individuals or social groups. Socioscientific issues are evaluated not only in terms of scientific accuracy or technological feasibility, but also in relation to justice, equity, access rights, and social consequences (Zeidler et al., 2005).

Therefore, it becomes important not only whether mathematical models "work," but also for whom, under what conditions, and with what social costs they operate. In areas such as health, organ transplantation, biotechnology, and resource allocation, models make visible questions regarding the criteria according to which limited resources should be distributed, which individuals should be prioritized, and which understanding of justice should guide decision-making processes. In this context, it was observed that during both intra-group and inter-group discussions, preservice teachers intensely negotiated not only the outcomes of the models, but also whether these models could produce fair or equitable consequences for different individuals and groups.

CLASSROOM DIALOGUES AND VIGNETTES

Cloning Discussions (Vignette): Practical Manifestations of Discussion Dimensions in SSI-Based Mathematical Modelling

In this study, a total of eight groups, most of which consisted of 3–5 pre-service elementary mathematics teachers, were formed on a voluntary basis during the 2025–2026 fall semester. One of the groups preferred to work individually. Before the implementation process, pre-service teachers were introduced to various socioscientific issues (e.g., nuclear energy, abortion, cloning, etc.) and asked to determine which topic they wished to study. The purpose of this process was to identify a problem situation that pre-service teachers were most interested in and through which natural discussion environments could emerge. Since the participants overwhelmingly preferred the topic of cloning, the implementation process was conducted around this issue. Prior to the main implementation, a pilot activity was conducted on Genetically Modified Organisms (GMOs) within the SSI context based on the modelling cycle proposed by Blum and Leiss (2007); the main implementation was then structured around the topic of cloning using the same modelling cycle framework. Data collection tools included observations of intra- and inter-group classroom interactions, video recordings, group reports submitted at the end of the process, and the researcher's field notes. The collected data were analyzed through content- and process-oriented approaches, and the discussion dimensions proposed in this section were developed based on the findings obtained from these analyses.

Assumptions and Model Construction in Cloning Vignettes

During the problem-understanding stage, preservice teachers accessed a wide range of scientific, ethical, economic, legal, and social information related to cloning. However, when they realized that they could not directly incorporate all of this information into the model, the need for simplification and assumption-making became evident. For the preservice teachers, this stage emerged not merely as a step included in the task instructions, but as a necessary precondition for constructing a model, and it became an important dimension of intra-group discussions. Statements voiced during group discussions, such as "How are we going to construct a model when there are so many variables?", "Wouldn't the formula become too complicated if we included everything?", and "I guess we need to assume some things as constant," made visible the tension preservice teachers experienced between real-life complexity and mathematical model construction. Therefore, questions such as "Which variables should we include in the model?", "Which information can we leave out?", "What should we assume to be constant in order to make the model work?", and "To what extent can we simplify the complexity of real life?" became central to intra-group discussions.

This situation was clearly observed in some assignments. For example, in the group examining the acceptability of therapeutic cloning in the context of organ and tissue production, the simplification stage was treated not only as a process of identifying variables, but also as a process of assigning relative weights to each variable. One group member's statement, "Scientific success is important, but I am not sure whether it can be considered acceptable if safety is low," shows that the variables were not merely listed, but that priority relationships among them were also discussed. Accordingly, variables such as scientific effectiveness, safety, ethical acceptability, legal appropriateness, cost, social benefit, and risk of rejection were selected, and the influence of these variables within the model to be constructed was represented through coefficients. In the same study, reproductive human cloning was excluded from the scope of the model in order to ensure its applicability, therapeutic cloning was taken as the basis, the relationships among variables were assumed to be partially linear, and the model was considered to be open to revision over time. In the same work, in order for the model to be applicable, reproductive human cloning was excluded from the scope, therapeutic cloning was taken as the basis, the relationships among variables were assumed to be linear to a certain extent, and the model was assumed to be updateable over time.

Similarly, in the model referred to as the Fair Clone-Organ Allocation Score (FCOAS), the participant established certain boundaries in order to make the model operational rather than attempting to incorporate the entirety of social and medical reality into the model. The participant's statement, "If I include everyone's economic condition, hospital resources, and family support, I cannot manage the model anymore; I first need to focus on the urgency of the patient's condition and the likelihood that the cloned organ will actually function," illustrates the necessary role of simplification in the model construction process. Within this framework, it was assumed that cloning costs would be covered by the state or health insurance, the patient's socioeconomic status would not be included in the model, the risk of rejection in therapeutic cloning would be very low, the contribution of urgency and expected life years would be linear, and each patient would experience only a single organ failure. These assumptions demonstrate that the model does not fully represent real life; rather, it was simplified in order to make a complex justice-related problem mathematically discussable. This situation also supports discussions emphasizing that, within modelling processes, the choices, priorities, and ways in which individual model constructors represent reality become highly influential (Doerr et al., 2017).

In the agricultural and animal cloning model, simplification was mainly carried out through genetic uniformity and epidemic risk. In this group, preservice teachers discussed that the identical or highly similar genetic structure of cloned animals might provide an advantage in terms of productivity, but that this similarity could pose a serious risk in the case of an epidemic disease. One preservice teacher's statement, "If they all have the same genetic structure [...] when a disease enters the herd, not just a few but almost all of them may be affected," laid the groundwork for constructing the model around the loss of genetic diversity and epidemic risk. Therefore, it was assumed that cloned animals could be affected at a high rate in the event of an epidemic disease, whereas in a natural herd, genetic diversity might prevent all individuals from being affected in the same way. In addition, by holding market conditions constant, the model was freed from macroeconomic complexity. This example shows that while preservice teachers reduced the number of variables, they also began to recognize which real-life dimensions the model left outside its scope.

The inclusion of dimensions such as economy, cost, and social benefit in the groups' models throughout the process may also be associated with perspectives suggesting that mathematical models serve to identify, make visible, and open social and economic inequalities to discussion (Blomhøj, 2009). This finding is also consistent with the aspect of critical mathematics education that addresses concepts such as social justice, mathematical competence, dialogue, and uncertainty together. Within this perspective, social justice is not treated as a fixed and unambiguous goal; rather, it is viewed as an open concept that must be continuously re-examined alongside processes of oppression, inequality, inclusion, and exclusion (de Abreu, 2020). From this perspective, the discussions carried out by pre-service teachers regarding which variables should be included in the model, which real-life conditions should be excluded, and which social priorities should be represented mathematically demonstrate that assumptions function not merely as technical simplifications, but also as critical choices with social consequences.

Multidimensional Model Development and Decision-Making Processes (SEE-STEP)

After the stages of understanding, simplification, and assumption generation, preservice teachers' work diverged in different directions throughout the process. While some groups addressed cloning in the context of therapeutic use, others developed models such as an environmental and biological risk model, an ethical-social acceptance model, kidney transplantation model, heart disease and stem cell treatment model, organ supply-demand balance model, animal/agricultural production model, and economic efficiency model. This diversification is important because it shows that the same socioscientific problem can be associated with different problem domains and represented through different mathematical models. It was also observed that multiple models were developed not only across groups but also within some groups.

When the models are evaluated through the SEE-STEP framework, it becomes clear that the models developed by preservice teachers were not limited only to scientific and technological dimensions; they also included ethical, social, economic, environmental, and political dimensions. However, each model did not represent the SEE-STEP components with the same scope and combination. Across the models, both the number of components included and the types of components emphasized varied. In models focused on health and organ/tissue production, the components of Science, Ethics/Morality, and Policy were particularly visible, whereas the Technology component became prominent in studies where laboratory processes, technical infrastructure, or success rates were explicitly incorporated into the model. In agricultural production and biological risk-oriented models, the components of Environment, Economy, Science, and Technology were more strongly represented. This differentiation shows that preservice teachers related the issue of cloning to different SEE-STEP domains depending on the problem context they selected. This general tendency becomes more visible when the variables included in the models developed by preservice teachers are examined. Below, the ways in which the variables used in some prominent models were related to SEE-STEP components are explained more concretely.

Due to having a family member waiting for an organ transplant, one participant preferred to work individually and developed a model called the Fair Clone-Organ Allocation Score (FCOAS). The primary aim of the model was to establish a decision-support structure for determining which patient should be prioritized in the allocation of a limited, costly, and time-intensive resource such as a cloned organ. In the model, the patient's medical urgency, expected quality-adjusted life years, cloning success rate, and laboratory cost/occupancy duration were evaluated together. Within this framework, the medical urgency score (U) was normalized as (u =

U/10); the expected quality-adjusted life years (Q) as (q = min(Q/50,1)); and the laboratory cost and occupancy duration (C) as (c = C/10). The cloning success rate was directly incorporated into the model as the coefficient (s).

The main function of the model was expressed as follows:

$$FCOAS = \frac{(0.60u + 0.40q) \cdot s}{c}$$

In this structure, the expression (0.60u + 0.40q) represents the balance between the patient's medical need and the expected benefit. The weight of 0.60 assigned to medical urgency positions saving the life of a patient in critical condition as the primary priority, whereas the weight of 0.40 assigned to expected quality-adjusted life years takes into account the long-term benefit of the organ. While the success coefficient (b) reflects the probability that the cloned organ can be successfully produced and adapted to the patient, placing the coefficient (m) in the denominator ensures that high costs and prolonged laboratory occupancy reduce the allocation score.

From the perspective of the SEE-STEP components, the variables of medical urgency, expected quality-adjusted life years, and cloning success rate are primarily associated with the Science dimension. The cloning success rate and laboratory occupancy time also make the Technology dimension visible, since they represent the production process and technical feasibility of the organ. Laboratory cost and occupancy time are related to the Economy component, while the question of to whom and according to which priority criteria limited organs should be allocated can be associated with the Ethics/Morality and Policy dimensions. In contrast, the Environment component was not explicitly included in the model; Sociology/Culture, on the other hand, was discussed indirectly rather than as a direct variable, particularly in relation to organ allocation, access to healthcare services, and social justice.

In this respect, the FCOAS model transformed the discussion of cloning from a general question of ethical acceptability into a more concrete problem of justice. In other words, the preservice teacher reformulated the question "Is cloning applicable?" into the question "To whom, according to which priority criteria, and based on which understanding of justice should the limited organs produced through cloning be allocated?" This transformation enabled the preservice teacher to mathematically structure a socioscientific problem based on personal experience and to establish a strong connection between ethical justification and mathematical modelling.

Among the models developed by the preservice teachers, the Therapeutic Cloning Acceptability Index Model (TCAI), which aimed to numerically evaluate the acceptability of therapeutic cloning in the context of organ and tissue production, emerged as particularly noteworthy. The reason this model is presented as an example in this section is that the decision regarding cloning was structured as a multidimensional decision problem in which variables such as scientific effectiveness, safety, ethical acceptance, legal appropriateness, social benefit, cost, and rejection risk were evaluated together.

In the model, scientific effectiveness (SE), safety (S), ethical acceptance (EA), legal appropriateness (H), and social benefit (P) were treated as variables increasing acceptability, whereas cost (C) and rejection/rejection risk (R) were considered variables decreasing acceptability. Based on these variables, the model was organized as follows to produce an interpretable acceptability index ranging between 0 and 100:

$$TCAI = 0.25SE + 0.20S + 0.15EA + 0.10L + 0.10SB + 0.10(100-C) + 0.10(100-R)$$

When examined in terms of the SEE-STEP components, the variables of scientific effectiveness (SE), safety (S), and rejection/rejection risk (R) in the model can be associated with the Science and Technology dimensions, since these variables are related to the medical success, patient safety, biotechnological feasibility, and reliability of organ/tissue production in therapeutic cloning. Ethical acceptance (EA) directly represents the Ethics/Morality dimension; legal appropriateness (L) corresponds to the Policy dimension; cost (C) represents the Economy dimension; and social benefit (SB) reflects the Sociology/Culture dimension. The Environment dimension, however, is not represented as an explicit variable in this model. Thus, the TCAI model does not reduce the acceptability of therapeutic cloning solely to scientific success; rather, it brings together different SEE-STEP components, including safety, ethical legitimacy, legal regulation, social benefit, and economic cost, within the same mathematical structure.

The model was interpreted through three temporal scenarios. In the 2024 scenario, representing present-day conditions, the acceptability score was calculated as approximately 65.3, suggesting that therapeutic cloning can currently be regarded not as a widespread clinical practice, but rather as a developing technology dependent on controlled and ethically supervised implementation. In the 2035 scenario, the increase of the score to approximately 77.3 was associated with improvements in safety, stronger ethical acceptance, advances in legal regulations, and reductions in cost. In the 2050 scenario, the score was calculated as approximately 84.5, indicating that therapeutic cloning may become more highly acceptable only when strong evidence of safety, ethical oversight, legal regulation, and equity of access are ensured. In this respect, the model enabled preservice teachers to engage in a decision-making process in which multiple variables were evaluated together, while also making visible which SEE-STEP components were represented strongly within the model and which component remained limited.

Another group, which initially opposed cloning, reconsidered its position after conducting an extensive literature review and developing a model based on this review. In this work, cloning was addressed in the context of animal production and the loss of genetic diversity. The preservice teachers discussed that cloning could increase production in the short term by enabling the reproduction of high-yield animals; however, they also noted that having an entire herd composed of animals with identical or highly similar genetic structures could create a serious biological risk. They reasoned that genetic diversity in natural herds serves a protective function against diseases, whereas in cloned herds composed of genetically similar individuals, this protective diversity may be reduced.Therefore, they discussed that if a highly contagious disease, such as foot-and-mouth disease or avian influenza, entered the herd, a virus affecting one individual could rapidly affect other genetically similar individuals as well, creating a "mass loss" scenario that could extend to the loss of the entire herd.

In this context, the group developed a model in which production efficiency increased as the cloning rate increased, while epidemic risk and biological vulnerability also increased due to genetic uniformity. The students interpreted the safe/optimal cloning rate as approximately 67% when the epidemic risk was 10%, and approximately 29% when the epidemic risk increased to 30%. The approximately 85% decrease in expected profit under the high-risk scenario in the case of full cloning demonstrated why genetic diversity was treated as a critical variable in the model. The group also proposed that cloned animals should not be housed as one large herd, but rather in small, spatially separated barn clusters. For example, they suggested that creating separate clusters of 100 animals could reduce the risk of a possible disease spreading to the entire population while helping to preserve the expected productivity gains from cloning.

When evaluated from the perspective of SEE-STEP, production efficiency, expected economic gain, and risk costs in the model can be associated with the Economy component. The variables of genetic uniformity, epidemic risk, and biological vulnerability make the Science and Environment dimensions visible, while cloning capacity and the biotechnological production process can be associated with the Technology component. The proposal to raise cloned animals in small and isolated clusters can be related to the Policy dimension in terms of biosecurity and sustainable production. In contrast, the Ethics/Morality and Sociology/Culture dimensions were not included as direct variables in the model; rather, they were indirectly opened to discussion in relation to food safety, the limits of human intervention, and societal attitudes toward biotechnological applications.

In this respect, the model demonstrates that cloning should be evaluated not only in terms of economic efficiency, but also in relation to biological vulnerability, loss of genetic diversity, epidemic risk, and long-term sustainability. It also shows that preservice teachers were able to relate different SEE-STEP components within the same problem situation through mathematical relationships and to structure a socioscientific issue in a multidimensional manner.

Another group, which initially opposed cloning, reconsidered its position after conducting an extensive literature review and developing a model based on this review. In this work, cloning was addressed in the context of animal production and the loss of genetic diversity. The preservice teachers discussed that cloning could increase production in the short term through the reproduction of high-yield animals; however, they also noted that having an entire herd composed of animals with identical or highly similar genetic structures could create a serious biological risk. In the model, the number of surviving clones at the end of time (t) was represented by S(t), while annual operational capacity (c), success rate (s), time (t), epidemic/genetic risk coefficient (r), and biological lifespan coefficient (l) were used as the main variables:

$$P(t) = (c \cdot s \cdot t) \times (1-r) \cdot l$$

The group argued that genetic diversity in natural herds serves a protective function against diseases, whereas in cloned herds composed largely of genetically identical individuals, this protective diversity may decrease. Therefore, they discussed that if a highly contagious disease, such as foot-and-mouth disease or avian influenza, entered the herd, a virus affecting one individual could rapidly affect other genetically similar individuals as well, potentially creating a "mass loss" scenario extending to the collapse of the entire herd.

In this context, the group developed a model in which production efficiency increased as the cloning rate increased, while epidemic risk and biological vulnerability also increased because of genetic uniformity. The students interpreted the safe/optimal cloning rate as approximately 67% when epidemic risk was 10%, and approximately 29% when epidemic risk increased to 30%. The approximately 85% decrease in expected profit under the high-risk scenario in the case of full cloning demonstrated why genetic diversity was treated as a critical variable in the model. The group also proposed that cloned animals should not be housed as one large herd, but rather in small and spatially separated barn clusters. For example, they suggested that creating separate clusters of 100 animals could reduce the risk of a possible disease spreading to the entire population while helping preserve the expected productivity gains from cloning.

When evaluated from the perspective of SEE-STEP, production efficiency, expected economic gain, and risk costs in this model can be associated with the Economy component, whereas genetic uniformity, epidemic risk, and biological vulnerability can be related to the Science and Environment components. Annual operational capacity and cloning success rate make the Technology dimension visible because they are associated with the biotechnological feasibility of cloning. Although the Ethics/Morality and Sociology/Culture dimensions were not included in the model as direct variables, they were indirectly opened to discussion in relation to food safety, the limits of human intervention, and societal attitudes toward biotechnological applications. In addition, the group's proposal, developed during the interpretation of the model results, to establish small and spatially isolated barn clusters can be considered not as a direct model variable, but rather as a discussion connected to the Policy dimension in the context of biosecurity and sustainable livestock policies. Thus, the model demonstrates that cloning should be evaluated not only in terms of economic efficiency, but also in relation to biological vulnerability, loss of genetic diversity, epidemic risk, and long-term sustainability.

On the other hand, classroom discussions also revealed that preservice teachers did not always make decisions solely based on the outcomes generated by the models they developed. Some groups argued, independently of their models, that cloning should be used only in limited situations, such as the preservation of endangered species. In contrast, other groups maintained that cloning should never be used under any circumstances, arguing that extinction processes in nature should be regarded as part of ecological balance. This view generated different reactions in the classroom. For example, one preservice teacher posed the counter-question, "Then if dinosaurs had not gone extinct, would we have to live together with them today?" thereby opening a discussion on whether natural extinction should always be viewed negatively. Another preservice teacher argued that bringing extinct species back could create new imbalances within ecosystems and that the aim of "protecting species" alone could not be considered a sufficient justification. These discussions indicate that preservice teachers treated mathematical models not as tools producing final decisions automatically, but as supportive discussion tools that needed to be evaluated together with ethical, environmental, and social justifications. In other words, the models produced by the groups functioned less as mechanisms that automatically determined decisions and more as thinking spaces in which different justifications became visible and open to discussion.

Another notable aspect of this process was that the modelling studies did not proceed in a linear structure following sequential steps. At times, preservice teachers returned to the understanding stage, reconsidered the variables they had identified, revised their assumptions after constructing the model, or felt the need to modify the structure of the model after interpreting the mathematical results. Therefore, the process displayed a cyclical structure involving returns, shifts, and revisions rather than a strictly linear progression from beginning to end. This non-linear progression demonstrates that preservice teachers were not merely producing formulas during the modelling process; they were also continuously re-evaluating and discussing the variables, assumptions, coefficients, and decision criteria they had selected. For example, some groups initially prioritized ethical acceptance but later felt the need to incorporate scientific success or cost variables into the model, while others reconsidered the effects of the coefficients they used on the final decision after interpreting the model outcomes. This situation indicates that mathematical modelling is not a closed, one-way, and linear procedure; rather, it is a dynamic thinking process requiring constant questioning, revisiting, restructuring, and justification. This characteristic became especially visible in socioscientific issue-based mathematical modelling problems. Indeed, the modelling literature also emphasizes that the modelling process follows a cyclical structure moving from real-life situations to mathematical models, and from solutions to interpretation and validation, while also involving returns to previous stages whenever necessary (Blum & Leiss, 2005; Doerr, 2007; Kaiser & Sriraman, 2006).

One of the most important aspects of this vignette from the perspective of mathematics education is that mathematical modelling functioned not as a tool reducing a socioscientific discussion to a single conclusion, but rather as a process deepening discussion and strengthening justification. At the beginning, preservice teachers tended to express more intuitive, emotional, or popular culture-based opinions about cloning. However, throughout the process, they encountered mathematical modelling practices such as selecting variables, constructing assumptions, determining coefficients, formulating equations, testing scenarios, making group decisions, and discussing the limitations of the model. This process also made visible the interdisciplinary nature of mathematical modelling intersecting with science, ethics, society, economics, and law. The science education literature likewise emphasizes that understanding the relationship between science, mathematics, and technology, as well as understanding the nature of science itself, occupies an important place in students' education (National Research Council, 1996). Thus, the issue of cloning evolved beyond being merely a biotechnological matter and became a cyclical and critical learning space in which science and mathematics were addressed together with ethical and social justification processes.

Social Interaction and Negotiation Processes

The group interactions and modelling studies conducted within the context of cloning demonstrated that preservice teachers' initially rigid judgments gradually became more negotiable and context-sensitive throughout the process. At the beginning, cloning was generally interpreted as "copying humans," "intervening in natural life," or "crossing ethical boundaries." However, following group interactions and literature reviews, preservice teachers began distinguishing between different types and purposes of cloning. During this process, some participants maintained their initial negative attitudes, whereas others developed more conditional evaluations in contexts such as organ and tissue production, stem cell therapies, or the preservation of endangered species.

For example, during one group discussion, a preservice teacher stated: "We know what happened to Dolly the sheep. Yes, it was a successful cloning, but she did not live long. The population is already increasing rapidly, so why do we even need cloning? Besides, I'm not sure I could tolerate another 'me'..." This remark caused laughter within the group. However, later in the process, the same participant reconsidered this view while discussing a model focused on reintroducing endangered species into populations through cloning. In this model, the group incorporated variables such as cloning capacity, success rate, time, biological lifespan, and epidemic/genetic risk, arguing that cloning should not merely be viewed as "copying a living being," but could also be discussed as a limited intervention capable of supporting species survival under certain conditions. During the presentation of the model, the same preservice teacher softened the initial opposition and stated: "I am still against human cloning, but if the issue is not creating another human being but saving a species on the verge of extinction, we cannot evaluate them in the same way."

In response, another group member argued: "But isn't the extinction of some species also part of natural balance? If we try to bring back every extinct species, we may disrupt the ecosystem." This comment emphasized that the model should not only be interpreted in terms of population increase, but also from the perspective of ecological balance. Following this discussion, the first participant repositioned the argument by stating: "Then the model should not only consider how many living beings can be produced, but also whether releasing them back into nature could harm the ecosystem."

This discussion demonstrates that preservice teachers' decisions regarding cloning were not solely shaped by their initial ethical reactions or by the numerical outcomes generated by the model. Instead, participants evaluated model results together with broader considerations such as species preservation, ecological balance, biological risks, and ethical responsibility. In this sense, social interaction functioned not merely as a process of sharing opinions, but as a negotiation space involving disagreement, persuasion, justification, and the repositioning of viewpoints.

On the other hand, one important finding of this study is that social interaction processes did not always result in consensus. In some cases, preservice teachers' personal experiences, ethical sensitivities, or dominant viewpoints shaped the direction of intra-group negotiation. In particular, it was observed that in some groups, preservice teachers with more dominant personalities had a stronger influence on variable selection, model type, or the interpretation of model outcomes. At the same time, some group members challenged dominant views and developed alternative variables or different interpretations. This indicates that social interaction in mathematical modelling is not merely a collaborative process, but also a dynamic negotiation space involving persuasion, opposition, repositioning, and intra-group power relations.

For example, in some group work, certain preservice teachers were observed to determine the direction of the model more strongly. During one group discussion, a preservice teacher stated, "I think we are extending the ethical part too much here. If cloning increases production and reduces hunger, we should think about that first... I mean, I think that is more important," attempting to shift the group's focus toward economic efficiency. In response, another group member argued, "But we cannot simply accept it as right just because production increases. If a disease occurs, the whole herd may disappear. We saw what happened during the pandemic..." emphasizing that biological risks should be incorporated more strongly into the model. As a result of this negotiation, the group revised the model that they had initially intended to construct solely around economic efficiency so that it also included epidemic risk and genetic diversity variables. This situation shows that intra-group social interactions functioned not merely as exchanges of opinions, but as negotiation processes that directly affected the mathematical structure of the model.

Another important point is that preservice teachers' personal experiences and individual life histories became clearly visible in the negotiation processes. This was particularly evident in the case of one preservice teacher whose father was waiting for a heart transplant. This participant approached cloning not merely as an academic problem situation, but as an issue directly connected to personal life. For this reason, the participant preferred to work individually rather than in a group and structured the model around patients waiting for heart transplantation. The following statement expressed by the participant during the process clearly reveals this orientation:

> *"My father is waiting for a heart transplant. That is why this topic did not feel like just an assignment to me. If I had worked with my friends, I would have had to keep the model within a shared topic, but I wanted to think about this problem around heart transplantation. If cloning is one day used in heart transplantation, it is very important to me for whom, under which equal conditions, according to which needs and characteristics, and how fairly it will be applied..."*

This statement shows how a socioscientific issue gained personal, ethical, and mathematical meaning for the preservice teacher. It also indicates that personal experiences directed preservice teachers toward deeper, more real-life-oriented, and solution-oriented engagement with the process. However, it was also observed that social interaction processes did not always produce shared consensus; in some cases, preservice teachers preferred to distance themselves to some extent from group negotiation in order to preserve their ethical sensitivities, lived experiences, and individual positions. In this respect, the SSI-based mathematical modelling process functioned not only as a space for producing shared ideas, but also as a social negotiation space in which individual experiences became visible, personal sensitivities were expressed, and, in some cases, individual positions were differentiated from group influence.

Justice, Equity, and Social Consequences

One of the most prominent dimensions emerging in SSI-based mathematical modelling discussions concerns the possible consequences that developed models may produce for different individuals or social groups. Socioscientific issues are evaluated not only in terms of scientific accuracy or technological feasibility, but also through questions of justice, equity, access rights, and broader social consequences (Sadler, 2004; Zeidler et al., 2005). Therefore, what becomes important is not merely whether mathematical models "work," but also for whom, under what conditions, and with what social costs they operate. Particularly in areas such as healthcare, organ transplantation, biotechnology, and resource allocation, the models developed make visible questions regarding according to which criteria limited resources should be distributed, which individuals should be prioritized, and which understanding of justice should guide decision-making processes. In this context, it was observed that during both intra-group and inter-group discussions, preservice teachers intensely negotiated not only the outcomes of the models, but also whether these models could produce fair or equitable consequences for different individuals.

In this respect, the work of the preservice teacher whose father was waiting for a heart transplant and who structured the model around the problem of organ allocation became one of the most significant examples highlighting issues of justice and equity in classroom discussions. Throughout the process, the participant occasionally shared personal experiences related to hospital procedures, waiting periods, and difficulties in accessing organ transplantation during interactions with the researcher or classroom presentations. These experiences generated emotional and ethical sensitivity within classroom discussions. The Fair Clone-Organ Allocation Score (FCOAS) model developed by the participant was designed as a decision-support structure intended to determine which patients should be prioritized for the allocation of a limited number of organs produced through cloning. In this model, cloned organs were not conceptualized as technologies that could be produced infinitely, but rather as costly and limited healthcare resources requiring time-intensive production processes. The central question of the model was formulated as: "To whom should the limited organs produced through cloning be allocated, and according to which priority criteria?" This question directly transformed cloning into a problem of justice.

Within the model, medical urgency, expected quality-adjusted life years, cloning success rate, and laboratory cost/occupancy duration were evaluated together. While these variables considered medical necessity and expected benefit on the one hand, they also incorporated system burdens such as limited laboratory capacity and financial cost on the other. One of the most striking justice-oriented features of the FCOAS model was the exclusion of the patient's economic status, social position, or profession from the model. Under the assumption of "complete equality and socioeconomic isolation," the participant assumed that cloning costs would be covered by the state or universal health insurance and attempted to construct the decision mechanism solely on medical data rather than variables such as wealth, profession, or social status. This choice may be interpreted as an equalizing attempt within the model against concerns that "wealthy and powerful individuals would gain easier access to the technology."

However, what became particularly significant during classroom discussions was not merely the technical structure of the model, but the ethical and social assumptions underlying the selected variables. In particular, the meaning of "priority" itself became a major topic of negotiation. Some preservice teachers argued that younger patients should be prioritized because of their potential for longer quality-adjusted life expectancy, whereas others stated that this approach could lead to age-based discrimination. One participant's question — "Aren't we automatically considering one person more valuable just because they are younger?... Are you thinking this way because we are young right now?..." — initiated a lengthy classroom debate. In contrast, another participant argued: "But if one person will survive only five more years with the same organ while another may live forty years, we also have to consider that," defending the efficient use of resources. Another preservice teacher added: "For example, a 30-year-old person may have children and a family, while there may also be an 18-year-old young person. How can one human life be considered more valuable than another simply because of age?" emphasizing that justice could not be established solely on the basis of expected lifespan. This model was discussed extensively throughout the class. As a result, preservice teachers began negotiating different understandings of justice, ethical priorities, and the ways in which human life is valued.

During these discussions, some preservice teachers also raised issues related to economic inequality. In particular, it was argued that because cloning technologies would initially be highly expensive, access to such technologies would likely be restricted to individuals with greater economic power. One participant's comment — "If this technology is going to be very expensive, then it simply means that only rich people will survive again" — led the model to be evaluated not only in medical terms, but also from the perspective of social inequality. Following this, some groups discussed how variables such as state-supported healthcare systems, priority rankings, and social access rights could be incorporated into the model more fairly. While some preservice teachers argued that economically disadvantaged individuals could systematically remain behind within the system, others suggested that distributing resources entirely equally might not always constitute the fairest solution. This finding is important because it demonstrates that preservice teachers did not interpret the concepts of equality and justice as synonymous; rather, they comparatively discussed different understandings of justice, ethical priorities, and social approaches to resource allocation. In this way, classroom negotiations made visible how issues such as access to healthcare, economic justice, and social equality could be integrated into mathematical modelling processes.

The organ supply-demand model shifted the justice discussion from the level of the individual patient to the level of the healthcare system. In this model, organ supply was represented through variables such as potential source, family/social consent, system efficiency, and cloning contribution. One particularly noteworthy result was that when the family/social consent rate was zero, total organ supply fell to zero regardless of how high the cloning contribution was. This result shows that technical capacity alone cannot create a fair and functioning system. When social trust, family consent, and institutional functioning are not ensured in relation to organ donation or the use of cloned organs, scientific progress cannot gain social traction. Therefore, the model opened the issue of equity to discussion not only through the question "How many organs can be produced?" but also through questions such as "How does society participate in this process, whose consent is taken into account, and how does the system distribute this resource?"

In the model developed in the context of stem cell treatment, justice became visible mainly through medical necessity, success potential, and data transparency. The model indicated that therapeutic cloning and stem cell applications may have high success potential in serious and widespread health problems such as kidney failure; however, it also emphasized that such applications should be conducted under strict ethical supervision and with data transparency. At this point, justice is not limited to access to treatment; which clinical outcomes are shared, whose long-term effects are monitored, and whether unsuccessful or risky results are made visible also become part of justice. In other words, data transparency is not only a matter of scientific reliability, but also a social responsibility in terms of patient protection and public accountability.

In the model focused on ethical, social, and religious acceptance, preservice teachers evaluated the general acceptability of cloning through the variables of social acceptance, ethical acceptance, and religious acceptance. This model did not focus on any specific area of application, such as organ transplantation, animal production, or the preservation of endangered species; rather, it opened discussion on the extent to which cloning could be accepted by society in general. In the model, the social acceptance rate was taken as 45%, the ethical acceptance rate as 35%, and the religious acceptance rate as 20%; these values were weighted, and the overall acceptance level was calculated as approximately 35.25%. This result led the group to interpret that cloning did not have a strong basis of acceptance at the general societal level.

During the interpretation of the model result, preservice teachers discussed in particular whether social and religious acceptance variables could be treated as fixed and universal values. One preservice teacher emphasized that acceptance rates may vary according to cultural context by stating, "Social acceptance or religious acceptance may not be the same in every country or every culture; some societies may evaluate cloning differently." In contrast, another preservice teacher expressed a more definite view: "I do not think any religion would accept cloning; especially the idea of copying a human being is already problematic from the start. That is probably why it is not officially permitted." In response to this view, another preservice teacher asked, "What if it is being done and we just do not know?" This question shifted the discussion beyond the level of religious acceptance toward issues of regulation, transparency, access to reliable information, and public control.

This discussion shows the difficulty of determining the variables of social acceptance, ethical acceptance, and religious acceptance used in the model as objective and fixed values. While preservice teachers attempted to numerically represent the general level of acceptance of cloning, they also recognized that such acceptance is closely related to culture, belief, level of knowledge, trust in official regulations, and the information disclosed to the public. From the perspective of justice and equity, an important aspect of the model is that it shows that cloning decisions should take into account not only scientific success or technical feasibility, but also the values, religious sensitivities, and ethical concerns of different social groups. At the same time, the model made visible that quantifying such qualitative variables involves subjective choices. In this respect, the model functioned not as a tool that renders social values entirely objective, but as a means of opening discussion on which values are included in the decision-making process and how these values influence the general acceptance of cloning.

In studies addressing the reintroduction of endangered species into populations through cloning, justice was considered not only in terms of present-day human benefit, but also from the perspectives of species preservation, ecological balance, and future generations. In these models, variables such as cloning capacity, success rate, budget, biological lifespan, and epidemic/genetic risk were evaluated together. The central social question emerging in this context was: "To which species should limited scientific and economic resources be allocated for preservation?" This question moved beyond a human-centered understanding of utility and brought issues of ecological justice and intergenerational responsibility into discussion. At the same time, the fact that some preservice teachers defended the view that "natural extinction may also be part of ecosystem balance" during classroom discussions demonstrated that cloning could not be regarded as an automatic solution to every conservation problem.

Overall, cloning is considered a powerful socioscientific issue that students may interpret as a moral problem within the context of genetic engineering (Sadler & Zeidler, 2004). For this reason, it is expected that preservice teachers would engage in discussions of justice and equity in their cloning models through issues such as organ allocation, economic access, social acceptance, and the distribution of risks. These discussions are closely related to the critical mathematics education perspective, which emphasizes that mathematical decision-making processes should be questioned in terms of their social consequences, equity, and inclusion–exclusion relations .In issues such as cloning, preservice teachers' decisions are not based solely on scientific knowledge or mathematical outcomes; personal experiences, cultural assumptions, religious sensitivities, social trust, family experiences, and perceptions of justice also become integrated into the modelling process. Indeed, Zeidler and Sadler (2008) emphasize that social and ethical issues allow students to attribute personal meaning to scientific concepts and make it possible for such issues to be addressed in closer relation to students' everyday lives. Within this framework, cloning models may be considered not only as products through which preservice teachers construct mathematical relationships, but also as representational tools through which they justify ideas concerning justice, equity, and social responsibility. The mathematical modelling process thus created an important learning environment in which issues such as justice, equity, access rights, social acceptance, and resource distribution — topics that often remain invisible in traditional mathematics classrooms — became open to discussion.

One of the most significant findings emerging under this dimension is that preservice teachers did not reduce justice to a single principle during the modelling process. In some models, justice was framed as "prioritizing individuals with the highest medical urgency"; in others, as "excluding economic status from the decision-making process"; in some, as "ensuring social trust and public acceptance"; in others, as "controlling risks without spreading them across the entire system"; and in yet others, as "protecting future generations and ecological balance." Therefore, the mathematical modelling studies conducted in the context of cloning demonstrated that justice is not a fixed or singular concept, but rather one that is continuously reshaped according to the selected variables, coefficients, assumptions, and problem contexts. This situation created an important learning space for preservice teachers regarding the idea that mathematical modelling cannot be considered independently from ethical and social consequences.

CONCLUSION AND IMPLICATIONS

In conclusion, this chapter demonstrates that the integration of socioscientific issues into mathematics education should not be regarded merely as a contextual expansion, but rather as an opportunity to reconceptualize mathematics as a more humanistic, critical, and public field of learning. The proposed discussion dimensions together with the SEE-STEP-based modeling and multidimensional decision-making processes create space for ethical evaluation, discussion of social consequences, negotiation practices, and the development of justified consensus across different perspectives. In this respect, the discussion dimensions developed in the chapter provide a holistic conceptual framework for examining more visibly and systematically the classroom interactions, decision-making processes, and modeling practices that emerge in SSI-based mathematical modeling processes. As Sadler and Zeidler (2005) argue, emotion-based reasoning in SSI contexts is not inferior to cognitive decision-making in terms of logical consistency and coherence; therefore, a strong mathematics education should not suppress emotions, but rather connect them with more explicit reasoning and justification processes. This argument suggests that SSI-based mathematical modeling provides a pedagogical ground that does not exclude students' emotional, ethical, and intuitive orientations, but instead brings them into more transparent and justifiable decision-making processes.

On the other hand, the models that emerged in this study should be evaluated, as emphasized by Barbosa (2006), not as neutral representations of an independent reality, but as more dynamic and social practices that arise in the process of investigating real social problems and that do not reduce modeling to the mere production of schematic representations. The more dynamic and humanized representation of modeling proposed by Rosa and Orey (2025) is also noteworthy in making visible the role of the individual model constructor, the transformative nature of pedagogy, and the action-oriented dimension that emerges when models are reflected back into the social context (Doerr et al., 2017). This situation demonstrates the importance of developing and evaluating the models emerging in SSI-based mathematical modeling more carefully and rigorously in terms of social awareness and action-oriented thinking processes. In addition, the use of tools such as dynamic mathematics software, simulations, data visualization tools, and artificial intelligence-supported technologies may contribute to sharing the models developed in classroom settings with educators, curriculum developers, educational policymakers, and broader social communities, thereby strengthening the transformative and social impact of modeling. From this perspective, the process may be regarded not only as a classroom learning activity, but also as a public learning practice that can contribute to developing more realistic, participatory, and justified decision-making environments regarding social problems.

In this SSI-based mathematical modeling study, discussing the assumptions students relied on, the risks they considered acceptable, the conceptions of justice they adopted, and the kinds of inequalities, responsibilities, and consequences their decisions might produce for different individuals or social groups constituted an important learning ground. Furthermore, the researcher's direct engagement with these processes contributed to making classroom discussions, decision-making processes, and students' ways of thinking more visible, thus providing an important pedagogical experience. At this point, Dewey emphasizes the central importance of inquiry in children's attitudes toward science, arguing that "ardent curiosity, fertile imagination, and love of experimental inquiry" are closely aligned with the attitude of the scientific mind; he also states that inquiry is not a one-time activity, but a continuous process in which outcomes become tools for further investigation (Dorier & Maass, 2020). In this respect, SSI-based mathematical modeling processes contribute not only to producing mathematical solutions, but also to creating more vibrant and participatory learning environments grounded in inquiry, discussion, rethinking, and social meaning-making processes. This emphasis strongly summarizes the pedagogical horizon of the chapter. The dimensions at the center of this framework—assumptions and model construction, multidimensional model development and decision-making, social interaction and negotiation, and justice, equity, and social consequences—constitute both the theoretical and pedagogical originality of SSI-based mathematical modeling.

REFERENCES

Abd-El-Khalick, F. (2003). Socioscientific issues in pre-college science classrooms. *The role of moral reasoning on socioscientific issues and discourse in science education*, 41–61. DOI: 10.1007/1-4020-4996-X_3

Anhalt, C. O., & Cortez, R. (2016). Developing understanding of mathematical modeling in secondary teacher preparation. *Journal of Mathematics Teacher Education, 19*(6), 523–545. DOI: 10.1007/s10857-015-9309-8

Apple, M. (2011). *Education and power*. Routledge., DOI: 10.4324/9780203143124

Atweh, B., Forgasz, H., & Nebres, B. (2001). *Sociocultural research on mathematics education: An international perspective*. Psychology Press., DOI: 10.4324/9781410600042

Barbosa, J. C. (2006). Mathematical modelling in classroom: A socio-critical and discursive perspective. *ZDM Mathematics Education, 38*(3), 293–301. DOI: 10.1007/BF02652812

Berkowitz, M. W., & Simmons, P. (2003). Integrating science education and character education. *The role of moral reasoning on socioscientific issues and discourse in science education*, 117–138. DOI: 10.1007/1-4020-4996-X_7

Bishop, A. (2020). Values in mathematics education. In *Encyclopedia of mathematics education* (pp. 893–896). Springer International Publishing., DOI: 10.1007/978-3-030-15789-0_160

Blomhøj, M. (2009). Different perspectives in research on the teaching and learning mathematical modelling. *Mathematical applications and modelling in the teaching and learning of mathematics*, 1.

Blum, W., & Ferri, R. (2009). Mathematical modelling: Can it be taught and learnt? 1, 45–58.

Blum, W., & Leiss, D. (2007). How do students and teachers deal with modelling problems. *Mathematical modelling (ICTMA 12): Education, engineering and economics*, 222-231. DOI: 10.1533/9780857099419.5.221

Caldeira, A. D. (2009). Mathematical modeling and environmental education. *Mathematical applications and modelling in the teaching and learning of mathematics*, 145.

Cotton, T., & Hardy, T. (2004). Problematising culture and discourse for mathematics education research. *Researching the socio-political dimensions of mathematics education*, 85–103.

Darragh, L. (2016). Identity research in mathematics education. *Educational Studies in Mathematics, 93*(1), 19–33. DOI: 10.1007/s10649-016-9696-5

de Abreu, G. (2020). Cultural diversity in mathematics education. In *Encyclopedia of mathematics education* (pp. 164–168). Springer International Publishing., DOI: 10.1007/978-3-030-15789-0_37

de Castro, R. S. (2024). Weaving the harmony of cultures into the infinite tapestry of mathematics. *Aracê, 6*(1), 219–230. DOI: 10.56238/arev6n1-013

de Oliveira Vidal, E., de Sousa, V. F., Alves, L. C. D. S. D., de Costa, D., Lopes, R. L. F., de Pontes, J. S., & de Luna, J. L. (2025). The socio-critical perspective of mathematical modeling: A proposal on prostate cancer. *Journal of Education and Learning, 14*(6), 57–72. DOI: 10.5539/jel.v14n6p57

Doerr, H. M. (2007). What knowledge do teachers need for teaching mathematics through applications and modelling? In *Modelling and applications in mathematics education: The 14th ICMI study* (pp. 69–78). Springer US., DOI: 10.1007/978-0-387-29822-1_5

Doerr, H. M., Ärlebäck, J. B., & Misfeldt, M. (2017). Representations of modelling in mathematics education. In *Mathematical modelling and applications: Crossing and researching boundaries in mathematics education* (pp. 71–81). Springer International Publishing., DOI: 10.1007/978-3-319-62968-1_6

Dorier, J. L., & Maass, K. (2020). Inquiry-based mathematics education. In *Encyclopedia of mathematics education* (pp. 384–388). Springer International Publishing. DOI: 10.1007/978-3-030-15789-0_176

Ernest, P., Sriraman, B., & Ernest, N. (2016). *Critical mathematics education: Theory, praxis, and reality*. IAP., DOI: 10.1108/978-1-68123-261-4

Eş, H., & Öztürk, N. (2021). An activity for transferring the multidimensional structure of SSI to middle school science courses: I discover myself in the decision-making process with SEE-STEP! *Research in Science Education, 51*(3), 889–910. DOI: 10.1007/s11165-019-09865-1

Eş, H., & Varol, V. (2019). Fen bilgisi öğretmenliği ve ilahiyat öğrencilerinin nükleer santral sosyo-bilimsel konusuyla ilgili informal argümanları [Informal arguments of pre-service science teachers and theology students regarding the socioscientific issue of nuclear power plants]. *Mersin Üniversitesi Eğitim Fakültesi Dergisi, 15*(2), 437–454. DOI: 10.17860/mersinefd.533013

Ferri, R. (2006). Theoretical and empirical differentiations of phases in the modelling process. *ZDM Mathematics Education, 38*(2), 86–95. DOI: 10.1007/BF02655883

Ferri, R. (2007). Modelling from a cognitive perspective: Individual modelling routes of pupils. In C. Haines, P. Galbraith, W. Blum, & S. Khan (Eds.), *Mathematical modelling. Education, engineering, and economics* (pp. 260–270). Horwood Publishing.DOI: 10.1533/9780857099419.5.260

Gates, P. (2020). Equity and access in mathematics education. In *Encyclopedia of mathematics education* (pp. 279–283). Springer International Publishing., DOI: 10.1007/978-3-030-15789-0_58

Gibbs, A. M. (2019). *Socio-critical mathematical modeling and the role of mathematics in society* (Doctoral dissertation, Florida Institute of Technology).

Gibbs, A. M., & Park, J. Y. (2022). Unboxing mathematics: Creating a culture of modeling as critic. *Educational Studies in Mathematics, 110*(1), 167–192. DOI: 10.1007/s10649-021-10119-z

Greer, B., & Mukhopadhyay, S. (2015). Honoring diversity in intercultural mathematics education. *Intercultural Education, 26*(4), 261–265. DOI: 10.1080/14675986.2015.1073876

Gutiérrez, R. (2013). The sociopolitical turn in mathematics education. *Journal for Research in Mathematics Education, 44*(1), 37–68. DOI: 10.5951/jresematheduc.44.1.0037

Gutstein, E. (2003). Teaching and learning mathematics for social justice in an urban, Latino school. *Journal for Research in Mathematics Education, 34*(1), 37–73. DOI: 10.2307/30034699

Herzig, A., & Steinthorsdottir, O. B. (2020). Cultural influences in mathematics education. In *Encyclopedia of Mathematics Education* (pp. 168–172). Springer International Publishing., DOI: 10.1007/978-3-030-15789-0_38

Högström, P., Gericke, N., Wallin, J., & Bergman, E. (2025). Teaching Socioscientific Issues: A Systematic Review. *Science & Education, 34*(5), 3079–3122. https://doi.org/. DOI: 10.1007/s11191-024-00542-y

Jorgensen, R., Gates, P., & Roper, V. (2014). Structural exclusion through school mathematics: Using Bourdieu to understand mathematics as a social practice. *Educational Studies in Mathematics, 87*(2), 221–239. DOI: 10.1007/s10649-013-9468-4

Kaiser, G. (2020). Mathematical modelling and applications in education. In *Encyclopedia of mathematics education* (pp. 553–561). Springer International Publishing., DOI: 10.1007/978-3-030-15789-0_101

Kaiser, G., Blomhøj, M., & Sriraman, B. (2006). Towards a didactical theory for mathematical modelling. *ZDM Mathematics Education, 38*(2), 82–85. DOI: 10.1007/BF02655882

Kaiser, G., & Sriraman, B. (2006). A global survey of international perspectives on modelling in mathematics education. *ZDM Mathematics Education, 38*(3), 302–310. DOI: 10.1007/BF02652813

Kieran, C., Forman, E., & Sfard, A. (2002). *Learning discourse: Discursive approaches to research in mathematics education.* DOI: 10.1007/0-306-48085-9

le Roux, K., & Swanson, D. (2021). Toward a reflexive mathematics education within local and global relations: Thinking from critical scholarship on mathematics education within the sociopolitical, global citizenship education and decoloniality. *Research in Mathematics Education, 23*(3), 323–337. DOI: 10.1080/14794802.2021.1993978

Lerman, S. (2001). Cultural, discursive psychology: A sociocultural approach to studying the teaching and learning of mathematics. *Educational Studies in Mathematics, 46*(1), 87–113. DOI: 10.1023/A:1014031004832

Lerman, S. (2006). Socio-cultural research in PME. In *Handbook of research on the psychology of mathematics education* (pp. 347–366). Brill., DOI: 10.1163/9789087901127_013

Louie, N., & Zhan, W. (2022). A socio-ecological framework for research in mathematics education. *Journal for Research in Mathematics Education, 53*(5), 365–371. Advance online publication. DOI: 10.5951/jresematheduc-2022-0003

Makramalla, M. (2022). Commentary: Mathematics education in a globally precarious reality–A simulated dialogue between two key players. *Research in Mathematics Education, 24*(2), 249–261. DOI: 10.1080/14794802.2022.2090423

Martin, D. (2019). Equity, inclusion, and antiblackness in mathematics education. *Race, Ethnicity and Education, 22*(4), 459–478. DOI: 10.1080/13613324.2019.1592833

Morgan, C. (2014). Social theory in mathematics education: Guest editorial. *Educational Studies in Mathematics, 87*(2), 123–128. DOI: 10.1007/s10649-014-9572-0

National Research Council. (1996). *National science education standards*. National Academic Press.

Orey, D. C., & Rosa, M. (2010). Ethnomodeling: A pedagogical action for uncovering ethnomathematical practices. *Journal of Mathematical Modelling and Application, 1*(3), 58–67.

Rosa, M., & Orey, D. (2025). *A didactic approach of sociocultural perspectives of mathematical modelling for classrooms: An ethnomodelling perspective*. REMATEC., DOI: 10.37084/REMATEC.1980-3141.2025.n54.e2025005.id772

Rosa, M., & Orey, D. C. (2015). Social-critical dimension of mathematical modelling. In G. A. Stillman, W. Blum, & M. S. Biembengut (Eds.), *Mathematical modelling in education research and practice* (pp. 385–395). Springer. https://doi.org/DOI: 10.1007/978-3-319-18272-8_32

Rosa, M., & Orey, D. C. (2021). An ethnomathematical perspective of STEM education in a glocalized world. *Bolema: Boletim de Educação Matemática, 35*(70), 840–876. https://doi.org/DOI: 10.1590/1980-4415v35n70a14

Sadler, T. D. (2004). Informal reasoning regarding socioscientific issues: A critical review of research. *Journal of Research in Science Teaching, 41*(5), 513–536. DOI: 10.1002/tea.20009

Sadler, T. D. (2009). Situated learning in science education: Socio-scientific issues as contexts for practice. *Studies in Science Education, 45*(1), 1–42. DOI: 10.1080/03057260802681839

Sadler, T. D., & Zeidler, D. L. (2004). The morality of socioscientific issues: Construal and resolution of genetic engineering dilemmas. *Science Education, 88*(1), 4–27. DOI: 10.1002/sce.10101

Sadler, T. D., & Zeidler, D. L. (2005). Patterns of informal reasoning in the context of socioscientific decision making. *Journal of Research in Science Teaching, 42*(1), 112–138. DOI: 10.1002/tea.20042

Sfard, A. (2008). *Thinking as communicating: Human development, the growth of discourses, and mathematizing*. Cambridge university press., DOI: 10.1017/CBO9780511499944

Simonneaux, L. (2007). Argumentation in socio-scientific contexts. In S. Erduran & M. P. JiménezAleixandre (Eds.), *Argumentation in science education: Perspectives from classroom-based research* (pp. 179–199). Springer. https://doi.org/10.1007/978-1-4020-6670-2_9.

Skovsmose, O. (2005). *Travelling through education: Uncertainty, mathematics, responsibility*. Brill., DOI: 10.1163/9789087903626

Skovsmose, O. (2013). *Towards a philosophy of critical mathematics education* (Vol. 15). Springer Science & Business Media., DOI: 10.1007/978-94-017-3556-8

Skovsmose, O. (2014). *Critique as uncertainty*. IAP., DOI: 10.1108/978-1-62396-755-0

Skovsmose, O. (2020a). Dialogic teaching and learning in mathematics education. In *Encyclopedia of Mathematics Education* (pp. 195–197). Springer International Publishing., DOI: 10.1007/978-3-030-15789-0_45

Skovsmose, O. (2020b). Mathematization as social process. In *Encyclopedia of Mathematics Education* (pp. 605–608). Springer International Publishing., DOI: 10.1007/978-3-030-15789-0_112

Skovsmose, O. (2022). Concerns of critical mathematics education and of ethnomathematics. *Revista Colombiana de Educación*, *86*(86), 365–382. DOI: 10.17227/rce.num86-13713

Solares-Rojas, A., Arellano-Aguilar, O., García González, M. M., López-Vargas, M. D. R., Coles, A., & Méndez Serrano, A. (2022). Mathematics education and social-environmental crises: An interdisciplinary proposal for didactic innovation with rural communities in Mexico. *Research in Mathematics Education*, *24*(2), 202–223. DOI: 10.1080/14794802.2022.2062781

Steffensen, L., & Kasari, G. (2023). Integrating societal issues with mathematical modelling in pre-service teacher education. *Education Sciences*, *13*(7), 721. DOI: 10.3390/educsci13070721

Stinson, D. W., & Bullock, E. C. (2012). Critical postmodern theory in mathematics education research: A praxis of uncertainty. *Educational Studies in Mathematics*, *80*(1), 41–55. DOI: 10.1007/s10649-012-9386-x

Valero, P. (2004). Socio-political perspectives on mathematics education. In Valero, P., & Zevenbergen, R. (Eds.), *Researching the socio-political dimensions of mathematics education: Issues of power in theory and methodology* (pp. 5–23). Springer., DOI: 10.1007/1-4020-7914-1_2

Valero, P. (2020). Political perspectives in mathematics education. In Lerman, S. (Ed.), *Encyclopedia of mathematics education* (pp. 663–666). Springer., DOI: 10.1007/978-3-030-15789-0_126

Vithal, R. (2004). Methodological challenges for mathematics education research from a critical perspective. In Valero, P., & Zevenbergen, R. (Eds.), *Researching the socio-political dimensions of mathematics education: Issues of power in theory and methodology* (pp. 227–248). Springer., DOI: 10.1007/1-4020-7914-1_18

Whitcombe, A. (1988). Mathematics creativity, imagination, beauty. *Mathematics in School*, *17*(2), 13–15.

Xenofontos, C., Fraser, S., Priestley, A., & Priestley, M. (2021). Mathematics teachers and social justice: A systematic review of empirical studies. *Oxford Review of Education, 47*(2), 135–151. DOI: 10.1080/03054985.2020.1807314

Zeidler, D. L. (2003). The role of moral reasoning on socioscientific issues and discourse in science education. In Zeidler, D. L. (Ed.), *The role of moral reasoning on socioscientific issues and discourse in science education* (pp. 201–216). Kluwer Academic Publishers., DOI: 10.1007/1-4020-4996-X

Zeidler, D. L. (2014). Socioscientific issues as a curriculum emphasis: Theory, research, and practice. In Lederman, N. G., & Abell, S. K. (Eds.), *Handbook of research on science education* (Vol. 2, pp. 697–726). Routledge.

Zeidler, D. L., & Keefer, M. (2003). The role of moral reasoning and the status of socioscientific issues in science education: Philosophical, psychological and pedagogical considerations. In Zeidler, D. L. (Ed.), *The role of moral reasoning on socioscientific issues and discourse in science education* (pp. 7–38). Springer., DOI: 10.1007/1-4020-4996-X_2

Zeidler, D. L., & Nichols, B. H. (2009). Socioscientific issues: Theory and practice. *Journal of Elementary Science Education, 21*(2), 49–58. DOI: 10.1007/BF03173684

Zeidler, D. L., & Sadler, T. D. (2008). Social and ethical issues in science education: A prelude to action. *Science & Education, 17*(8–9), 799–803. DOI: 10.1007/s11191-007-9130-6

Zeidler, D. L., Sadler, T. D., Simmons, M. L., & Howes, E. V. (2005). Beyond STS: A research-based framework for socioscientific issues education. *Science Education, 89*(3), 357–377. DOI: 10.1002/sce.20048

KEY TERMS AND DEFINITIONS

Mathematical Modelling: A cyclical process of representing, interpreting, and analyzing real-world situations through mathematical structures such as variables, relationships, assumptions, and quantitative representations. The process typically includes understanding the problem, simplifying, mathematizing, interpreting, validating, and revising models.

SEE-STEP Framework: A multidimensional analytical framework used to examine socioscientific reasoning and decision-making through the dimensions of Sociology/Culture, Environment, Economy, Science, Technology, Ethics/Morality, and Policy.

Socio-Critical Mathematics Education: A perspective in mathematics education that examines mathematics in relation to democracy, power, inequality, justice, and social participation. It emphasizes the critical examination of how mathematical knowledge and models influence society and public decision-making.

Sociocultural Mathematics Education: An approach that views mathematics learning as a socially and culturally mediated process shaped through language, interaction, participation, classroom discourse, and students' lived experiences and cultural resources.

Socioscientific Issues (SSI): Open-ended, controversial, and multidimensional social problems that involve scientific knowledge together with ethical, social, political, environmental, and economic considerations. SSI require evidence-based reasoning, public decision-making, and the evaluation of multiple perspectives rather than a single correct answer.

SSI-Based Mathematical Modelling: An interdisciplinary modelling approach in which socioscientific issues are addressed through mathematical reasoning, quantitative analysis, and multidimensional decision-making processes. This approach integrates ethical, social, cultural, scientific, and mathematical considerations within modelling activities.

Chapter 3

Escape Rooms as Socio-Cultural Learning Environments in Primary School Mathematics:

Designing Collaborative and Digitally Mediated Practices

Emel Çilingir Altıner
http://orcid.org/0000-0002-8085-553X
Çukurova University, Turkey

Ahmet Oğuz Akçay
Eskişehir Osmangazi University, Turkey

ABSTRACT

This chapter examines how escape-room-based digital tasks function as socio-cultural learning environments supporting collaborative mathematical participation among fourth-grade students. Nine student groups worked collaboratively on Code.org tasks involving angles, direction, geometric shapes, and spatial reasoning. Data were collected through video recordings, field notes, and focus group interviews and analyzed using inductive thematic analysis. Findings revealed four interconnected dimensions of collaborative activity: participation structures, iterative experimentation, artefact-supported regulation, and planning-oriented reasoning. Despite using the digital platform for the first time, students successfully coordinated actions, predicted directional outcomes, and constructed geometric shapes collaborative-

DOI: 10.4018/979-8-2600-3485-9.ch003

ly. The chapter demonstrates how escape rooms can be intentionally designed to support spatial reasoning, shared dialogue, and participation-rich mathematics learning environments.

INTRODUCTION

Socio-cultural perspectives on mathematics education conceptualize learning as a socially mediated activity in which participation, discourse, and identity are constituted through interaction with others and with cultural tools (Bakhtin, 1981; Lave & Wenger, 1991; Sfard, 2008; Vygotsky, 1978; Wenger, 1998). From this standpoint, mathematics classrooms are not neutral containers for cognition but interactional and ideological settings that define what qualifies as valid mathematical activity, whose contributions are recognized, and how learners come to see themselves as competent mathematical actors (Cobb et al., 2009; Gutiérrez, 2008). Designing learning environments that intentionally broaden participation and sustain dialogic engagement is, therefore, central to equity-oriented mathematics education. In education, gamification and game-based learning approaches are increasingly being adopted to enhance student participation, motivation, and engagement in learning processes (Ratinho et al., 2023; Ruiz et al., 2024). In recent years, research on the use of escape rooms as an effective pedagogical tool has gained momentum, particularly in abstract and challenging disciplines such as mathematics and computer programming, where traditional teaching methods are insufficient (Ugo et al., 2025). However, while digital escape rooms have been widely examined from motivational and cognitive perspectives, their potential as socio-cultural learning environments that structure participation, discourse, and identity development in mathematics classrooms remains underexplored.

Socio-Cultural Participation and Identity in Mathematics Classrooms

Socio-cultural theories frame learning as participation in socially organized practices rather than as the accumulation of decontextualized knowledge (Lave & Wenger, 1991; Wenger, 1998). In mathematics education, this lens directs attention to how students enter shared activity, negotiate meanings, and become recognized as legitimate contributors to mathematical work (Cobb et al., 2009; Sfard, 2008). Participation, therefore, extends beyond task completion: it includes explaining, questioning, justifying, coordinating with peers, and responding to others' ideas in ways that sustain collective reasoning. From this perspective, classrooms operate as participation structures that distribute access to authority, visibility, and recognition.

These structures influence who speaks, which contributions are taken up, and what counts as mathematically valuable activity (Gutiérrez, 2008). When participation is reduced to brief teacher-led response patterns or procedural reproduction, opportunities for equitable engagement narrow. By contrast, intentionally designed learning environments can expand pathways for involvement and support broader forms of competence. Within such environments, students are not only participants but also positioned in particular ways through interaction. Positioning refers to how individuals are recognized, authorized, or constrained as contributors within classroom discourse (Cobb et al., 2009). For example, students may be positioned as explainers, questioners, verifiers, or passive observers depending on the participation structures established by instructional design. These positioning processes shape students' access to learning opportunities and influence how authority and responsibility are distributed within collaborative mathematical activity.

Learning mathematics can be understood as becoming fluent in specialized forms of communication through which meanings are jointly produced (Sfard, 2008). Such communication is inherently multimodal: students draw on spoken language, gesture, symbolic notation, visual representations, and shared manipulation of tools. A dialogic orientation further emphasizes that understanding develops through encounters among multiple voices, perspectives, and interpretations (Bakhtin, 1981). As students compare, challenge, and reconcile ideas, they refine both concepts and methods, strengthening their capacity for collective reasoning.

Identity development is closely intertwined with participation and positioning processes. Mathematical identity concerns how students view themselves and are positioned by others as capable participants in mathematical practices (Cobb et al., 2009; Wenger, 1998). Classrooms that distribute responsibility, validate diverse contributions, and support multiple solution pathways are more likely to foster positive identity trajectories. Recent research highlights the importance of identity-safe environments in which learners' linguistic, cultural, and interactional resources are treated as assets rather than deficits (Boaler, 2002; Gutiérrez, 2008). In such environments, students can participate through different communicative styles and problem-solving approaches while remaining accountable to shared mathematical goals. This supports not only conceptual growth but also the social and affective conditions necessary for sustained engagement. Digital technologies can reinforce these dynamics when used as mediational tools for joint activity. Platforms that require shared planning, iterative testing, and collective interpretation of feedback can promote coordinated participation and dialogic reasoning. From a socio-cultural standpoint, these tools are not neutral delivery systems; rather, they shape interactional possibilities and discourse patterns in consequential ways (Wertsch, 1998). When intentionally integrated into collaborative learning environments, digital tools can function as mediational spaces that support distributed responsibility and sustained mathematical interaction.

Considered collectively, socio-cultural perspectives suggest that equitable mathematics learning depends on the deliberate design of environments where participation, positioning, discourse, and identity are developed in tandem. By sharing authority and focusing on deep tasks, these settings empower students to see themselves as mathematically capable. This theoretical framing provides a foundation for examining instructional designs -such as escape rooms- that intentionally structure collective activity and support inclusive participation in mathematics learning.

Escape Rooms in Education: Pedagogical Foundations

Game-based learning and gamification have gained increasing attention in contemporary education as approaches that promote active participation, motivation, and collaborative problem solving (Vidergor, 2021). Among these approaches, educational escape rooms have emerged as structured, team-based learning environments in which participants work collaboratively to solve interconnected challenges within a limited time frame. These environments typically require players to decode clues, interpret information, and coordinate actions to achieve a shared goal, such as unlocking sequential tasks or completing a mission scenario (Ugo et al., 2025; Vidergor, 2021).

The pedagogical foundations of educational escape rooms draw upon a confluence of multiple learning theories. Flow theory posits that learners experience deep engagement when task difficulty is appropriately calibrated to their existing competencies, maintaining a balance between challenge and competence (Csikszentmihalyi, 1990; Vidergor, 2021). In escape room settings, this balance is achieved through progressive task sequencing, where early tasks establish familiarity and later tasks require increasingly complex reasoning (Pinero Charlo, 2020). Like problem-based learning (PBL), escape rooms involve interpreting problems and iteratively testing strategies, reflecting established stages of mathematical discovery (Pinero Charlo, 2020).

From a design perspective, escape rooms can also be interpreted through the lens of cognitive load theory, which emphasizes the importance of aligning task demands with learners' cognitive capacities (Sweller et al., 2011; Ugo et al., 2025). When narrative elements and puzzle structures are carefully integrated, learners are better positioned to direct their cognitive resources to meaningful problem-solving rather than extraneous or superfluous information. Beyond these cognitive perspectives, escape rooms also provide opportunities to develop communication, leadership, and collaborative decision-making skills, which are broadly acknowledged as indispensable competencies for twenty-first-century learning (Kim et al., 2024).

The use of escape room activities, which have emerged in recent years as an innovative active learning tool, is becoming increasingly widespread in primary school settings. Although studies directly focused on the elementary school level are relatively limited (Taraldsen et al., 2022), existing research clearly demonstrates that this pedagogical tool has a transformative effect on students' motivation and collaboration skills. For example, it has been found that digital escape rooms implemented in primary school settings not only increase students' intrinsic and extrinsic motivation but also significantly strengthen group cohesion and a sense of teamwork (Vidergor, 2021). Similarly, these game-based environments, which require students to work together in groups to discover clues and complete tasks within a limited time frame, enrich the learning process by inherently necessitating communication and collaboration-based teamwork (Kahila et al., 2020).

Beyond motivation and teamwork, escape rooms are also a highly effective tool for developing problem-solving skills in elementary school students. Especially when integrated into subjects like science education, these activities encourage students to make logical inferences, overcome challenges using clues, and solve complex problems step by step, resulting in significantly higher problem-solving skill scores compared to traditional teaching methods (Huang et al., 2020). At this point, 'primary school teachers' play a key role in designing and effectively implementing the process. Indeed, to prevent students from becoming frustrated and to keep their interest consistently engaged, teachers must balance the difficulty level of the puzzles with the time spent on them in a way that aligns with the children's developmental and cognitive levels, while also serving as active guides throughout the process (Yllana-Prieto et al., 2025). As a result, these gamified experiences not only enable the learning of challenging academic content in an engaging format but also equip young students with the fundamental social and cognitive competencies they will need in the future.

Importantly, although escape rooms are frequently examined through motivational and cognitive frameworks, their collaborative and interactional nature suggests strong alignment with socio-cultural theories of learning. The shared constraints and collective goals embedded in escape room design create conditions that require negotiation, explanation, and coordination among participants. These features position escape rooms as promising environments for supporting sustained participation and collaborative reasoning in classroom settings.

Escape Rooms in Mathematics Education

Traditional mathematics education is often criticized for its teacher-centered, abstract theories, rote learning, and repetition-based structure; this structure can cause negative affective states in students such as high levels of math anxiety, shyness, and low motivation (Aydın & Ata, 2024; Fuentes-Cabrera et al., 2020; Pais et al., 2023; Zapata et al., 2024). Active learning strategies such as gamification and digital escape rooms offer a strong alternative for overcoming the negative barriers that students often develop toward the complex nature of mathematics. Escape rooms place students at the center of the learning process by concretizing abstract mathematical concepts and transforming mathematics from a passive memorization process into an interactive journey of exploration and discovery. Specific escape room applications in mathematics education vary widely, from primary school to higher education. A study conducted at the primary school level (3rd grade) used Educational Digital Escape Games to teach basic mathematical concepts such as time measurement, money, and weighing. It found that this application significantly increased academic achievement and intrinsic motivation compared to traditional classroom activities (Işık & Liman Kaban, 2025). In another primary school study, escape rooms were positioned as a tool for "horizontal mathematization" that is the process of transforming real-life problems (e.g., using scales to balance weights) into mathematical symbols and equations was examined (Pinero Charlo, 2020). Compared to conventional teaching, escape rooms significantly improved secondary students' performance and engagement in linear equations while lowering anxiety levels (Fuentes-Cabrera et al., 2020). Similarly, 3D Augmented Reality (AR) based and physical/digital hybrid escape room activities for high school students have been reported to improve problem-solving skills and conceptual understanding (Zapata et al., 2024). In higher education, escape room activities designed to address the demotivation of life sciences students in analytical mathematics courses resulted in more positive attitudes toward learning mathematics. In addition, the interactive environment contributed to making the course more engaging (Pais et al., 2023).

A detailed examination of the outcomes and effects on student motivation reveals that the game aspect of escape rooms and the narrative elements they contain act as a driving force on intrinsic motivation (Aydın & Ata, 2024). Students viewing mathematics not just as a subject but as a means to an end, such as "escaping the room" or "solving the mystery," increases their interest in the subject and necessitates collaboration with their teammates (Vidergor, 2021). Quantitative and qualitative analyses confirm a strong positive correlation between students' intrinsic motivation and their performance on academic achievement tests (Işık & Liman Kaban, 2025). However, limitations such as restricted lesson times (time constraints), classroom management, the ineffectiveness of games with weak educational dimensions, technical deficiencies, and sometimes increased cognitive load are also listed in the literature as challenges faced by educators (Aydın & Ata, 2024; Ugo et al., 2025). In conclusion, math escape rooms use gamified learning to break down barriers, fostering higher achievement and intrinsic motivation through active participation. These findings suggest that escape rooms may function not only as motivational tools but also as social participation structures that organize collaborative mathematical activity.

Escape Rooms as Socio-Cultural Mathematics Learning Environments

Educational escape rooms can be conceptualized as intentionally designed socio-cultural learning environments that organize participation, structure discourse, and distribute responsibility across learners. From a socio-cultural perspective, learning environments are not neutral settings in which tasks are completed individually; rather, they function as mediational spaces that coordinate interaction, shape communication patterns, and regulate access to shared resources (Wertsch, 1998; Lave & Wenger, 1991). Within mathematics classrooms, escape rooms provide structured opportunities for students to engage in collaborative reasoning through time-bound challenges, role assignments, and shared problem-solving goals.

Unlike traditional classroom routines that often position students as individual responders to teacher prompts, escape rooms create collective problem-solving contexts in which responsibility is distributed across group members. The presence of shared constraints-such as time limits, locked sequences, and task progression-encourages learners to coordinate actions, negotiate strategies, and justify reasoning in relation to group decisions. These features transform mathematical tasks into socially mediated activities in which meaning is constructed through interaction rather than transmitted through instruction.

One distinctive feature of escape rooms is their capacity to generate sustained dialogic interaction. The requirement to decode clues, verify intermediate results, and agree on solution pathways naturally produces talk that includes explanation, questioning, and argumentation. Such discourse practices align closely with socio-cultural views of learning, which emphasize that knowledge emerges through communicative participation (Sfard, 2008; Bakhtin, 1981). As students interpret clues and reconcile differing interpretations, they encounter multiple perspectives that support conceptual refinement and collaborative sense-making.

Escape rooms also create structured opportunities for positioning learners in varied roles within group activities. Role-based participation, such as coordinating, explaining, verifying, or executing actions, enables students to assume diverse responsibilities that extend beyond solving individual problems. These roles influence how authority is distributed within groups and shape the peers' recognition of competence (Cobb et al., 2009). When learners take responsibility for explaining reasoning or monitoring progress, they become active contributors to the collective learning process rather than passive recipients of information.

Another defining characteristic of escape rooms is their reliance on mediational tools, particularly digital platforms that organize interactional sequences. In digitally mediated escape rooms, tools such as interactive puzzles, visual feedback systems, and programmable tasks do not merely serve as passive supports; rather, they function as structuring elements that shape the nature, timing, and direction of student interaction. These tools act as shared reference points for group discussion while simultaneously constraining and affording particular forms of participation. For example, immediate feedback mechanisms can redirect students' attention, prompt re-evaluation of strategies, and encourage iterative cycles of hypothesis testing and revision. From a socio-cultural standpoint, such tools act as cultural artifacts that shape both the content and form of mathematical activity (Wertsch, 1998). However, extending this view, digitally mediated environments also invite a socio-material interpretation in which technology operates not only as a tool but as an active participant in the learning process. In this sense, digital platforms co-structure interaction by sequencing tasks, regulating access to information, and influencing decision-making pathways. Thus, the escape room environment emerges as a hybrid interactional space where human actors and digital systems jointly produce learning processes. The integration of these tools into classroom practice enables learners to externalize reasoning processes and coordinate collective problem-solving efforts, and iteratively evaluate strategies within a dynamically structured environment.

Importantly, escape rooms can support identity-safe participation when tasks are designed to allow multiple entry points and varied forms of contribution. In group-based escape room settings, students may contribute through explaining patterns, organizing materials, testing solutions, or coordinating group actions. This multiplicity of roles reduces reliance on single forms of competence and encourages recognition of diverse strengths. Such design features align with equity-oriented mathematics education, which emphasizes the importance of broadening participation and supporting positive mathematical identity development (Gutiérrez, 2008; Boaler, 2002).

The narrative structure commonly embedded in escape room design further contributes to sustained engagement. Story-lines that frame tasks as parts of a mystery or mission create meaningful contexts that support collaborative sense-making. Narratives provide continuity across tasks and establish shared goals that encourage collective responsibility. When students perceive their work as contributing to a larger objective, they are more likely to maintain engagement and coordinate efforts among group members.

From an instructional design perspective, escape rooms can be understood as sequences of interconnected tasks that progressively increase complexity while maintaining continuity in participation structures. Early tasks often function as entry points that establish group interaction norms, while later tasks require deeper reasoning and coordination. This progression supports the gradual development of collaborative competence and enables students to refine their strategies over time.

In mathematics education specifically, escape rooms provide a promising context for integrating reasoning, representation, and communication. Mathematical puzzles embedded within escape room sequences require students to interpret patterns, construct models, test conjectures, and justify conclusions. These processes align with contemporary views of mathematics learning that emphasize reasoning and communication as central components of mathematical proficiency (Sfard, 2008; Cobb et al., 2009).Viewing escape rooms through a socio-cultural lens shifts their interpretation from engaging classroom activities to structured learning environments that intentionally organize participation, discourse, and identity development. Within such environments, responsibility is distributed across learners, dialogic interaction is encouraged, and mediational tools support collaborative reasoning. These features enable students to participate as active contributors to shared mathematical practices rather than as isolated problem solvers. Conceptualizing escape rooms in this way highlights their potential as instructional models for designing identity-safe and participation-rich mathematics classrooms.

Current Study

This chapter investigates how escape-room-based coding activities function as socio-cultural learning environments that support collaborative mathematical engagement among primary school students. While escape rooms are often used as motivational or game-based instructional tools, their potential as structured environments for organizing participation, discourse, and mediated problem-solving remains underexplored in primary mathematics contexts. In particular, there is limited empirical evidence documenting how students coordinate their actions, negotiate mathematical ideas, and utilize mediational tools during collaborative digital problem-solving tasks. The present study addresses this gap by examining a classroom-based escape-room implementation conducted with fourth-grade primary students. The design of the escape-room tasks emphasized collective reasoning, distributed responsibility, and the use of digital and written artifacts to support problem-solving. Rather than focusing solely on task completion or correctness, this study conceptualizes learning as a socially mediated process in which students construct meaning through interaction with peers and tools.

Guided by socio-cultural perspectives on learning, the study aims to explore how collaborative participation unfolds within digitally mediated mathematics tasks and how students organize their actions when confronted with complex coding challenges. Special attention is given to the role of discourse, experimentation, planning, and artifact use in shaping collective mathematical reasoning.

Accordingly, the study is guided by the following research questions:

1. How do students organize participation and distribute responsibility during collaborative escape-room tasks?
2. What discursive and interactional strategies do students use while negotiating mathematical ideas during problem solving?
3. How do students use mediational tools, including digital platforms and written artifacts, to support collective reasoning?
4. What patterns of experimentation and planning emerge as students attempt to solve mathematical coding tasks collaboratively?

METHOD

Research Design

This chapter draws on a design-based classroom implementation conducted with fourth-grade primary school students. Design-based approaches are particularly appropriate for examining how instructional environments function as mediational spaces that shape interaction, discourse, and collaborative engagement (Brown, 1992; Cobb et al., 2003). In this study, the escape room was conceptualized not merely as an instructional activity but as a socio-cultural learning environment designed to structure participation, distribute responsibility, and organize collaborative reasoning.

The escape room consisted of one progressively structured digital task hosted on the Code.org platform. While such tasks are commonly used to support computational thinking development (Wing, 2006), in this study, they were treated as cultural artifacts that mediated interaction by shaping talk, regulating access to actions, and coordinating group participation (Brennan & Resnick, 2012; Wertsch, 1998). From a socio-cultural perspective, the analytic focus shifted from individual performance to collective engagement, examining how students explained, justified, checked, debugged, and coordinated strategies while working toward shared solutions.

Participants and Context

The implementation was conducted in a public primary school with fourth-grade students. A total of 31 students participated in the study. Students were organized into heterogeneous groups based on academic achievement, gender distribution, and observed social interaction patterns to support equitable participation. Each group consisted of three to four students, and all members of a given group engaged collaboratively through a single computer. Before the implementation, students were informed about the activity procedures and expectations, and brief orientation sessions were conducted to familiarize them with collaborative working norms.

The escape room experience was conducted during regular lesson hours and lasted approximately 120 minutes, during which students completed tasks. During the implementation, the teacher adopted a facilitative role rather than providing direct solutions. The teacher supported students by clarifying instructions and encouraging group discussion when needed, but did not intervene in the problem-solving process or provide procedural guidance. This approach allowed students to assume responsibility for decision-making, negotiation, and revision of their ideas, thereby preserving the collaborative and student-centered nature of the activity. Such facilitative positioning aligns with socio-cultural perspectives that emphasize guided participation and the distributed nature of learning within peer groups (Vygotsky, 1978; Lave & Wenger, 1991).

Design of the Escape Room Learning Environment

The escape room learning environment was structured around three interconnected tasks designed to increase in complexity and encourage collaborative mathematical reasoning.

Task: Drawing Geometric Paths Using Angles

In the task, students constructed geometric paths using directional commands and angle-based turns. Students collaboratively explored how different turning angles (e.g., common turn (45°), quarter (90°), half (180°), three-quarter (270°), and full turns (360°) influenced movement direction and shape formation within the Code.org environment. Students collaboratively determined how turning angles influenced the direction of movement and experimented with commands to produce specific shapes. This task supported the development of foundational spatial reasoning skills related to directionality, rotation, and geometric shape construction. Each task functioned as a "challenge room" within the escape room structure, where groups were required to complete the task before progressing to the next stage. Working collaboratively on a shared computer, students tested movement and rotation commands and immediately observed the resulting output. When the expected shape was not produced, groups revisited their reasoning, discussed alternative strategies, and revised their commands accordingly. These tasks were designed not only to reach correct solutions but to elicit discussion around mathematical ideas such as angle size, direction, and efficient coding strategies (e.g., the use of loops), thereby making students' reasoning processes visible during collaborative interaction.

Role-based participation structures were integrated into the design to support distributed responsibility. Students were assigned rotating roles such as solver, checker, explainer, and coordinator. These roles were intended to encourage collaborative engagement and to ensure that responsibility for reasoning was shared among group members.

Data Collection

Multiple data sources were used to document interactional processes and participation patterns during the implementation. During the escape room experience, classroom interactions were documented through video recordings, screen recordings, and researcher field notes. These records captured students' verbal interactions, gestures, collaborative decisions, and moments of breakdown and repair.

Following the implementation, students wrote short reflective notes describing their experiences. In addition, semi-structured focus group interviews were conducted with selected students to explore their perceptions of collaboration, engagement, and learning. The interview protocol was developed by the researchers and reviewed by two field experts to ensure content validity. Revisions were made based on expert feedback before final implementation.

Data Analysis

Data analysis followed an inductive thematic approach informed by socio-cultural constructs, including mediation, positioning, and participation (Cobb et al., 2009; Sfard, 2008; Wertsch, 1998). Video recordings, field notes, and interview transcripts were analyzed iteratively to identify recurring interactional patterns and discourse structures. Initial coding focused on identifying instances of explanation, justification, negotiation, coordination, and repair during collaborative activity. Codes were then grouped into broader categories representing participation patterns and positioning dynamics. Emerging themes were refined through repeated comparison across data sources.

To enhance analytic reliability, coding decisions were reviewed collaboratively by the researchers, and disagreements were resolved through discussion until consensus was reached.

Findings were reported in the form of short classroom vignettes illustrating:

- how mathematical meanings were negotiated in situ
- how roles were adopted, redistributed, and coordinated within groups

- how students engaged in iterative experimentation through testing, revising, and refining their solutions
- how digital and written tools supported collective decision-making processes
- how planning and predictive reasoning emerged as students anticipated outcomes before executing commands

These vignette-based representations enabled detailed examination of interactional processes and highlighted the socio-cultural nature of collaborative mathematical activity.

FINDINGS

This section presents the findings derived from the analysis of video recordings, researcher field notes, and focus group interviews conducted with nine student groups during the escape-room implementation. The analysis focused on identifying recurring patterns in students' participation, interaction, and use of mediational tools while collaboratively solving coding-based mathematical tasks.

An inductive thematic analysis was employed to examine how students coordinated their actions, negotiated mathematical ideas, and responded to challenges encountered during the tasks. Through iterative coding and comparison across groups, four interrelated themes were identified that represent distinct yet overlapping dimensions of collaborative mathematical activity. These themes capture variations in participation structures, collective experimentation processes, artifact-supported reasoning, and planning-oriented discourse.

The themes presented below are supported by illustrative examples drawn from classroom observations and student reflections, demonstrating how mathematical understanding developed through shared activity, negotiation, and tool-mediated interaction.

Theme 1: Participation Structures in Collaborative Coding

Analysis of Groups 1 and 2 revealed distinct participation structures that shaped how students coordinated their actions during the coding-based shape construction task. Although both groups worked collaboratively, differences emerged in how responsibility was distributed and how decisions were initiated. In particular, participation patterns varied between moments where a single student initiated and guided the sequence of coding actions and moments where students jointly negotiated solutions by proposing ideas, questioning each other's reasoning, and reaching shared decisions. These differences influenced not only the organization of the task but also the depth of interaction and the extent to which reasoning processes were collectively constructed.

In Group 1, participation was relatively distributed across members, with students jointly monitoring progress and evaluating alternative solutions. They frequently paused to review previously inserted commands and collaboratively identify potential errors. Responsibility for regulating the problem-solving process shifted dynamically among group members, as students inspected outputs together and negotiated revisions rather than relying on a single dominant participant.

Classroom Vignette 1: Revisiting Commands through Collective Inspection

During one phase of the activity, Group 1 noticed that the drawn shape did not align with their expectations. Instead of attributing the error to an individual, students paused and collectively reviewed the sequence of commands displayed on the screen. One student pointed to an earlier movement, while another suggested modifying the direction. After a brief discussion, the group replaced the command and re-executed the sequence. The revised output produced a more accurate representation of the intended shape, prompting visible agreement among members.

This interaction illustrates how distributed participation supported collaborative regulation of mathematical activity. The group treated errors as shared challenges, encouraging collective reasoning and joint decision-making. Focus group responses reinforced this observation. Students described reviewing their previous actions together and discussing possible corrections before continuing. One student stated:

"When we made a mistake, we stopped and looked at it together. First, we checked what we had done."

This reflection indicates that collaborative inspection functioned as a shared strategy for maintaining progress.

In contrast, participation in Group 2 was more centrally organized around a leading student who frequently initiated decisions and directed group attention. This leadership pattern shaped the structure of interaction by establishing a primary source of suggestions while other members evaluated or confirmed proposed actions. Although participation remained collaborative, authority appeared more concentrated in comparison to Group 1.

Classroom Vignette 2: Leadership-Guided Decision Making

During the middle phase of the task, Group 2 encountered difficulty when the drawn output deviated from the expected shape. A leading student quickly suggested inserting a new movement command to correct the direction. The student explained the proposed step while pointing at the screen, and other group members responded by observing and confirming the suggestion. After executing the revised command, the output improved, leading the group to proceed with the next stage.

This episode demonstrates how leadership functioned as an organizing mechanism that structured participation. While responsibility for proposing solutions was concentrated, error correction still involved collective review. Focus group responses supported this pattern. Students reported that one member frequently initiated suggestions while the group evaluated the effectiveness of proposed actions. One student explained:

> *"Generally, he/she suggested it, and we gave it a try. But when it didn't work out, we looked into it together again."*

This statement suggests that leadership did not eliminate collaboration but influenced how decisions were introduced into group discussion.

Taken together, the comparison of Groups 1 and 2 highlights how participation structures varied across groups, shaping the organization of mathematical reasoning. While Group 1 demonstrated distributed regulation, Group 2 illustrated leadership-centered coordination. These findings indicate that collaborative mathematical activity can emerge through multiple participation structures rather than a single uniform pattern.

Theme 2: Iterative Collective Experimentation

Analysis of Groups 5, 6, and 7 revealed a recurring pattern of collective experimentation in which students relied on iterative cycles of testing, evaluating, and revising commands to construct the intended geometric shapes. Across these groups, participation was characterized by shared idea generation and immediate testing of alternative solutions. For example, students frequently referred to specific angle values while testing alternative solutions. One student stated, *"We need to turn 45 degrees instead of 90 degrees; otherwise, the shape won't form a corner,"* illustrating how mathematical reasoning guided the revision of commands. Such interactions indicate that experimentation was not random but informed by an emerging understanding of angle size and directional change.

Students reported that each group member contributed suggestions, and newly proposed ideas were rapidly tested to determine their effectiveness. Rather than engaging in extended planning prior to action, these groups adopted an exploratory approach in which experimentation functioned as a primary strategy for resolving uncertainty. This pattern suggests that reasoning was distributed across members through repeated cycles of action and feedback.

Video records indicated that students frequently removed previously inserted commands and replaced them with new alternatives after observing incorrect outputs.

These rapid testing cycles allowed groups to refine their strategies progressively without prolonged delays. Such behavior reflects an adaptive problem-solving process supported by immediate visual feedback from the digital environment.

Classroom Vignette 3: Testing Alternative Suggestions

During one sequence, Group 5 members proposed multiple suggestions regarding the next movement step. Rather than debating extensively, the group inserted one suggested command and executed the sequence. When the output did not match the expected shape, another student proposed an alternative command. The group quickly replaced the earlier step and retested the sequence. This process was repeated several times until the shape began to resemble the intended design.

This interaction demonstrates how experimentation functioned as a shared regulatory mechanism. Responsibility for proposing and testing alternatives shifted across members, enabling the group to maintain progress despite repeated errors. Focus group responses further supported this interpretation. A student from Group 5 explained:

> *"Everyone shared their ideas, and we tried to sketch out the shape by experimenting. Whenever a different suggestion was made, we immediately tested it to make sure it was correct so we wouldn't repeat the same mistake."*

Similar reflections were reported in Groups 6 and 7, where students described testing multiple alternatives to prevent repeating earlier mistakes. These statements indicate that iterative experimentation was perceived as a collective responsibility rather than an individual action.

Finally, Groups 5, 6, and 7 demonstrated a shared pattern of iterative collective experimentation, in which trial-and-revision cycles supported the gradual refinement of mathematical actions. This pattern highlights the importance of exploratory interaction as a socially mediated pathway to problem solving.

Theme 3: Discursive Planning and Predictive Reasoning

Analysis of Groups 3 and 9 revealed patterns of planning-oriented reasoning in which students engaged in predictive thinking before executing commands. Unlike groups that relied primarily on rapid testing cycles, these groups demonstrated a shift toward deliberate planning and forecasting possible outcomes.

In Group 3, students frequently paused to discuss possible alternatives before inserting commands. Rather than immediately testing each suggestion, group members engaged in short episodes of reasoning in which they evaluated potential directions and anticipated the resulting movement. These discussions allowed students to coordinate their actions and reduce unnecessary repetitions.

Classroom Vignette 4: Predicting Outcomes before Execution

During one stage of the activity, Group 3 paused before inserting the next command. Students discussed which direction would best align with the previously drawn segment. One student suggested a possible movement, while another questioned whether the angle would produce the desired shape. The group briefly compared alternative possibilities before selecting a command. After executing the sequence, the resulting movement closely matched their expectations.

This interaction illustrates how predictive reasoning functioned as a planning mechanism. Rather than relying solely on trial-and-error, students used discussion to anticipate outcomes and guide decision-making. Focus group responses supported this interpretation. One student explained:

"We talked it over and thought it through first, then we did it. We didn't just jump right in."

This reflection indicates that verbal reasoning supported the coordination of mathematical actions.

In Group 9, a similar shift toward planning was observed, particularly after repeated unsuccessful attempts. Initially, students engaged in rapid cycles of trial-and-revision. However, when the resulting output produced an unintended straight-line shape, the group decided to adopt a more deliberate strategy. At this point, students introduced paper-based planning as a tool to anticipate outcomes before executing commands.

Classroom Vignette 5: Transitioning from Trial to Paper-Based Planning

During an early phase of the task, Group 9 repeatedly tested commands by inserting and removing blocks. When the output produced an incorrect straight-line figure, the group paused and decided to plan their actions before continuing. Students began sketching possible movements on paper and discussed which angles might produce the intended shape. They predicted how different angle values would influence the resulting pattern and selected commands accordingly.

Following this planning phase, the group executed the revised sequence and observed improved results. This transition from repeated testing to structure planning demonstrates how students adapted their strategies in response to persistent difficulties.

Focus group responses confirmed this shift. One student reported:

"We tried erasing and redrawing it over and over, but it didn't work. Then we decided to plan it out. We tried to predict what the image would look like depending on the angle we drew it from."

This statement illustrates how planning emerged as a response to earlier errors and supported more efficient problem-solving.

Theme 4: Artifact-Mediated Regulation of Learning

Analysis of Groups 4, 8, and 9 revealed the strategic use of external artifact as mediational tools that supported planning, monitoring, and decision-making processes. Unlike groups that relied solely on verbal interaction or repeated testing, these groups incorporated written documentation to regulate their problem-solving processes and reduce repetitive errors. In some groups, written notes became critical tools for regulating collective reasoning. For instance, one student recorded attempted angle values and noted which commands were unsuccessful. During the discussion, a group member explained, *"We wrote 90 here, but it didn't work, so we tried 45,"* demonstrating how written records supported comparison of alternative strategies and prevented repetition of ineffective commands.

In Group 4, the introduction of written recording emerged directly from student initiative. Following repeated unsuccessful attempts, Mustafa proposed writing down previously tested commands to prevent repetition. The group adopted this suggestion and assigned Mustafa responsibility for documenting each attempted sequence. The written record enabled students to review earlier steps and make informed decisions about subsequent actions. This practice transformed the problem-solving process into a more structured and systematic activity.

Classroom Vignette 6: Initiating Written Tracking to Avoid Repetition

During one phase of the activity, Group 4 experienced repeated difficulty producing the intended shape. Mustafa suggested recording the tested commands on paper so that unsuccessful sequences would not be repeated. The group agreed and began documenting each attempt. Before testing new commands, students frequently consulted the written list to determine whether similar sequences had already been attempted. This process reduced redundant actions and allowed the group to focus on generating new alternatives. Focus group responses supported this observation. One student explained:

"Mustafa suggested writing it down. We wrote it down on paper so we wouldn't have to do the same thing again."

This statement indicates that written documentation functioned as a collectively constructed strategy that supported group coordination.

A similar pattern emerged in Group 8, where one student assumed responsibility for recording tested commands and identifying which sequences were successful or unsuccessful. Unlike passive note-taking, this role involved active evaluation of outcomes. Students reported returning to written notes after errors occurred and using these records to guide future decisions.

Classroom Vignette 7: Using Written Records to Identify Effective Strategies

During the activity, Group 8 maintained a written list of tested commands. After each attempt, the group reviewed whether the sequence produced the intended movement. The student is responsible for recording marked unsuccessful attempts and identifying sequences that produced partial success. When the group encountered difficulties, they referred to these notes rather than restarting from the beginning. This process enabled the group to refine its strategy systematically. A student described this process as follows:

"The notes he/she took throughout the process were very helpful to us. Whenever we made a mistake, we went back to those notes and used them as a guide."

This reflection highlights how written documentation functioned as a shared memory system that supported evaluation and strategic filtering.

Group 9 demonstrated an additional variation of artifact-mediated regulation through the use of paper-based planning after repeated unsuccessful attempts. Initially, students engaged in rapid cycles of trial-and-revision. However, when the resulting output repeatedly produced an incorrect straight-line pattern, the group decided to sketch possible movements before continuing. This shift toward external planning enabled students to predict outcomes and coordinate their actions more effectively. Planning behaviors became particularly visible when repeated attempts produced unexpected results. In one group, students paused to discuss directional outcomes before executing commands. A student commented, *"If we don't turn 180 degrees, the shape will go straight,"* indicating the use of predictive reasoning based on angle relationships and movement direction.

Classroom Vignette 8: Shifting from Repetition to Paper-Based Planning

During an early phase of the task, Group 9 repeatedly inserted and removed commands without achieving the desired outcome. After recognizing the limitations of repeated testing, students decided to plan their actions on paper. They sketched possible movements and discussed how different angles might influence the resulting shape. This planning phase allowed the group to anticipate outcomes before executing commands, leading to more accurate results. One student explained:

"We tried erasing and redrawing it over and over, but it didn't work. Then we decided to plan it out. We tried to predict what the image would look like de-

pending on the angle from which we drew it. We jotted these down and put our notes into practice."

This statement demonstrates how external representations supported predictive reasoning and improved problem-solving efficiency.

Cross-Theme Synthesis: Multiple Pathways of Mediated Mathematical Participation

Across the four themes identified in this study, students demonstrated diverse yet interconnected pathways for coordinating mathematical activity within collaborative coding environments. Rather than following a single uniform pattern, groups employed different participation structures, discursive strategies, and mediational tools that collectively supported problem solving.

The findings indicate that locally negotiated interactional structures shaped participation in collaborative coding tasks. In some groups, responsibility was distributed among members, allowing students to monitor and revise their actions jointly. In others, leadership-based coordination emerged, providing direction and continuity during decision-making processes. These variations suggest that collaborative engagement does not depend on a single model of participation but evolves through group-specific social dynamics.

Across multiple groups, iterative experimentation emerged as a central mechanism for refining mathematical actions. Students frequently described testing alternative solutions, evaluating outputs, and adjusting commands through repeated cycles of action and feedback. This process highlights the importance of experimentation as a socially mediated strategy rather than an individual trial-and-error behavior.

At the same time, several groups demonstrated a shift toward planning-oriented reasoning, particularly when repeated attempts produced unexpected results. Students engaged in predictive discussion and, in some cases, introduced paper-based planning to anticipate outcomes before execution. These moments illustrate how students transitioned from reactive strategies to more deliberate forms of reasoning.

Perhaps most notably, the introduction of written artifacts in several groups highlighted the powerful role of external representations in organizing collective thinking. Students used written records to track attempted solutions, identify ineffective strategies, and guide future actions. These artifacts functioned as shared memory systems that extended cognitive processes beyond individual recall, supporting systematic decision-making and reducing redundant errors.

Together, these findings illustrate that collaborative mathematical activity in digitally mediated environments is not solely determined by task structure or technological design. Instead, it emerges through the interaction of participation patterns, discourse practices, and mediational tools that students construct and adapt during problem-solving processes. Such diversity in interactional pathways reflects the fundamentally socio-cultural nature of mathematical learning, in which knowledge develops through shared activity and the coordinated use of cultural resources.

DISCUSSION

The central claim advanced in this chapter is that escape rooms can be intentionally designed to support collaborative reasoning and identity-safe participation by expanding legitimate ways of contributing and sustaining dialogic interaction in which students negotiate strategies, reconcile competing interpretations, and position themselves and others as competent contributors. In this sense, escape rooms function not merely as engaging instructional activities but as structured environments for participation that foreground learning as a collective activity rather than an individual accomplishment.

The findings of the present study provide empirical support for this claim by demonstrating how students organized participation, negotiated mathematical ideas, and coordinated actions across multiple collaborative groups. Across the nine student groups, participation emerged as a socially negotiated process in which responsibility was distributed through roles, leadership behaviors, and peer-supported decision-making. In several groups, students alternated between proposing solutions, checking commands, and coordinating next steps, illustrating how responsibility for reasoning became shared rather than centralized. These findings align with socio-cultural perspectives that conceptualize learning as participation in socially organized practices rather than individual knowledge acquisition (Lave & Wenger, 1991; Wenger, 1998). Beyond these general patterns of collaborative participation, it is important to consider the specific affordances of the escape room format that may have shaped these outcomes. Unlike standard collaborative coding tasks, the escape room introduced a narrative progression, time-bound challenges, and a shared goal of "escaping," which collectively intensified students' engagement and coordination. The sequential unlocking of tasks required groups to sustain joint attention and continuously negotiate strategies in order to progress. In this sense, the escape room functioned not only as a task environment but also as a structuring mechanism that organized interaction around urgency, interdependence, and goal-oriented action. These features appear to have amplified collaborative behaviors such as shared decision-making, iterative testing, and distributed responsibility, suggesting that the observed interaction patterns were shaped not only by group work itself but also by the distinctive design of the escape room context.

The emergence of iterative collective experimentation, observed particularly in Groups 5, 6, and 7, further illustrates how learning unfolded through cycles of testing, revising, and refining ideas. Students frequently described "trying again," "deleting incorrect commands," and "testing new suggestions," indicating that experimentation functioned as a collaborative reasoning strategy rather than isolated trial-and-error behavior. Such patterns reflect Vygotsky's (1978) view of learning as a socially mediated process in which problem solving evolves through interaction and shared regulation. However, this pattern was not equally evident across all groups. In some cases, groups experienced difficulty sustaining iterative engagement, tending to rely on repeated trial without collective reflection or strategic discussion. This contrast suggests that while the escape room environment can support collaborative experimentation, the quality of interaction remains contingent on group dynamics and participation structures.

Another important finding concerns the role of artifact-mediated regulation, particularly evident in Groups 4, 8, and 9, where students used written notes and sketches to track successful and unsuccessful attempts. These written artifacts functioned as external memory systems that supported systematic reasoning and reduced repetitive errors. The use of paper-based planning alongside digital coding tools highlights the multi-modal nature of mathematical thinking and reinforces the importance of mediational tools in structuring collective cognition (Wertsch, 1998; Sfard, 2008).

The findings also revealed the emergence of planning and predictive reasoning, particularly in situations where repeated attempts produced unexpected results. Students began to anticipate possible outcomes, particularly when discussing turning angles and movement directions required to construct specific geometric shapes. In several groups, students predicted how changes in angle size would affect the direction of movement, demonstrating emerging spatial reasoning related to rotation and geometric representation. These moments illustrate a shift from reactive experimentation to strategic reasoning, suggesting that collaborative environments can support the development of higher-level thinking processes in primary mathematics contexts. The findings also highlight the development of mathematical concepts related to angles, direction, and geometric shape construction. Students' discussions frequently involved reasoning about turning degrees, predicting directional outcomes, and coordinating movements to produce desired geometric patterns. These interactions demonstrate how spatial reasoning skills can be strengthened through collaborative and digitally mediated environments. Such findings suggest that collaborative digital environments can support not only social participation but also the development of foundational geometric and spatial reasoning skills in primary mathematics.

An additional point of significance concerns students' prior experience with digital tools. All participating students engaged with the Code.org platform for the first time during the implementation. Despite this initial unfamiliarity, students demonstrated the ability to coordinate actions, interpret feedback, and collaboratively refine their strategies. This finding suggests that collaborative escape-room environments can support students' entry into unfamiliar digital contexts by distributing responsibility across group members and encouraging peer-supported learning. From a socio-cultural perspective, this reinforces the view that competence emerges through mediated participation rather than individual technical expertise (Vygotsky, 1978; Wertsch, 1998).

Equally important was the role of the classroom teacher, who adopted a facilitative rather than directive position throughout the implementation. Instead of providing procedural solutions, the teacher offered clarification and motivational prompts that encouraged students to reconsider their strategies and explore alternative solutions. This facilitative stance created opportunities for students to negotiate meaning collaboratively and assume ownership of their problem-solving processes. Such instructional positioning is consistent with socio-cultural models of guided participation, in which support structures enable learners to engage actively within shared problem-solving environments (Lave & Wenger, 1991).

In summary, socio-cultural and digitally mediated escape rooms offer a pedagogical alternative that centers the student as an active participant rather than a passive recipient of information. In contrast to traditional instructional approaches that position the teacher as the sole source of knowledge, escape-room-based learning environments distribute authority across learners and encourage shared responsibility for reasoning (Zapata et al., 2024). These environments create conditions that support motivation, reduce hesitation associated with making errors, and foster collaborative confidence among peers. As students work collectively to solve challenges, mathematics becomes an exploratory process shaped by dialogue, experimentation, and mutual support (Işık & Liman Kaban, 2025; Zapata et al., 2024).

CONCLUSION

This chapter demonstrates that escape-room-based mathematics activities can function as structured socio-cultural learning environments that support collaborative reasoning, distributed responsibility, and identity-safe participation among primary school students. The findings reveal that students' mathematical engagement extended beyond task completion to include collective experimentation, artifact-supported reasoning, and planning-oriented dialogue. Importantly, the results show that even students with no prior experience using digital coding platforms were able to participate meaningfully in collaborative problem-solving processes. This highlights the potential of well-designed escape-room environments to support entry into new learning contexts while promoting confidence and shared responsibility.

By conceptualizing escape rooms as mediational environments rather than merely motivational tools, this chapter contributes to socio-cultural mathematics education literature and provides empirically grounded design insights for educators. The findings suggest that when participation structures, collaborative dialogue, and mediational tools are intentionally aligned, escape-room-based learning environments can foster inclusive, engaging, and mathematically meaningful classroom experiences. Digital escape rooms provide a unique context for students to solve mathematical puzzles, develop strategies within a group, use digital materials effectively, and collaborate under time pressure.

Nevertheless, several limitations should be acknowledged. The study was conducted with a single group of fourth-grade students within a limited time frame, which may constrain the generalizability of the findings. In addition, variations in group dynamics and interaction patterns suggest that the effectiveness of escape-room-based learning may depend on contextual and social factors. Future research could explore longer-term implementations across diverse classroom settings and examine how different design features of escape rooms influence the quality of collaborative engagement and learning outcomes.

REFERENCES

Aydın, C. S., & Ata, R. (2024). Use of digital games in mathematics education in Turkey: A systematic review study. *Inonu University Journal of the Faculty of Education, 25*(2), 570–596. DOI: 10.17679/inuefd.1415803

Bakhtin, M. M. (1981). *The dialogic imagination: Four essays (M. Holquist* (Emerson, C., & Holquist, M. Eds. & Trans.). University of Texas Press.

Boaler, J. (2002). Exploring the nature of mathematical activity: Using theory, research and working hypotheses to broaden conceptions of mathematics knowing. *Educational Studies in Mathematics, 51*(1-2), 3–21. DOI: 10.1023/A:1022468022549

Brennan, K., & Resnick, M. (2012-April). New frameworks for studying and assessing the development of computational thinking. *Proceedings of the 2012 Annual Meeting of the American Educational Research Association,* Vol. 1, Vancouver, 13-17 April 2012, 25 p. https://scratched.gse.harvard.edu/ct/files/AERA2012.pdf

Brown, A. L. (1992). Design experiments: Theoretical and methodological challenges in creating complex interventions in classroom settings. *Journal of the Learning Sciences, 2*(2), 141–178. DOI: 10.1207/s15327809jls0202_2

Cobb, P., Confrey, J., diSessa, A., Lehrer, R., & Schauble, L. (2003). Design experiments in educational research. *Educational Researcher, 32*(1), 9–13. DOI: 10.3102/0013189X032001009

Cobb, P., Gresalfi, M., & Hodge, L. L. (2009). An interpretive scheme for analyzing the identities that students develop in mathematics classrooms. *Journal for Research in Mathematics Education, 40*(1), 40–68. DOI: 10.5951/jresematheduc.40.1.0040

Csikszentmihalyi, M. (1990). *Flow: The psychology of optimal experience.* Harper & Row.

Fuentes-Cabrera, A., Parra-González, M. E., Lopez-Belmonte, J., & Segura-Robles, A. (2020). Learning mathematics with emerging methodologies-The escape room as a case study. *Mathematics, 8*(9), 1586. DOI: 10.3390/math8091586

Gutiérrez, R. (2008). A "gap-gazing" fetish in mathematics education? Problematizing research on the achievement gap. *Journal for Research in Mathematics Education, 39*(4), 357–364. https://www.jstor.org/stable/40539302. DOI: 10.5951/jresematheduc.39.4.0357

Huang, S.-Y., Kuo, Y.-H., & Chen, H.-C. (2020). Applying digital escape rooms infused with science teaching in elementary school: Learning performance, learning motivation, and problem-solving ability. *Thinking Skills and Creativity, 37*, 100681. DOI: 10.1016/j.tsc.2020.100681

Işık, H. K., & Liman Kaban, A. (2025). Escaping traditional learning: The role of educational digital escape games on achievement and motivation in primary math education. *Journal of Educational Technology and Online Learning, 8*(1), 136–151. DOI: 10.31681/jetol.1603015

Kahila, J., Parkki, T., Gröhn, A., Karvinen, A., Telimaa, E., Riikonen, P., Tiitta, R., Haantio, P., Keinänen, A., Kerkkänen, T., Jormanainen, I., Penttinen, S., & Tedre, M. (2020-November). Escape room game for CT learning activities in the primary school. *Koli Calling 20: Proceedings of the 20th Koli Calling International Conference on Computing Education Research*. 9, 1-5. Association for Computing Machinery. DOI: 10.1145/3428029.3428063

Kim, C., Na, H., Zhang, N., & Bai, C. (2024). Escape rooms for education: A meta-analysis. *International Journal of Instruction, 17*(4), 219–234. DOI: 10.29333/iji.2024.17413a

Lave, J., & Wenger, E. (1991). *Situated learning: Legitimate peripheral participation*. Cambridge University Press., DOI: 10.1017/CBO9780511815355

Pais, S., Sousa, A., & Pires, A. (2023-October). Using an escape room activity to enhance the motivation of undergraduate life science students in mathematics classes: A case study. *Proceedings of the 17th European Conference on Games Based Learning, 17* (1), 490-496. DOI: 10.34190/ecgbl.17.1.1431

Pinero Charlo, J. C. (2020). Educational escape rooms as a tool for horizontal mathematization: Learning process evidence. *Education Sciences, 10*(9), 213. DOI: 10.3390/educsci10090213

Ratinho, E., Martins, C., & Machado, J. (2023). The role of gamified learning strategies in student's motivation in high school and higher education: A systematic review. *Heliyon, 9*(8), e19033. DOI: 10.1016/j.heliyon.2023.e19033

Ruiz, J. J. R., Sanchez, A. D. V., & Figueredo, O. R. B. (2024). Impact of gamification on school engagement: A systematic review. *Frontiers in Education, 9*, 1466926. DOI: 10.3389/feduc.2024.1466926

Sfard, A. (2008). *Thinking as communicating: Human development, the growth of discourses, and mathematizing*. Cambridge University Press. DOI: 10.1017/CBO9780511499944

Sweller, J., Ayres, P., & Kalyuga, S. (2011). *Cognitive load theory*. Springer. DOI: 10.1007/978-1-4419-8126-4

Taraldsen, L. H., Haara, F. O., Lysne, M. S., Jensen, P. R., & Jenssen, E. S. (2022). A review on use of escape rooms in education – touching the void. *Education Inquiry, 13*(2), 169–184. DOI: 10.1080/20004508.2020.1860284

Ugo, C. C., López-Pernas, S., Gordillo, A., & Tukiainen, M. (2025). Escape rooms for learning programming: A systematic literature review. *Review of Education, 13*(3), 1-26, e70123. DOI: 10.1002/rev3.70123

Vidergor, H. E. (2021). Effects of digital escape room on gameful experience, collaboration, and motivation of elementary school students. *Computers & Education, 166*, 104156. DOI: 10.1016/j.compedu.2021.104156

Vygotsky, L. S. (1978). *Mind in society: The development of higher psychological processes* (Cole, M., John-Steiner, V., Scribner, S., & Souberman, E., Eds.). Harvard University Press. DOI: 10.2307/j.ctvjf9vz4

Wenger, E. (1998). *Communities of practice: Learning, meaning, and identity*. Cambridge University Press. DOI: 10.1017/CBO9780511803932

Wertsch, J. V. (1998). *Mind as action*. Oxford University Press. DOI: 10.1093/acprof:oso/9780195117530.001.0001

Wing, J. M. (2006). Computational thinking. *Communications of the ACM, 49*(3), 33–35. DOI: 10.1145/1118178.1118215

Yllana-Prieto, F., González-Gómez, D., & Jeong, J. S. (2025). The escape room and breakout as an aid to learning STEM contents in primary schools: an examination of the development of pre-service teachers in Spain. *Education 3-13*, 53(1), 15-31. DOI: 10.1080/03004279.2022.2163183

Zapata, M., Ramos-Galarza, C., Valencia-Aragón, K., & Guachi, L. (2024). Enhancing mathematics learning with 3D augmented reality escape room. *International Journal of Educational Research Open, 7*, 1–11, 100389. DOI: 10.1016/j.ijedro.2024.100389

Chapter 4
Bridging Ethnomathematics and School Mathematics:
Concept Mapping Across Program–Textbook–Task Levels

Fatih Kaleci
http://orcid.org/0000-0001-6823-3773
Necmettin Erbakan University, Turkey

Betül Tekerek
http://orcid.org/0000-0001-7066-6885
Faculty of Education, Kahramanmaraş Sütçü İmam University, Turkey & Faculty of Educational Sciences, Ankara University, Turkey

ABSTRACT

This chapter,bridges Ethnomathematics and Culturally Responsive Mathematics Education (CRME) through concept mapping and systematic document analysis of Türkiye Yüzyılı Maarif Modeli (TYMM) mathematics program and its Grade 5–6 textbooks—the complete set published during the reform's initial implementation year. Employing the C–T–M–E model and a 0–3 depth-of-integration rubric, the chapter codes 53 cultural elements across four textbook volumes (Cohen's κ=0.926). Results reveal that tokenistic (D1=34.0%), contextual (D2=43.4%), and transformative (D3=22.6%) integration coexist across the corpus, with a pronounced cross-grade gradient (Grade 5 D3=15.8%; Grade 6 D3=40.0%) and rare power-identity signaling (E5=9.4%). The theoretical framework integrates ethnomathematics, CRME, culturally sustaining pedagogy, critical mathematics education, teachers' textbook

DOI: 10.4018/979-8-2600-3485-9.ch004

use, and ethnomodeling–task design research. Practical deliverables include concept maps, a Sankey flow diagram, a depth rubric, and three task transformation patterns for curriculum designers and teachers.

INTRODUCTION

Across educational systems, school mathematics is positioned as foundational alongside literacy; however, achieving mathematical proficiency—conceptual understanding, procedural fluency, strategic competence, adaptive reasoning, and productive disposition—remains a persistent challenge for many learners (National Research Council [NRC], 2001). Contemporary standards similarly emphasize that proficiency requires sustained engagement with problem solving, reasoning and proof, communication, connections, and representation, rather than a narrow focus on routine procedures (NCTM, 2000). Yet in many contexts, mathematics continues to be experienced by students as an impersonal, technique-driven subject where success is equated primarily with speed and accuracy, restricting opportunities for sense-making, justification, modeling, and meaningful participation (Bishop, 1991; NRC, 2001). A cultural perspective helps explain why technique-oriented representations can be educationally limiting. Bishop (1991) conceptualizes mathematics as a cultural product and highlights how curriculum traditions can foreground "technique-oriented" approaches at the expense of understanding and critical engagement with mathematics in daily and societal life. Educating learners mathematically entails not only teaching mathematics but also supporting learners in understanding what mathematics is for, how it is justified, and how it relates to values and decisions in society.

Ethnomathematics and Culturally Responsive Mathematics Education (CRME) offer complementary frameworks for addressing this problem. Ethnomathematics (D'Ambrosio, 2001) foregrounds mathematical ideas and practices developed within cultural groups, treating everyday practices—measurement, trade, craft, design—as legitimate sites of mathematical knowing/doing. Building on this orientation, ethnomodeling provides a methodological bridge from community practices to school mathematics (Rosa & Orey, 2013, 2016). CRME, extended by culturally sustaining pedagogy (Paris, 2012; Paris & Alim, 2017) and critical mathematics education (Skovsmose, 1994; Gutstein, 2006), goes further; it asks not only whether cultural resources support mathematical sense-making but whether curriculum and materials position learners as legitimate, rights-bearing mathematical agents who can use mathematics to examine and critique social reality (Gutiérrez, 2008, 2013; Frankenstein, 1983).

Despite strong theoretical arguments for culturally grounded mathematics education, a persistent practical gap remains: cultural integration is increasingly demanded in curricula and materials, yet the presence of cultural references does not automatically guarantee meaningful mathematical work or equitable participation. Textbooks are central in this gap because they operationalize curriculum into explanations, representations, tasks, and assessments that structure students' opportunities to learn (OTL). However, the relationship between textbook design and learning opportunities is mediated by teacher use—how teachers interpret, adapt, and enact textbook tasks fundamentally shapes the opportunities that students actually experience (Remillard, 2005; Brown, 2009; Remillard & Heck, 2014). This means that culturally grounded task design must be examined not only as a designed artifact but as an enacted curriculum, with implications for both how textbooks are written and how teachers are supported in activating their cultural integration potential.

This chapter responds to that need through a document-based, concept-mapping approach that systematically codes culturally embedded units across four Grades 5–6 mathematics textbook volumes-the complete set of Türkiye's *Türkiye Yüzyılı Maarif Modeli* (TYMM) mathematics program-aligned middle school mathematics textbooks available at the time of analysis (Ministry of National Education (MoNE) (2024b). In Türkiye, the Ministry of National Education (MoNE) centrally produces and distributes one official textbook set per grade and subject for all state schools, with no competing publisher editions. The chapter develops a theoretical framework that integrates four literature strands that have not previously been systematically combined for this type of analysis:

(1) teachers' textbook use and curriculum enactment research;
(2) culturally sustaining pedagogy and critical mathematics education;
(3) textbook-task enactment and opportunity-to-learn; and
(4) ethnomodeling–task design bridges. These strands are woven into the C–T–M–E model and inform the findings, discussion, and toolkit presented in subsequent sections.

BACKGROUND: THEORETICAL ANCHORS FOR BRIDGING ETHNOMATHEMATICS AND CRME

Ethnomathematics as Cultural–Mathematical Practice

Ethnomathematics, in D'Ambrosio's foundational framing, is not a "supplement" to school mathematics but an epistemological and educational stance that treats mathematical knowledge as historically and culturally produced (D'Ambrosio, 1985).

A widely used practical definition describes ethnomathematics as "the mathematics practiced by cultural groups," recognizing mathematical activity embedded in occupational communities, rural and urban practices, professional groups, and children's everyday reasoning (D'Ambrosio, 2001). Crucially, ethnomathematics distinguishes between context as decoration and context as epistemic resource: cultural practices should be treated as sites for mathematizing—identifying quantities and relations, articulating constraints, representing patterns, justifying methods, and generalizing. From a curriculum–textbook perspective, ethnomathematics provides:

(i) a selection logic (what counts as culturally meaningful mathematical activity),
(ii) a translation problem (how to move from informal/community practice to school-formal mathematics without erasing meaning), and
(iii) a quality criterion (whether tasks support genuine reasoning, modeling, representation, and critique rather than tokenism).

Figure 1 illustrates this cycle: cultural practice serves as the entry point for mathematizing, which proceeds through problem formulation, mathematical representation, validation and interpretation, and generalization and critique before feeding back into community knowledge as a renewed cultural resource.

Figure 1. The ethnomathematics cycle

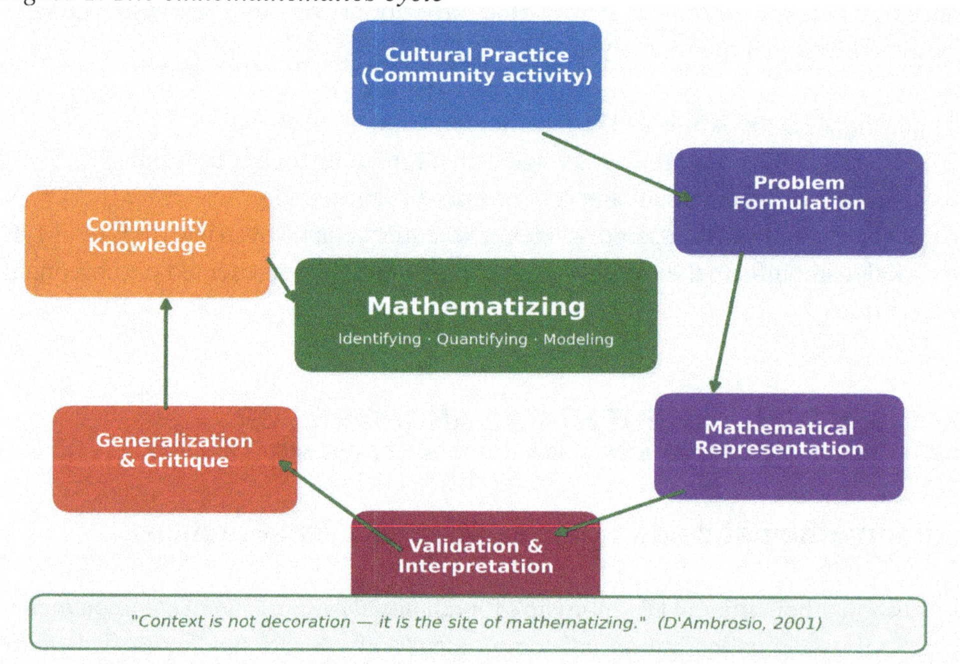

The diagram shows the six-stage iterative process through which cultural practices are transformed into mathematical knowledge and returned to community life. The central node, *Mathematizing* (identifying, quantifying, modeling), is the operative link between cultural practice and formal mathematical representation. Directed arrows indicate the cyclical, non-linear character of ethnomathematical inquiry. The foundational claim at the base—"Context is not decoration — it is the site of mathematizing" (D'Ambrosio, 2001)—summarizes the epistemological distinction between cultural context as decoration (D1) and cultural context as epistemic resource (D3) that organizes the C–T–M–E depth analysis throughout this chapter.

Mathematical Enculturation: From Technique to Meaning, Values, and Purpose

Bishop's (1991) mathematical enculturation perspective frames mathematics education as an induction into a mathematical culture: not merely learning procedures, but internalizing symbol systems, norms of justification, and value-laden purposes for using mathematics. This viewpoint makes visible the "hidden curriculum" of mathematics—what textbooks implicitly teach about what mathematics is, what counts as evidence, and whose knowledge is valued. Bishop's central critique of technique-oriented curriculum traditions—where mathematics is represented as an action-based discipline built through procedural rules at the expense of understanding—directly informs the chapter's depth-of-integration rubric. A culturally grounded curriculum should include societal uses and values attached to mathematical explanations, and a framing of mathematics as a universal human activity occurring in all societies (Bishop, 1991, 1997). Figure 2 maps this spectrum: the horizontal gradient from *technique* (rules and procedures) through *justification* (why it works) and *societal use* (what it is for) to *critical participation* (questioning and creating) corresponds directly to the depth-of-integration levels D1 through D3 in the C–T–M–E model, making enculturation depth an observable and codable property of individual textbook tasks.

Figure 2. Mathematical enculturation spectrum (Bishop, 1991; Bishop, 1997)

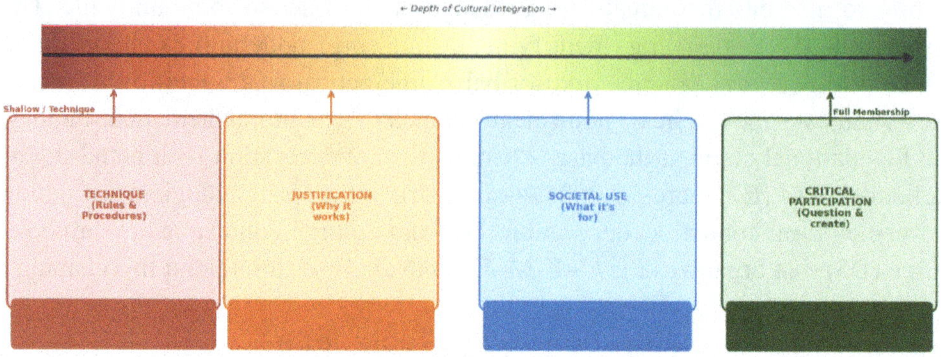

The gradient represents the continuum from technique-oriented traditions (rule-following) toward full cultural membership (critique, justification, societal purpose). The lower row maps each enculturation level onto its C–T–M–E code: C1–C3 with Decor T-role produces D1; C2 with Directive T-role produces D2; C4 with Modeling Agent T-role produces D3.

Ethnomodeling: From Community Practice to Task Design

Ethnomodeling is a methodological and theoretical bridge between ethnomathematics and mathematical modeling. Rosa and Orey (2013) conceptualize ethnomodeling as situated at the intersection of cultural anthropology, mathematics, and mathematical modeling—aiming to understand and represent mathematical phenomena within cultural practices while enabling translation into school-formal representations. Under ethnomodeling, culture functions as a driver of: (i) problem formulation, (ii) mathematization, (iii) validation/interpretation, and (iv) communication and critique. In their subsequent work, Rosa and Orey (2016) developed ethnomodeling into a glocal approach that respects locally situated mathematical practices (emic perspective) while translating them into school-formal representations accessible to broader mathematical communities (etic perspective). This glocal framing has direct implications for task design: a glocally oriented task begins with an authentic local practice, requires students to mathematize it (formulate a model), validate the model against contextual constraints, and generalize the mathematical structure (the complete ethnomodeling cycle that corresponds to D3 in the present study's depth classification).

The task design implications of ethnomodeling have been developed most concretely by Rosa and Orey (2015), who proposed a trivium curriculum framework organizing mathematical tasks around literacy (reading mathematical reality), matheracy (using mathematics to critique and transform reality), and technoracy (using technological tools to model mathematical phenomena). This three-level framework maps onto the D1/D2/D3 depth structure in the present study: literacy-level tasks correspond to D1–D2 (reading and interpreting mathematical representations of cultural reality), while matheracy-level tasks correspond to D3 (using mathematics to critique and transform understanding of cultural phenomena). More recently, Fouze and Amit (2018) demonstrated that explicitly designed ethnomathematics-based tasks—grounded in local folklore, traditional games, and cultural artifacts—can significantly increase student engagement, mathematical reasoning quality, and cultural identity affirmation, providing empirical validation for the task design principles implied by ethnomodeling theory. These developments establish that ethnomodeling is not merely a theoretical framework but an empirically tested design science for mathematics tasks.

Culturally Responsive Mathematics Education: From Equity to Cultural Sustaining and Critical Dimensions

CRME builds on culturally relevant and responsive pedagogy traditions by treating culture not only as a resource for meaning-making but also as a central axis of equity. Ladson-Billings's (1995) culturally relevant pedagogy foregrounds three broad aims—academic success, cultural competence, and critical consciousness. Gay's (2018) culturally responsive teaching emphasizes that instruction becomes more effective when filtered through students' cultural experiences and communication repertoires. Aguirre and Zavala (2013) provide a materials lens for document analysis by integrating mathematical thinking, language, culture, and social justice.

Culturally Sustaining Pedagogy: Maintaining and Extending Cultural Wealth

Paris (2012) introduced culturally sustaining pedagogy (CSP) as an extension and revision of culturally relevant pedagogy, arguing that cultural responsiveness is necessary but insufficient if its goal is only to bridge from students' culture to academic content. Rather than leveraging students' cultural practices to ease entry into dominant academic culture, CSP aspires to maintain and revitalize those practices, languages, and ways of knowing as inherently valuable. Paris and Alim (2017) extended this framework in a collection demonstrating how culturally sustaining teaching looks in practice across diverse school contexts, from indigenous language revitalization to hip-hop literacies.

For mathematics textbook analysis, CSP introduces a criterion beyond D'Ambrosio's ethnomathematics and Gutiérrez's equity dimensions: *does this task sustain the cultural practice from which the mathematical problem is drawn, or does it appropriate the cultural setting as a motivational shell while erasing its meaning?* This question is operationalized in the present study through the C4 (Value/ethical dimension) sub-code and the distinction between Decor T-role (cultural erosion) and Modeling Agent T-role (cultural sustaining). A task that positions Ottoman architectural ornamentation as a visual backdrop for shape identification (C3 Decor) appropriates cultural material while erasing its constructive logic; a task that requires students to reverse-engineer the mathematical construction rules of that ornamentation (C3 Modeling Agent) sustains the cultural practice by making its mathematical knowledge explicit and valued. CSP thus provides an additional normative criterion for the depth-of-integration rubric: D3 is not only mathematically superior but culturally sustaining in Paris's sense.

Critical Mathematics Education: Mathematics as a Tool for Social Critique

Critical mathematics education, rooted in Freire's (2014) critical pedagogy, frames mathematical learning as a tool for reading and writing the world—understanding mathematical structures that shape social reality and using mathematics to critique and transform those structures. Skovsmose (1994) developed this tradition as *critical mathematics education* (CME), arguing that mathematical competence must include the ability to examine the mathematical formatting of social reality: how decisions are justified through mathematical models, who benefits from those models, and what is obscured or erased by mathematical framing. Frankenstein (1983) provided one of the earliest curriculum applications, demonstrating how adult education mathematics courses could be organized around social justice data and questions, developing both mathematical skills and critical consciousness simultaneously.

Gutstein's (2006) K–12 application distinguishes "reading the world with mathematics" (using mathematical analysis to understand social phenomena) from "writing the world with mathematics" (using mathematics to act toward social change). Gutstein and Peterson (2005) compiled practical classroom examples demonstrating both dimensions.

For the present study, CME and CSP together provide the theoretical grounding for the E5 (power-identity signaling) sub-code—the equity sub-code that most directly corresponds to Gutiérrez's (2008) "changing the game" dimension. A task that invites students to examine data about personal data violations and propose protection measures is not merely "culturally responsive" in the sense of connecting to students' lived experience; it engages students in mathematical critique of social reality — reading the world with mathematics (Gutstein, 2006). When such tasks are absent or rare, the materials sustain cultural references at the level of national pride and heritage while failing to invite students to use mathematics to examine, critique, or transform social structures.

Figure 3 maps the relationship among these three frameworks: CRME establishes equity as an observable design feature of opportunity structures (E2+E3+E4), CSP extends this by treating community knowledge as a mathematical asset to be sustained rather than merely leveraged (C1+C3 with Modeling Agent T-role), and CME goes furthest by positioning mathematics as a tool for social critique and transformation (C4+E5). The Gutiérrez (2008) equity axis running across all three frameworks—from "playing the game" (access and achievement) to "changing the game" (identity and power)—shows that D1 integration remains confined to the dominant axis, while D3 integration requires crossing into the critical axis where CSP and CME converge.

Figure 3. CRME, culturally sustaining pedagogy, and critical mathematics education: Equity dimensions

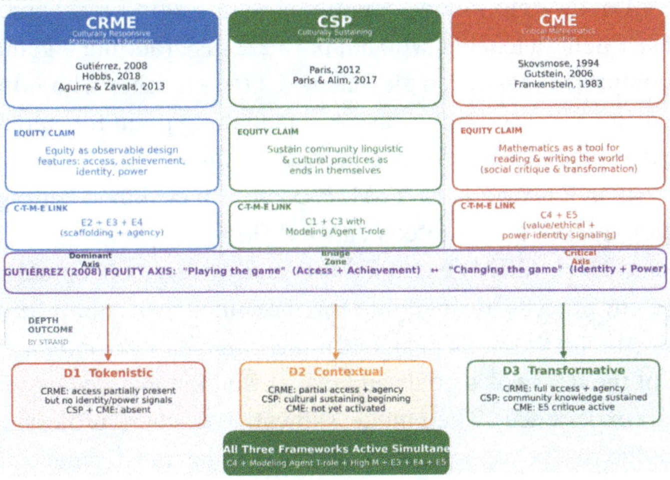

The three-column framework maps each tradition's equity claim, key scholars, and C–T–M–E code linkage across Gutiérrez's (2008) axis from "playing the game" (access) to "changing the game" (identity + power). D3 requires all three frameworks simultaneously: C4 with Modeling Agent T-role, high M demand, and E3+E4+E5 signals. In the TYMM corpus, only 5 units (9.4%) reach the E5 threshold, reflecting the virtue-oriented rather than critically-oriented character of the TYMM value framework.

Teachers' Textbook Use: From Design to Enactment

Research on teachers' use of curriculum materials consistently shows that textbook design is necessary but insufficient for determining what opportunities students actually encounter. Remillard's (2005) foundational review identified four forms of curriculum use—following, interpreting, adapting, and subverting—and argued that teacher–material interaction is participatory: both teacher and materials are transformed through use. Teacher interpretation mediates between the designed curriculum and the enacted curriculum.

Brown (2009) developed the concept of the teacher-tool relationship to describe how teachers interact with curriculum resources. Brown distinguishes three modes of use: *offloading* (following the curriculum as a script), *adapting* (modifying the curriculum for specific contextual needs), and *improvising* (drawing from the curriculum as a resource while departing substantially from its structure). For cultural integration in mathematics tasks, this typology has direct implications: a teacher who offloads a D1 task (following a national data routine mechanically) cannot produce D3 enactment; a teacher who adapts a D2 task (adding a genuine research question to a comparison task) can elevate its depth; a teacher who improvises with D3 materials (using a privacy rights task as a launching point for student-generated inquiry) can extend the task's cultural and mathematical reach. Brown's framework therefore suggests that textbook C–T–M–E depth is a designed ceiling for cultural integration potential, not a guaranteed enacted floor.

Remillard and Heck (2014) extended this analysis by conceptualizing the curriculum enactment process as involving three distinct curriculum forms: the *written curriculum* (the textbook as designed), the *intended curriculum* (the teacher's interpretation of the textbook's goals and tasks), and the *enacted curriculum* (what students actually experience). Their model shows that teachers' professional contexts, content knowledge, and pedagogical beliefs mediate the transformation from written to enacted curriculum. For the present study, this three-form curriculum model is directly relevant: the C–T–M–E analysis operates on the written curriculum layer, establishing the designed depth of cultural integration; the findings and toolkit are intended to influence both the written curriculum (through textbook author recommendations) and the intended curriculum (through teacher planning tools) in ways that increase the likelihood of D3 enactment.

International comparative work by Pepin and Haggarty (2001) examined how teachers in England, France, and Germany use mathematics textbooks and found systematic national differences in the degree to which textbooks are followed versus adapted—reflecting both cultural norms of teacher autonomy and the structural features of textbooks themselves. In contexts where textbooks provide detailed procedural guidance with little room for teacher decision-making (high "determination"), teachers tend to follow materials more closely, which means that D1 task design is more likely to remain D1 in enactment. This finding is particularly relevant to the TYMM context: given that Türkiye's centralized single-textbook system may be associated with higher textbook determination than systems with multiple competing publishers, the design depth of TYMM textbooks is likely to have particularly strong consequences for the enacted depth of cultural integration across all state schools.

Figure 4 visualizes this three-stage mediation: the written curriculum (textbook as designed) establishes a D3 ceiling of 22.6% that teachers can at best maintain but more typically reduce through one of Brown's (2009) three interaction modes—offloading the task as a script, adapting it to contextual needs, or improvising with it as a flexible resource. The task degradation risk identified by Stein et al. (2007) is marked explicitly: a D3 task whose ethical constraint is stripped or whose open formulation stage is pre-answered by the teacher is enacted at D1–D2 regardless of its written depth, which means that improving textbook cultural integration depth is a necessary but not sufficient condition for improving students' actual opportunity to learn culturally grounded mathematics.

Figure 4. Teachers' textbook use: From written to enacted curriculum (Brown, 2009; Remillard & Heck, 2014)

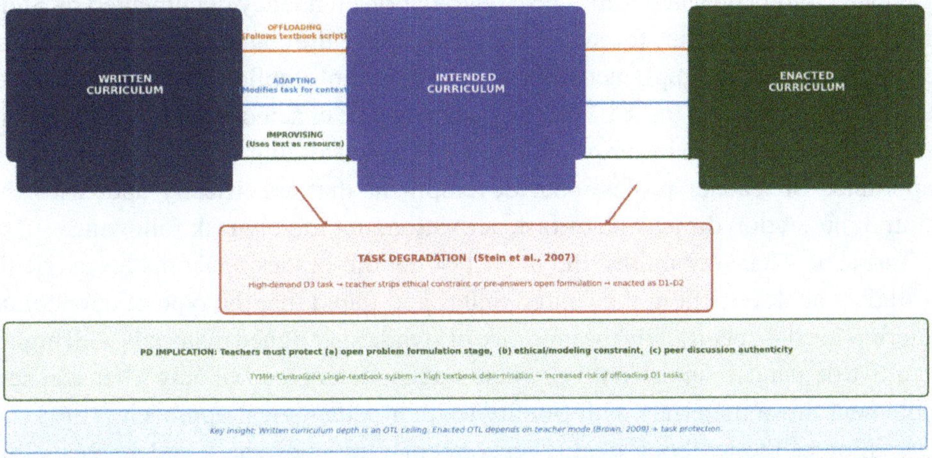

The diagram shows the three curriculum forms—written, intended, and enacted—connected through teachers' three interaction modes (Brown, 2009): offloading, adapting, and improvising. The task degradation box (Stein et al., 2007) marks the key risk for D3 tasks: high-demand culturally grounded tasks are reduced in cognitive complexity when teachers strip the ethical modeling constraint or pre-answer the open problem formulation stage. In the TYMM context, the centralized single-textbook system increases the likelihood of offloading, making the written D3 ceiling of 22.6% an upper bound rather than a reliable predictor of enacted depth.

Textbook-Task Enactment and Opportunity-to-Learn

Opportunity to learn (OTL) refers to the learning opportunities made available through curriculum, instruction, and assessment (Carroll, 1963; Husén, 1967). Schmidt et al.'s (1997) analysis across 45 countries showed that differences in textbook cognitive demand account for substantial achievement variation. Their "splintered vision" concept describes how shallow curricula fragment learning by spreading across many low-depth topics. This applies directly here: a corpus dominated by D1 tokenistic integration provides the appearance of cultural responsiveness without genuine OTL for culturally grounded mathematical reasoning.

Stein et al. (2007) provided a comprehensive analysis of how curriculum influences student learning, synthesizing research on the textbook-to-classroom pathway. Their model identifies three stages of curriculum influence: *written curriculum* (as designed and published), *intended curriculum* (as interpreted and planned by teachers), and *enacted curriculum* (as experienced by students). At each stage, the cognitive demand and cultural integration potential of tasks can be either maintained or reduced, rarely elevated. This "task degradation" tendency documented by Stein and colleagues—in which teachers frequently reduce the cognitive demand of high-demand tasks during implementation—has important implications for the present study: even D3 tasks in the TYMM textbooks may be enacted at D1 level if teachers strip the ethical constraint or provide too much scaffolding. This underscores the importance of teacher professional development that specifically addresses the cultural integration dimension of task activation, not merely task following.

Tarr et al. (2008) examined the impact of middle-grades mathematics curricula on student achievement in the United States and found that the type of curriculum materials used—specifically the presence of standards-aligned materials with higher cognitive demand—significantly predicts student outcomes, but only when teachers implement those materials with fidelity to their pedagogical approach. This finding reinforces the textbook quality–enactment–outcomes chain and suggests that improving textbook cultural integration depth (from D1 to D3) is a necessary but not sufficient condition for improving culturally situated mathematical reasoning: teacher enactment support is the complementary necessary condition. For the present study, this literature motivates the teacher-facing components of the toolkit (Section 7)—specifically the "remove-context" diagnostic test and the three upgrade patterns, which are designed to be operable during lesson planning rather than requiring restructuring of the full instructional sequence.

Within the textbook-task enactment strand, Haggarty and Pepin's (2002) comparative analysis of mathematics textbooks in England, France, and Germany examined not only textbook content but how textbooks' structural features—their organization of tasks, their scaffolding design, their representation repertoires—enable or constrain different forms of enactment. Their finding that textbooks carry implicit pedagogical messages about what mathematics is and how it should be learned resonates with Bishop's (1991) "hidden curriculum" concept and with the present study's concept map approach: the TYMM curriculum concept map (Figure 5) makes visible the explicit structural connections between the curriculum's value framework and the task design features that should operationalize those values. When those connections are present in the written curriculum but absent from the enacted task (as in the D1 token bridge pattern), the implicit message of the textbook is that cultural references are decorative rather than mathematically serious.

Figure 5 maps this five-stage chain: curriculum intention produces a written OTL design (textbook), teacher interpretation transforms the written into an intended curriculum, classroom enactment produces the enacted curriculum, and students finally experience the actual OTL. The declining red arc across the bottom of the figure operationalizes Stein et al.'s (2007) task degradation finding: because high-demand tasks are consistently reduced in cognitive complexity at the enactment stage, the written D3 ceiling of 22.6% functions as an upper bound rather than a reliable predictor of the OTL students actually receive. Schmidt et al.'s (1997) "splintered vision" critique—that shallow distribution of cultural references across many surface topics without depth constitutes a fragmented OTL—applies with equal force to the TYMM corpus: a corpus dominated by D1 tokenistic integration provides the appearance of culturally grounded mathematics without the depth that constitutes authentic opportunity to learn culturally sustained mathematical reasoning.

Figure 5. Textbook-task enactment and opportunity-to-learn chain (Schmidt et al., 1997; Stein et al., 2007)

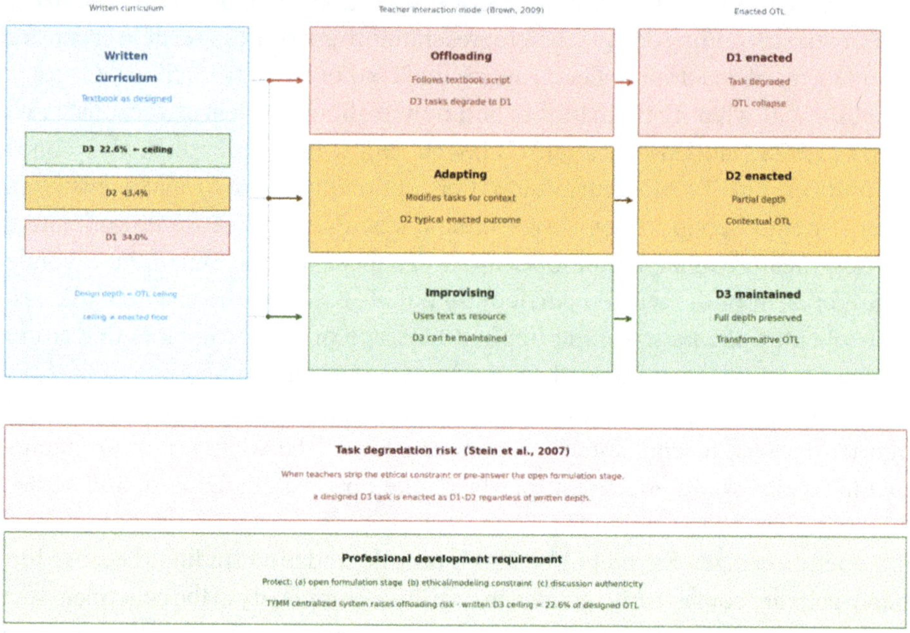

The five-node chain shows how curriculum policy intention flows through textbook design (written OTL), teacher interpretation (intended OTL), classroom enactment (enacted OTL), and student experience (experienced OTL). The declining arc visualizes cumulative attenuation of designed depth through the enactment chain. The written D3 ceiling of 22.6% is an upper bound: teacher interaction modes and task degradation risk (Stein et al., 2007) further reduce it. In the TYMM centralized single-textbook system, this ceiling represents the maximum culturally transformative OTL available nationally.

Shield and Dole (2013), in their analysis of mathematics textbooks' potential to promote deep learning, found that cultural and real-world contexts tend to be introduced at surface level—primarily motivating computation rather than generating mathematical inquiry. Their framework for assessing "deep learning potential" anticipates several features of the C–T–M–E model, particularly the distinction between tasks where context is incidental and tasks where context structures mathematical thinking.

Xenofontos and Andrews (2014), in their comparative study across national contexts, demonstrated that the depth of cultural integration is not simply a function of intention but of the structural role assigned to cultural context in national curriculum frameworks. Their findings suggest that nations differ systematically in whether cultural contextualization functions as motivational (paralleling D1) or epistemically active (paralleling D3) a macro-level parallel to the micro-level T-role distinction in the C–T–M–E model.

Pöhler and Prediger (2015), examining cultural context in German mathematics materials, documented a "scaffolding without rigor" pattern in which culturally grounded tasks provide representational support without increasing mathematical demand or expanding student agency. This pattern is precisely the E2-without-E4 gap in the TYMM E-axis analysis. Together, these three comparative studies confirm that the TYMM patterns are widespread design tendencies, not TYMM-specific failures. The C–T–M–E model advances prior comparative work by adding the functional dimension ("what does culture do?") and the equity dimension ("does scaffolding support genuine agency?").

Turkish Mathematics Education Context: TYMM and Reform Trajectory

Türkiye's mathematics education reform trajectory provides institutional backdrop for interpreting the TYMM corpus. The Turkish school system has undergone several curriculum reform cycles since 2005, responding to both international assessment pressures (PISA and TIMSS) and domestic educational policy priorities. Prior curricula emphasized a constructivist-influenced learning approach (2005, 2013, 2018 cycles), and Turkish mathematics education scholarship has examined how successive reform cycles have translated into textbook design, teacher practice, and student outcomes (Bingölbali et al., 2016; Işık & Kar, 2012).

TYMM (Türkiye Yüzyılı Maarif Modeli), introduced for the 2024–2025 academic year, represents the most comprehensive educational reform in Türkiye in the past decade. Unlike previous curricula, TYMM explicitly integrates a holistic character development framework (Erdem–Değer–Eylem: Virtue–Value–Action) with disciplinary learning objectives and positions culture, values, and cross-disciplinary literacies as integral to—not supplementary to—the academic program (MoNE, 2024a). In mathematics, the reform introduces a skills-based architecture and explicitly names "Kültür Okuryazarlığı" (Culture Literacy, OB5) as a cross-disciplinary process component embedded in learning outcomes (MoNE, 2024b). Türkiye operates a centralized single-textbook system: MoNE directly produces and distributes one official textbook set per grade and subject for all state schools, with no competing publisher editions. The four volumes analyzed constitute 100% of officially available TYMM lower-secondary mathematics materials at the analyzed grade levels (MoNE, 2025c).

Why Textbooks? Opportunities to Learn and Empirical Traces

Textbooks are a major mechanism through which the intended curriculum is translated into OTL in classrooms. Cross-national textbook research has shown that textbooks communicate norms about what students should do with content—functioning as a policy-to-practice medium at scale (Valverde et al., 2002). The Mathematical Tasks Framework classifies tasks by cognitive demand and shows how materials can over-concentrate opportunities in lower-demand categories, thereby limiting meaning-making and reasoning (Stein et al., 2000). For culturally contextual tasks specifically, this literature provides an analytic bridge: tasks can be evaluated not only for cultural presence but for the mathematical work they demand, the equity signals they carry, and—given the teachers' use literature reviewed.

The C–T–M–E Bridge Model and Depth-of-Integration Logic

The C–T–M–E model is the analytic heart of this chapter. It operationalizes the bridge between Ethnomathematics and CRME—drawing on ethnomodeling (Rosa & Orey, 2013, 2016), culturally sustaining pedagogy (Paris, 2012), critical mathematics education (Skovsmose, 1994; Gutstein, 2006), equity/access design (Hobbs, 2018; Gutiérrez, 2008), and OTL (Stein et al., 2000)—into a relational diagnostic framework that can be applied to individual textbook tasks. This section (2.10) presents the model's theoretical architecture and sub-code definitions; the operationalization of these definitions as a coding instrument (including decision rules for boundary cases) is detailed in Section 2.10.7 and its methodological application is described in Section 4.5. Empirical frequency distributions are reported in the Findings (Section 5). Figure 6 provides a comprehensive visual overview of the complete model.

Figure 6. The C–T–M–E bridge model: A comprehensive overview

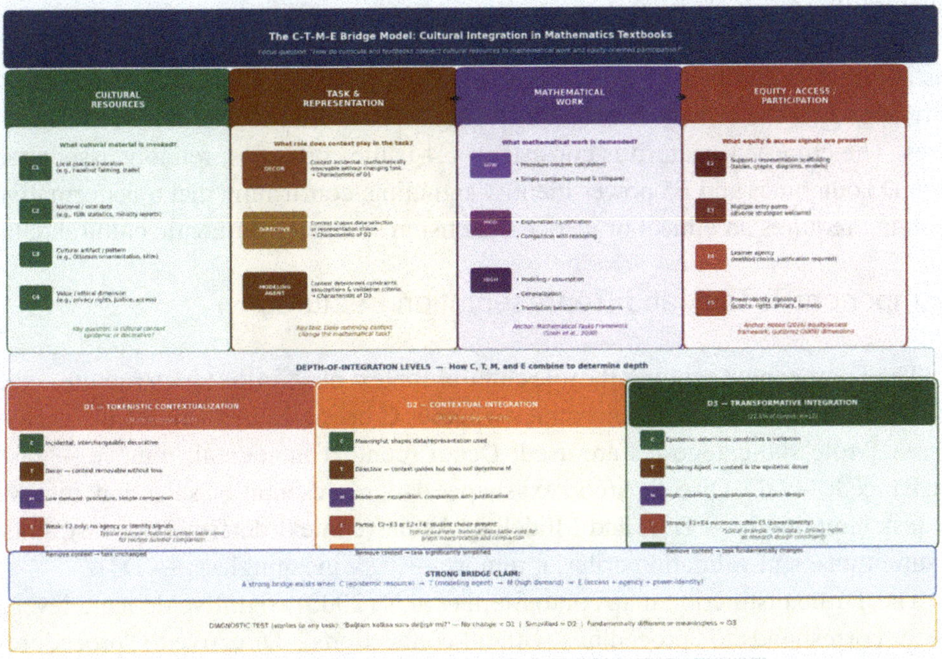

Note: Top panel: four coupled analytic components. Detail panels: sub-codes with operational definitions, analytic questions, and theoretical anchors. Bottom panel: depth levels D1/D2/D3 with characteristic configurations, corpus frequencies, and typical examples. Banner rows: the strong bridge claim and the diagnostic test.

Component C: Cultural Resources

The C component captures the cultural material invoked in a textbook task. Four sub-codes are used: *C1 (Local practice/vocation)*—community economic practices (hazelnut farming, carpet weaving, local trade); *C2 (National/local data)*—public institutional data (TÜİK statistics, ministry reports); *C3 (Cultural artifact/pattern)*—architectural forms, craft patterns, heritage objects; *C4 (Value/ethical dimension)*—privacy rights, justice, environmental responsibility, accessibility. C is meaningful when it can be mathematized: when the cultural material introduces constraints, variables, or validation criteria that shape the mathematical task. C is not meaningful when it provides only a narrative label interchangeable with any other setting.

From a CSP perspective (Paris, 2012), C1 and C3 codes relate most directly to sustaining community mathematical knowledge, while C4 (value/ethical dimension) corresponds to the critical dimension—using mathematics to examine social structures. The empirical pattern is consistent: C4 is the code most reliably associated with D3 outcomes and E5 power-identity signaling, confirming that transformative bridging requires an ethical or justice dimension alongside authentic cultural data.

Component T: Task and Representation Features

The T component captures the role cultural context plays in the task structure—the functional assignment that determines whether C becomes mathematically active. Three T-role sub-categories are used: Decor (context incidental; remove → task unchanged; → D1), Directive (context shapes data/representation selection; remove → task simplified; → D2), and Modeling Agent (context determines constraints, assumptions, and validation criteria; remove → task meaningless; → D3).

The T-role distinction maps onto Stein et al.'s (2000) cognitive demand levels: Decor corresponds to "procedures without connections"; Directive to "procedures with connections"; and Modeling Agent to "doing mathematics," where students formulate, test, and validate models in response to culturally situated constraints. T-role is also the design feature most susceptible to task degradation during enactment (Brown, 2009; Remillard & Heck, 2014): a Modeling Agent task becomes Directive or Decor when a teacher provides excessive scaffolding, strips an ethical constraint, or pre-formulates the research question the task was designed to leave open. Teacher professional development should therefore explicitly address how to protect the Modeling Agent T-role.

Component M: Mathematical Work

The M component captures the mathematical practices demanded by the task, grounded in the Mathematical Tasks Framework (Stein et al., 2000). Six M sub-codes are organized into three cognitive demand levels: Low demand—procedure (routine calculation) and simple comparison; Moderate demand—explanation/justification and comparison with reasoning; High demand—modeling/assumption, generalization, and translation between representations.

The key insight is that M-code type alone does not determine depth: the T-role is decisive. The same comparison task (moderate M) remains D1 when C is Decor, reaches D2 when C is Directive, and achieves D3 when C is Modeling Agent. This T–M interaction is what prior binary (present/absent) cultural integration analyses miss. From an ethnomodeling perspective (Rosa & Orey, 2015), high-demand M corresponds to the matheracy level—using mathematics to critically analyze cultural phenomena—while low-demand M corresponds to literacy-level engagement. The D3 design challenge is to achieve matheracy-level work without sacrificing cultural authenticity.

Component E: Equity, Access, and Participation

The E component captures equity-relevant design features. Five sub-codes are used: E2 (support for explanation/representation—scaffolding, tables, graphs, diagrams; near-universal in corpus, n=50/53), E3 (multiple entry points—diverse strategies acceptable, n=15), E4 (learner agency—method choice, question formulation, n=28), E4+ / collaboration (peer discussion invitation, n=22), and E5 (power-identity signaling—justice, rights, privacy, fairness, n=5).

The E-axis draws on three theoretical streams: Hobbs (2018) for equity as observable opportunity structure (E2 = representational access; E3+E4 = agentic participation); Paris (2012) for sustaining community mathematical knowledge (E3+E4 with C1/C3); and Gutstein (2006) and Skovsmose (1994) for critical consciousness (E5 = reading the world with mathematics). The hierarchy encodes a progression from passive access (E2) through agency (E3+E4) to critique (E5). E5's rarity in the corpus (9.4%) is analyzed in Section 6.3.

Depth-of-Integration Logic

D1 — Tokenistic Contextualization (34.0%, n=18): C incidental (Decor T); M low; E weak (E2 only). Cultural material provides motivational packaging without reorganizing mathematical work. Consistent with D'Ambrosio's (2001) critique of "context as decoration" and with Gutiérrez's (2008) concern about cultural integration that only "plays the game."

D2 — Contextual Integration (43.4%, n=23): C meaningful (Directive T); M moderate; E partial (E2+E3 or E2+E4). Cultural context contributes to mathematical sense-making. Corresponds to Shield and Dole's (2013) contextual integration level and to Rosa and Orey's (2015) literacy-level mathematical engagement.

D3 — Transformative Integration (22.6%, n=12): C epistemic (Modeling Agent T); M high; E strong (E3+E4, often E5). Cultural context is the epistemic driver of the mathematical inquiry. Corresponds to the matheracy level of Rosa and Orey's (2015) trivium framework, to Gutstein's (2006) "reading and writing the world with mathematics," and to Paris's (2012) culturally sustaining pedagogy standard for valuing and activating community mathematical knowledge.

The Strong Bridge Claim and Diagnostic Test

The C–T–M–E model makes a central bridge claim: a strong bridge exists when C (epistemic resource) → T (modeling agent) → M (high demand) → E (access + agency + power-identity). Each coded unit is subjected to the "context-removal diagnostic test"—*"Bağlam kalksa soru değişir mi?" (Does removing the context change the mathematical task?)*—with three outcomes: No change → D1; Task significantly simplified → D2; Task fundamentally changed or meaningless → D3. This single diagnostic question provides a rapid, reliable, theoretically grounded instrument for depth classification, operable in both formal coding and teacher lesson planning contexts.

Operationalization: Codebook Decision Rules

The C–T–M–E model is operationalized through a codebook with six key decision rules for boundary cases:

- D1 vs. D2: "Does the cultural context determine which specific data values, representations, or strategies are used?" → Yes = D2; No = D1.
- D2 vs. D3: "Does removing the cultural context fundamentally change or destroy the task structure?" → Yes = D3; No = D2.
- E2 vs. E4: "Does the task require students to make a genuine mathematical decision (choose method, formulate question, select representation)?" → Decision required = E4; Support only = E2.
- C2 vs. C4: "Is the cultural resource a data source alone, or does it carry a values/rights/ethical dimension shaping the inquiry constraints?" → Values/rights shaping constraints = C4; Data source only = C2.
- C3 Decor vs. C3 Modeling Agent: "Does the task require engaging with the mathematical construction logic of the artifact, or merely identifying shapes displayed in it?" → Construction logic = Modeling Agent; Identification only = Decor.
- Inter-rater boundary (D1/D2): "Would any numerical dataset serve equally well?" → Yes = D1; No, this specific dataset matters = D2.

MAIN FOCUS OF THE CHAPTER

Case Context: TYMM, Centralized Textbooks, and the Scope of Analysis

This chapter is situated within Türkiye's 2024 curriculum reform (TYMM). At the time of this study, the officially published TYMM-aligned lower-secondary mathematics textbooks are available exclusively for Grades 5 and 6 (two volumes per grade: Books 1 and 2). Grades 7 and 8 textbooks had not yet been published. The four volumes analyzed therefore constitute the complete and comprehensive corpus of TYMM lower-secondary mathematics textbooks currently available—not a sample from a larger set. Additionally, Türkiye operates a centralized single-textbook system: MoNE produces and distributes one official textbook set per grade and subject for all state schools, with no competing publisher editions.

Research Questions

RQ1 (Distribution and forms): How and where do the written curriculum (TYMM mathematics, grades 5–6) and officially approved textbooks invoke cultural resources relevant to ethnomathematics and CRME?

RQ2 (Depth of integration): To what extent do these invocations reflect tokenistic, contextual, or transformative integration—what are the dominant C–T–M–E configurations and depth patterns across grade levels and content domains?

RQ3 (Alignment and gaps): How aligned are curriculum intentions and textbook instantiations regarding cultural integration and equity/access—where do strong bridges appear and where do systematic ruptures occur?

METHODOLOGY

Research Design

This chapter employs a qualitative document analysis design, strengthened by a systematic content analytic coding procedure and a concept-mapping synthesis stage (Bowen, 2009). The design is two-stage: Stage A involves evidence extraction and coding using the C–T–M–E framework (detailed in Section 2.10); Stage B involves relational synthesis through concept mapping, converting coded evidence into propositional relations and cross-links that enable alignment/gap diagnosis across curriculum and textbook layers (Novak & Cañas, 2008). The theoretical framework integrating teachers' textbook use (Remillard & Heck, 2014; Brown,

2009), OTL/enactment (Stein et al., 2007), CSP (Paris, 2012), and CME (Gutstein, 2006) enriches the interpretive layer—findings are interpreted not only as design artifacts but as designed OTL with enactment implications.

Data Sources and Corpus

The curriculum corpus includes Türkiye's TYMM common framework text and the Lower Secondary Mathematics Curriculum (MoNE, 2024a, 2024b). The textbook corpus consists of all four officially approved TYMM mathematics textbook volumes for Grades 5 (MoNE, 2024c; MoNE, 2024d) and Grades 6 (MoNE, 2025a; MoNE, 2025b) — (two volumes per grade)—the complete set of TYMM lower-secondary mathematics textbooks in circulation at the time of analysis, centrally produced and nationally distributed with no competing editions. Sampling foregrounds Statistical Investigation and Geometric Quantities as primary analytic domains, with Algebra/Numbers and Data–Probability as comparative domains, reflecting the high likelihood of culturally grounded practices in measurement, spatial design, and data reasoning (D'Ambrosio, 2001; Rosa & Orey, 2013). The four textbook volumes together contain six thematic units, as summarized in Table 1.

Table 1. Textbooks aligned with TYMM and their thematic contents

Theme	5th Grade Book 1	5th Grade Book 2	6th Grade Book 1	6th Grade Book 2
Geometric Shapes				
Numbers and Quantities	✓	✓	✓	
Geometric Quantities	✓			✓
Statistical Investigation Process		✓	✓	
Algebraic Thinking through Operations		✓		✓
From Data to Probability		✓	✓	

Reliability and Trustworthiness

Inter-rater reliability was established through a double-coding protocol conducted with an expert mathematics educator (a specialist in mathematics curriculum research and textbook analysis with extensive experience in qualitative content analysis of Turkish mathematics materials) on a pilot sub-sample of approximately 10–15% of coded units. Cohen's kappa was $\kappa=0.926$ (observed $P_0=0.967$, expected $P_E=0.348$), with weighted kappa $\kappa\_w=0.962$ and $\kappa\_w^2=0.902$. All three values exceed $\kappa\geq0.80$

(Landis & Koch, 1977; Cohen, 1960). All six disagreements occurred at the D1–D2 boundary (zero D1↔D3 discrepancies). Disagreements were resolved through codebook refinement prior to full corpus coding.

Concept Mapping Procedure

Concept mapping is used as an analytic transformation of coded evidence into relational structures (Novak & Cañas, 2008). Focus question: "How do curricula and textbooks connect cultural resources to mathematical work and equity-oriented participation?" Propositions are generated in a controlled grammar (e.g., "cultural practice supports representation choice"; "value/ethical dimension drives modeling constraints"). Cross-links diagnose alignment or rupture.

Coding Procedure

Each textbook unit identified as culturally embedded in Stage A was coded independently by two raters using the C–T–M–E framework. Coding proceeded in four passes corresponding to the four model components (C, T, M, E), followed by an integrative depth-level assignment (D1/D2/D3). The context-removal diagnostic test—"Does removing the cultural context change the mathematical task?" ("Bağlam kalksa soru değişir mi?")—was applied to every unit as the primary depth-classification instrument. Boundary cases were resolved using six codebook decision rules developed iteratively during the pilot coding phase (see Section 2.10.7 for the full codebook with operational definitions): (1) D1 vs. D2: does the cultural context determine which specific data values, representations, or strategies are used? (2) D2 vs. D3: does removing the context fundamentally destroy the task structure? (3) E2 vs. E4: is genuine student decision-making required? (4) C2 vs. C4: does the resource carry a values/rights/ethical dimension that shapes inquiry constraints? (5) C3 Decor vs. C3 Modeling Agent: does the task require engaging with the artifact's mathematical construction logic? (6) Inter-rater D1/D2 boundary: would any numerical dataset serve equally well? All six rules were applied consistently across the 53 coded units following pilot agreement and codebook refinement prior to full corpus coding.

Limitations

Document-based inference is constrained by artifacts: textbooks represent designed opportunities but do not capture how teachers interpret or enact materials in practice (Remillard, 2005). Sampling introduces potential bias when multiple approved textbook sets exist, and "cultural integration" and "equity/access" are inherently interpretive constructs. Concept mapping necessarily simplifies complex textual phenomena. Generalizability is analytic rather than statistical: the codebook, depth rubric, and mapping workflow can be applied to other curricula but results should be generalized cautiously.

FINDINGS

Three patterns organize the corpus findings. First, Decor T-role dominance in cultural artifact tasks drives D1 tokenistic integration (34.0%), concentrated in 5th grade Book 1. Second, national data tasks consistently produce D2 contextual integration (43.4%), which is the modal outcome across both grades. Third, a concentrated D3 transformative corridor (22.6%) emerges in 6th grade Book 1 statistical investigation tasks where C4 (value/ethical dimension) combines with Modeling Agent T-role and high mathematical demand. These three patterns recur across all four layers of analysis reported below.

This section reports concept-mapped findings in four layers: (i) the written curriculum map (macro–meso), (ii) corpus-level distribution of cultural elements across textbook volumes and thematic units (RQ1), (iii) C–T–M–E sub-level profiles (RQ2), and (iv) alignment/gap patterns between curriculum intentions and textbook instantiations (RQ3). Findings are anchored in the full corpus of 53 coded units and illustrated through concrete examples drawn from the textbook evidence base.

Curriculum Concept Map — TYMM Macro–Meso Layer

Figure 7 makes four key propositions visible. (1) The Virtue–Value–Action (Erdem–Değer–Eylem) framework "incorporates" Culture Literacy (Kültür Okuryazarlığı: OB5) as a formal process component—establishing that culture literacy is integral, not supplementary. (2) The Köprü Kurma (Bridging) principle branches into an Application Pathway (risking D1 tokenistic integration) and a Societal-Literacy Pathway (enabling D3 transformative integration). (3) The C–T–M–E chain is the meso-level bridge mechanism. (4) The strong bridge claim anchors the map: C (epistemic) → T (modeling agent) → M (high demand) → E (access + agency).

Figure 7. TYMM curriculum concept map (macro–meso layer)

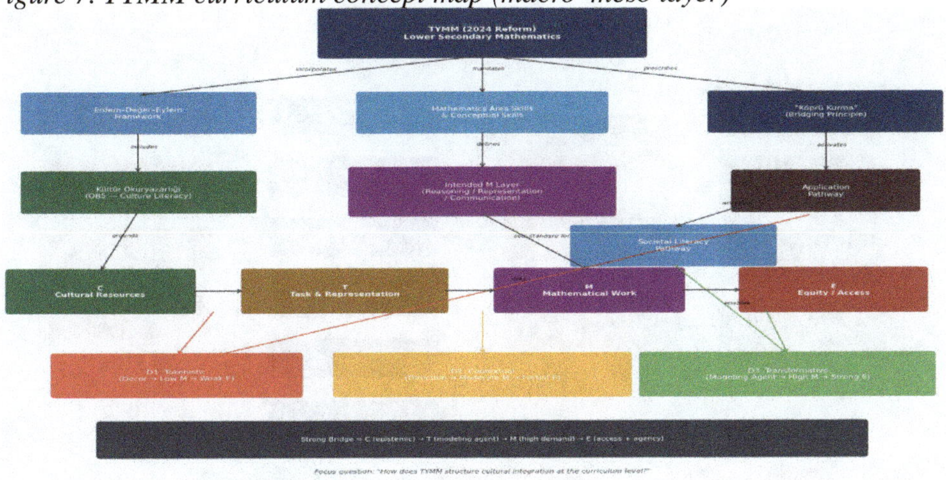

Note: Nodes represent curriculum components and C–T–M–E elements. Red dashed = tokenistic risk pathway; green = transformative enabling pathway.

C–T–M–E Textbook Concept Map — Meso–Micro Layer

Figure 8 makes the corpus-level structure empirically visible: C3/C4 → Decor → D1 (the token bridge rupture); C2 → Directive → D2 (the modal outcome); C4 → Modeling Agent → D3 (the transformative corridor). E4+E5 co-occurs with Generalization and Modeling M-codes, confirming that deep equity signals co-occur with high mathematical demand—not with low-demand tasks with discussion add-ons.

Figure 8. C–T–M–E textbook concept map (meso–micro layer)

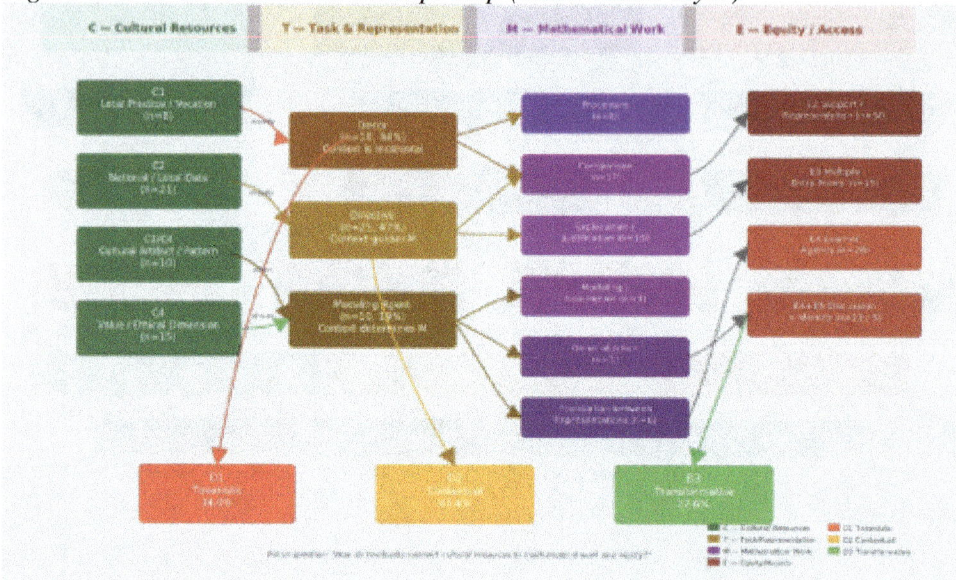

Note: Frequency counts from the 53-unit corpus in parentheses. Depth level outcomes at the base show the empirical distribution.

Sankey Flow Diagram

Figure 9 makes the D1-dominance problem structurally visible. The largest single flow is Cultural Heritage (C3) through Decor to D1; the narrowest but most significant flow is Values/Ethics (C4) through Modeling Agent to D3. The central design opportunity: redirect the large Decor→D1 flow by reassigning cultural artifacts from decorative to modeling-agent T-roles, activating their mathematical construction logic. The enactment implication is that this T-role reassignment requires both textbook redesign and teacher support for activating modeling cycles rather than following prescribed steps.

Figure 9. Sankey flow: Curriculum intentions cultural resource types task representation role mathematical work demand depth level outcomes (n=53)

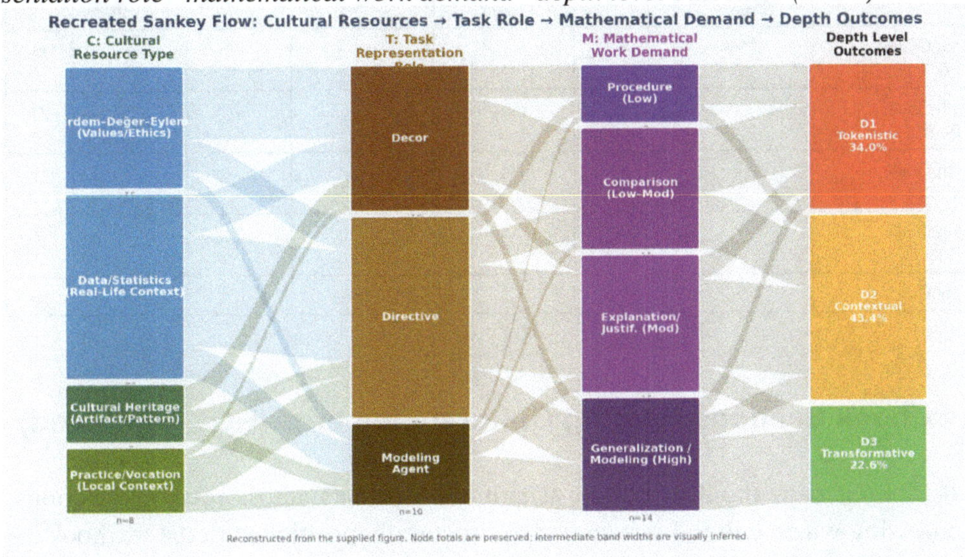

Note: Band widths proportional to n.

Distribution of Cultural Elements Across Textbook Volumes and Themes (RQ1)

Table 2 presents the number of cultural elements identified in each textbook volume by thematic unit, addressing RQ1 (distribution and forms). Three patterns from Table 2 warrant emphasis. First, the corpus contains 53 cultural elements in total, unevenly distributed across volumes: 5th grade Book 2 carries the largest share (n=27, 50.9%), while 6th grade Book 2 is the sparsest (n=2, 3.8%). Second, the Statistical Investigation theme dominates across the two grades that include it (5th grade Book 2: n=16; 6th grade Book 1: n=8; combined n=24, 45.3%). This confirms the theoretical expectation that data reasoning creates a natural scaffold for cultural contextualization. Third, Geometric Shapes appears exclusively in 5th grade Book 1 (n=6) and 6th grade Book 2 (n=—, only Geometric Quantities), suggesting that the cultural integration of spatial/geometric content is concentrated in 5th grade introductory materials.

Table 2. Number of cultural elements by theme and volume

Textbooks	Geom. Shapes	Numbers	Geom. Quant.	Statistical Inv.	Algebraic	Data-Prob.	Total
5th Grade Book 1	6	3	2	—	—	—	11
5th Grade Book 2	—	2	—	16	4	5	27
6th Grade Book 1	—	4	—	8	—	1	13
6th Grade Book 2	—	—	—	—	2	—	2
Total	6	9	2	24	6	6	53

Textbook Sections Hosting Cultural Elements (RQ1 continued)

Table 3 maps the distribution of cultural elements across textbook sections, addressing where cultural resources are structurally positioned in the textbook architecture. Table 3 reveals a key structural pattern: Activity sections host the largest share of cultural elements across 6th grade materials (6.1: n=9), while 5th grade Book 2 distributes cultural elements more broadly across Examples (n=10), Activities (n=7), and Assessment/Evaluation Questions (n=4). Performance Tasks carry only n=2 cultural elements across the full corpus—one in each Book 2—yet the kappa analysis and vignette evidence (Section 5.6) indicate that these two performance tasks are among the corpus's strongest D3 instances. The *Hazır mıyız?* (Are We Ready?) and *Bunları Bilelim* (Things to Know) sections appear exclusively in 5th grade Book 1, and explanatory text sections drop to zero by 6th grade. This structural shift indicates that the textbook architecture itself becomes more inquiry-oriented across grade levels, with cultural elements increasingly embedded in open-ended activity and performance formats rather than introductory or explanatory text.

Table 3. Sections of the textbooks where cultural elements are identified and their frequencies

Textbook Section	5th Grade Book 1	5th Grade Book 2	6th Grade Book 1	6th Grade Book 2
Are We Ready?	2	1	—	—
Let's Start	2	2	2	—
Explanatory Text	3	2	—	—
Things to Know	1	—	—	—

continued on following page

Table 3. Continued

Textbook Section	5th Grade Book 1	5th Grade Book 2	6th Grade Book 1	6th Grade Book 2
Activity	1	7	9	2
Example	—	10	—	—
Performance Task	—	1	1	—
Problem	2	—	—	—
Assessment and Evaluation Questions	—	4	—	—
Progress Monitoring Tests	—	—	1	—
Total	11	27	13	2

C–T–M–E Sub-Level Frequency Analysis (RQ2)

Table 4 presents the frequencies of C–T–M–E sub-level codes and depth levels across all four textbook volumes, providing the empirical backbone for the depth-of-integration analysis.

Table 4. Frequencies of C–T–M–E sub-level codes and depth levels

Component	Sub-Level	5th Gr. Bk 1	5th Gr. Bk 2	6th Gr. Bk 1	6th Gr. Bk 2
C: Cultural Resources	C1: Local practice / vocation	1	2	4	1
	C2: National / local data	1	14	6	—
	C3: Cultural artifact / pattern	6	2	1	1
	C4: Value / ethical dimension	3	9	3	—
T: Task / Representation	Decor	6	9	2	1
	Directive	2	16	6	1
	Modeling agent	3	2	5	—
M: Mathematical Work	Procedure	2	2	2	—
	Explanation / justification	4	7	3	1
	Comparison	2	12	3	—
	Modeling / assumption	1	1	1	—
	Generalization	1	1	2	1

continued on following page

Table 4. Continued

Component	Sub-Level	5th Gr. Bk 1	5th Gr. Bk 2	6th Gr. Bk 1	6th Gr. Bk 2
	Translation between representations	—	4	2	—
E: Equity / Access	E2: Support for explanation/ representation	10	26	12	2
	E3: Multiple entry points	1	8	6	—
	E4: Learner agency	4	14	10	—
	Collaboration / peer discussion invitation	2	9	10	1
	E5: Power-identity signaling	—	2	3	—
Depth Level	D1 — Tokenistic	6	9	2	1
	D2 — Contextual	2	15	5	1
	D3 — Transformative	3	3	6	—

Depth Profile Analysis: Cross-Grade and Cross-Domain Patterns

Table 5 makes the depth-of-integration gradient empirically precise. Aggregating across volumes:

Table 5. Aggregated depth-level profile by grade (n=53)

	Total n	D1 n (%)	D2 n (%)	D3 n (%)
Grade 5 (Books 1+2)	38	15 (39.5%)	17 (44.7%)	6 (15.8%)
Grade 6 (Books 1+2)	15	3 (20.0%)	6 (40.0%)	6 (40.0%)
Full corpus	53	18 (34.0%)	23 (43.4%)	12 (22.6%)

Four depth-level findings stand out. First, the cross-grade gradient is pronounced and theoretically meaningful: Grade 5 D3=15.8% vs. Grade 6 D3=40.0%. Grade 6 Book 1 alone contributes 6 of the corpus's 12 D3 instances, confirming its status as the primary transformative corridor. Grade 6 Book 2, by contrast, produces zero D3 units despite containing 2 cultural elements—both in the Algebraic Thinking theme with local practice (Kayısı/Malatya cebirsel ifade, Narman Peri Bacaları örüntüsü) that functions decoratively.

Second, the D2 category (contextual integration, 43.4%) is the most frequent overall, concentrated in 5th grade Book 2 (n=15) and 6th grade Book 1 (n=5). This suggests that the textbook system consistently achieves a baseline of sense-making contextualization but rarely reaches fully transformative integration. Third, D1 (tokenistic, 34.0%) is disproportionately concentrated in 5th grade Book 1 (n=6/11, 54.5%), the volume with the highest share of cultural artifact references and explanatory text. This confirms the pattern that cultural artifacts used as visual backgrounds or introductory examples—without accompanying mathematization demands—produce tokenistic integration. Fourth, power-identity signaling (E5) appears only in 5th grade Book 2 (n=2) and 6th grade Book 1 (n=3), with zero instances in the other two volumes. This highly uneven distribution demonstrates that the equity dimension of the TYMM values framework is operationalized only in specific statistical investigation contexts, not yet embedded as a systematic design feature across domains.

C-Level Analysis: From Artifact to Value/Ethical Dimension

The C-level sub-codes in Table 4 reveal a volume-specific typology of cultural resources. Cultural artifact/pattern dominates in 5th grade Book 1 (n=6/11), reflecting the heavy use of Ottoman architectural ornamentation (Osmanlı mimari süslemeleri) in the Geometric Shapes theme. National/local data dominates in 5th grade Book 2 (n=14/27), driven by TÜİK-derived datasets, ministry reports and national statistics throughout the Statistical Investigation theme. Local practice/vocation is strongest in 6th grade Book 1 (n=4/13), reflecting TÜBİTAK innovation contexts, hazelnut production (Giresun fındık üretimi) and apricot agriculture (Malatya kayısı tarımı) Value/ethical dimension is notably absent from 6th grade Book 2 (n=0), confirming that ethics-augmented cultural resources—the corridor most reliably producing D3 outcomes—are confined to the Statistical Investigation theme in 6th grade Book 1.

T-Level Analysis: Decor Dominates Grade 5, Modeling Agent Rises in Grade 6

The T-level sub-codes show a clear gradient. Decor is most prevalent in 5th grade Book 1 (n=6/11, 54.5%)—the volume with the most tokenistic profile. Directive peaks in 5th grade Book 2 (n=16/27, 59.3%), indicating that this volume consistently provides cultural context that guides but does not fully determine mathematical work. Modeling agent rises markedly in 6th grade Book 1 (n=5/13, 38.5%) and is absent from 6th grade Book 2 (n=0). This gradient—from Decor through Directive to Modeling Agent—tracks the depth-level gradient across volumes and provides the proximate mechanism through which D3 outcomes are produced: when cultural context functions as a modeling agent, it structures the mathematical problem formulation rather than merely providing a setting.

M-Level Analysis: Comparison Dominates, Generalization and Representation Translation Are Rare

The M-level sub-codes reveal the mathematical work distribution. Comparison is the most frequent sub-code (total n=17), concentrated in 5th grade Book 2 (n=12). While comparison can reach moderate demand (D2) when it requires justification and interpretation, the corpus shows that many comparison tasks are purely descriptive ("which is greater?") without requiring explanation of why—producing D1 or borderline D2 outcomes. Explanation/justification appears at moderate frequency (n=15 total) and is the primary M-code associated with D2 units. Generalization (n=5 total) and translation between representations (n=6 total) are the rarest M-codes, yet they are disproportionately associated with D3 outcomes. The three D3 units in 5th grade Book 1 all involve generalization (carpet area–perimeter relationship,vOlympic rings triangle construction, geometric tool determination from a cubic picture). This demonstrates that *M-code diversity* is a stronger predictor of D3 than C-code type alone.

E-Level Analysis: Support for Explanation Ubiquitous; Agency and Power-Identity Scarce

The E-level sub-codes present the equity dimension of the corpus. "Support for explanation/representation" (E2) is by far the most frequent sub-code (n=50 instances across all volumes), present in virtually every coded unit. This indicates that the textbooks systematically provide tablo, grafik, şema (table, graph, diagram) supports—but this alone does not constitute equity in the Hobbs (2018) sense; it is necessary but not sufficient for access to robust mathematical engagement. Learner agency (E4: n=28) is present at a meaningful level, concentrated in 5th grade Book 2 (n=14) and 6th grade Book 1 (n=10). Invitation to collaboration/peer discussion (n=22) is also substantial in these two volumes, supporting D2 and D3 outcomes when combined with high M demand. Power-identity signaling (E5: n=5) is the rarest E sub-code (5th grade Book 2: n=2; 6th grade Book 1: n=3; all others: zero). These five instances—concentrated in privacy/data protection, environmental justice, and accessibility rights tasks—represent the corpus's most genuinely equity-expanding moments, where cultural context positions students as citizens with rights and responsibilities rather than merely as consumers of mathematical techniques.

Qualitative Evidence: Examples from Each C–T–M–E Level

The following examples, drawn directly from the textbook corpus (Tables 5–8 of the evidence base), illustrate each major C–T–M–E level with concrete textbook evidence.

Examples for Cultural Resources (C-Level)

The four C-level sub-categories are illustrated below.

Local practice/vocation: A task in 6th grade Book 2 (p. 64) contextualizes algebraic expression writing in Malatya's apricot production sector. By emphasizing Türkiye's global leadership in apricot production and foregrounding the relationship between two specific farmers' outputs, the task requires students to formulate an algebraic expression representing one farmer's production as a function of the other's, and to discuss their reasoning with peers.

National/local data: A task in 5th grade Book 2 (p. 95) presents the graph titled "Primary Problems of Provinces in 2022," published by the Ministry of Environment, Urbanization, and Climate Change, and expects students to interpret the data and compare provincial priorities. The graph is reproduced below.

Figure 10. Primary problems of provinces (Ministry of Environment, 2023; 5th grade Book 2, p. 95)

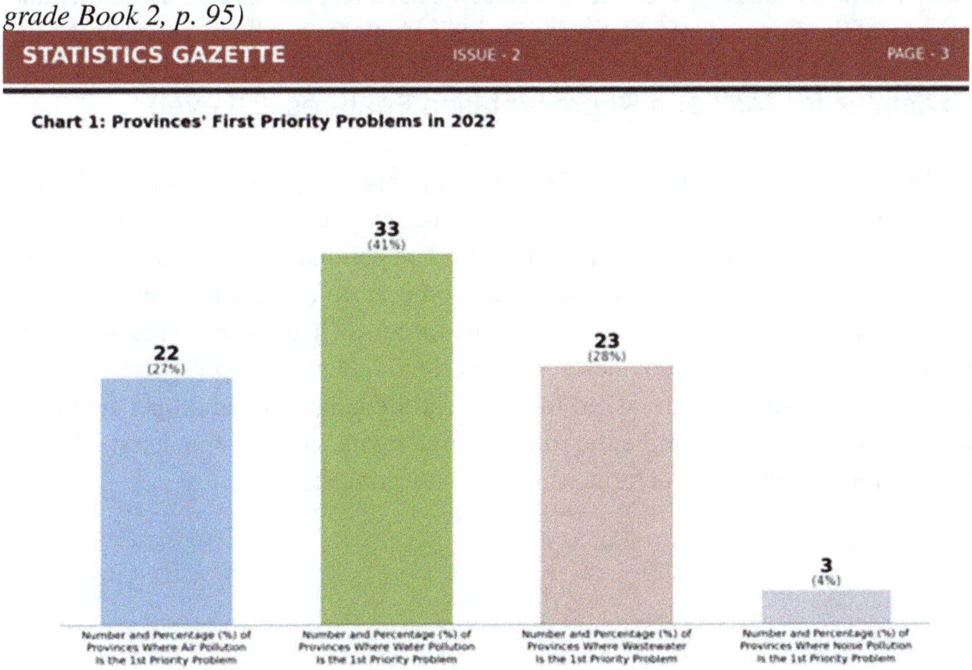

Cultural artifact/pattern: In 5th grade Book 1 (p. 14), visuals of significant examples from Turkish architecture—ornaments from the Süleymaniye Mosque, the Sultanahmet Mosque, and Topkapı Palace—are shared. Students are asked to identify the geometric shapes they contain and the mathematical tools that can be used to create these shapes.

Figure 11. An example of Süleymaniye Mosque, Sultan Ahmed Mosque, and Topkapı Palace respectively (5th grade Book 1, p. 14)

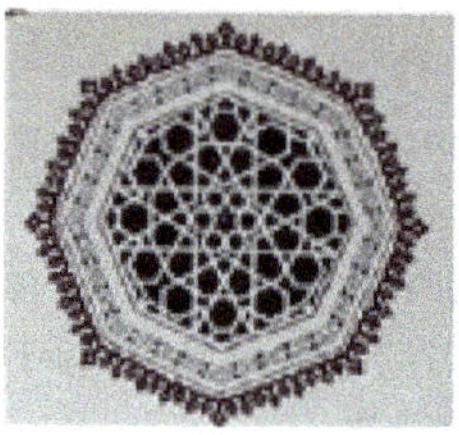

Value/ethical dimension: In 6th grade Book 1 (p. 74), after introducing the e-Government (E-Devlet) platform and explaining its privacy measures protecting personal data in digital environments, students are asked to prepare survey questions regarding the platform's data protection practices and discuss results using collected data. The task explicitly instructs that students must respect the confidentiality and privacy of the personal information of the individuals from whom they collect data.

Examples for Task and Representation Features (T-Level)

Decor: In 5th grade Book 1 (p. 106), the national tree-planting initiative ("Millî Ağaçlandırma Günü") provides a table of saplings planted per year, followed by questions about comparing natural numbers. The national event context functions decoratively—removing it would leave the comparison task unchanged.

Directive: In 5th grade Book 2 (p. 36), the success of Türkiye's women's national volleyball team in 2023 provides heights at the moment of a spike (shown in the figure below). Students are asked to represent the heights as improper fractions, compare them, and determine the roles of numerators and denominators—the national sporting context structures which data are compared.

Figure 12. Volleyball spike height context (5th grade Book 2, p. 36).

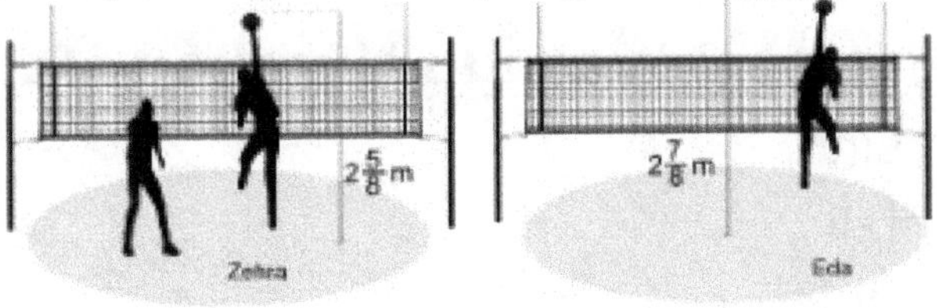

Modeling agent: In 6th grade Book 1 (p. 63), a TÜİK population projection excerpt situates questions about interpreting temporal change, determining whether the data are categorical, proposing research questions, and selecting the most appropriate visualization. Here the cultural resource (TÜİK report) is the epistemic driver: the modeling cycle begins with and is structured by this culturally situated data source.

Examples for Mathematical Work (M-Level)

Procedure: In 6th Grade Book 1 (p. 212), a Javelin (Cirit)—a traditional equestrian sport played in Anatolia—scenario asks students to determine the experimental probability, as a percentage, of an athlete catching a thrown javelin in mid-air (5 out of 50), a task that includes a clear cultural reference but mathematically remains procedural, requiring only a single-step fraction-to-percentage conversion.

Explanation/justification: In 5th grade Book 2 (p. 178), meteorological data on average rainy days in June across several Turkish provinces is presented alongside three friends' predictions on a probability spectrum (see Figure 9). Students must identify which friend's reasoning they agree with, demonstrate their own predictions, and justify which city warrants carrying an umbrella.

Figure 13. Meteorology data table and probability spectrum (5th grade Book 2, p. 178)

Provinces	Average Number of Rainy Days
Mersin	2
Denizli	5
Eskişehir	4
Rize	14
Bursa	6
Ankara	9

(Retrieved from general web sources.)

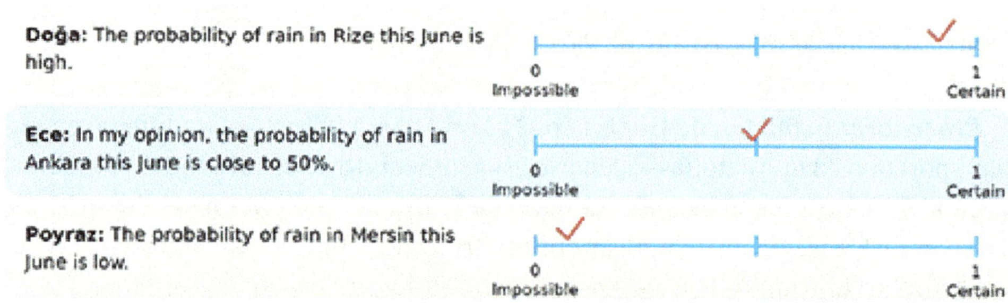

Comparison: In 5th Grade Book 1 (p. 115), students are presented with a table showing the number of teams that applied to TEKNOFEST—a major national aviation, space, and technology festival in Türkiye—across its first four years and are asked to compare the figures by year; while the context offers a contemporary national technology frame, the mathematical demand remains limited to straightforward comparison rather than deeper interpretation of trends or patterns.

Modeling/assumption: In 5th grade Book 2 (p. 149), a brief introduction to Al-Khwarizmi connects algorithm logic to a scenario where one child guides another to draw the Turkish flag based on verbal instructions. Students must discuss how instructions should be phrased and in what order steps should proceed—the modeling task is structuring an algorithm, and the cultural context (Al-Khwarizmi heritage + Turkish flag) shapes the assumptions of what "a good algorithm" looks like.

Generalization: In 6th grade Book 2 (pp. 77–78), following an introduction to Narman Peri Bacaları (Narman Fairy Chimneys, see Figure 10), a travel agency tour scenario asks students to find the relationship between the number of weeks and buses operated, to write the total for the 10th week, and to calculate and express algebraically the pattern rule for the nth week.

Figure 14. Narman Peri Bacaları and weekly bus schedule table (6th grade Book 2, pp. 77–78)

Table: Number of Buses Departing by Week

Week	Number of Buses Departing Weekly	Total Number of Buses Departing to the Fairy Chimneys
Week 1	8	8
Week 2	5	13
Week 3	5	18
Week 4	5	23
Week 5	5	28
:	:	:
Week n	5	

Translation between representations: In 6th grade Book 1 (p. 113), students are expected to conduct research on characteristics of children in Türkiye using

TÜİK Child Portal data, present findings through data visualization, select a data type based on interest, formulate a research question, decide on visualization tools, compare findings, derive conclusions, and identify errors in the dataset with potential explanations. This full inquiry cycle—spanning data selection, visualization choice, analysis, and validation—exemplifies translation between representations in a culturally grounded research context.

Examples for Equity/Access/Participation (E-Level)

Multiple entry points: In 6th grade Book 1 (p. 88), rural and urban population data of Türkiye for six historical time points (1927–2012) is presented alongside two open-ended questions: explain the change, then discuss the reasons. The historical data provides multiple analytical angles—demographic, economic, social—enabling different student starting points (see Figure 11).

Figure 15. Rural-urban population data 1927–2012 (6th grade Book 1, p. 88)

Year	Rural	Urban	Rural (%)	Urban (%)
1927	10,342,391	3,305,879	75.8%	24.2%
1940	13,474,701	4,346,249	75.6%	24.4%
1960	18,895,089	8,859,731	68.1%	31.9%
1980	25,091,950	19,645,007	56.1%	43.9%
2000	23,797,743	44,006,184	35.1%	64.9%
2012	17,178,953	58,448,431	22.7%	77.3%

Support for explanation/representation: In 5th grade Book 1 (p. 95), TÜİK tourist data for Istanbul is provided in a structured table with explicit column headers (see Figure 12). The table format reduces the representational barrier to number-reading tasks, illustrating E2 support—though in this case M (place value procedure) remains low, exemplifying the common gap between E2 support and deep mathematical engagement.

Figure 16. TÜİK tourist data table (5th grade Book 1, p. 95)

Countries	Number of Tourists
Germany	97,577
United Kingdom	49,961
Iran	88,056
Russia	148,016
Turkmenistan	9,496

Learner agency (choice, determining methods): In 5th grade Book 2 (p. 46), national traffic accident fault data is presented alongside a diagnostic branched tree. Students must follow the correct path through the tree and are asked to compare their final exit with classmates—the branching structure gives each student a partially distinct path, creating choice and differentiated engagement.

Invitation to collaboration/peer discussion: In 5th grade Book 2 (p. 59), an environmental pollution map from the Ministry is presented, followed by six sub-questions building from locating the student's own province to formulating a research question and discussing whether it qualifies as a valid research question with peers (see Figure 17). The progression from local to general creates genuine discussion scaffolding.

Figure 17. Ministry of Environment air pollution map with research question scaffolding (5th grade Book 2, p. 59)

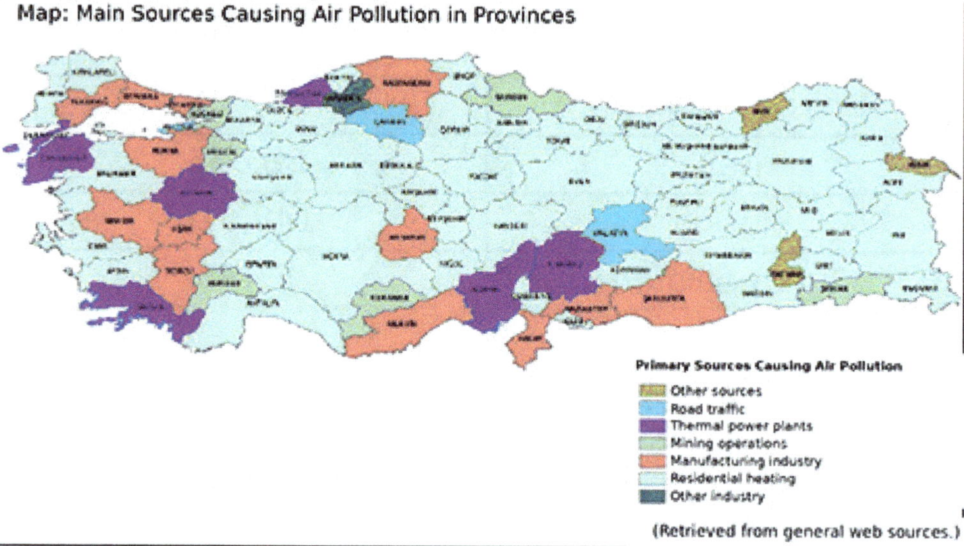

Power-identity signaling (justice/privacy/rights): In 6th grade Book 1 (p. 100), after explaining personal data and its protection, the Personal Data Protection Authority (KVKK) is introduced as an institution providing recourse for citizens whose data may be compromised. Students are asked to compare ihlal (violation) categories from a KVKK data table and propose protection measures. The task positions students as rights-holding citizens who can engage with and critique data protection mechanisms—the most direct E5 (power-identity signaling) instance in the corpus.

Depth Level Examples (RQ2)

Table 6 presents one concrete textbook example for each of the three depth levels (D1, D2, D3), showing how the C–T–M–E coding and the context-removal diagnostic apply to actual tasks. Each row reports the source task, its C–T–M–E reading, the outcome of the diagnostic test, and a brief note on how teacher enactment could raise or protect its depth. The purpose is to ground the abstract depth-level definitions in Section 2.10.5 with concrete corpus evidence.

Table 6. Depth level examples from the TYMM textbooks

Depth	Source	C–T–M–E Reading and Enactment Implications
D1 Tokenistic	*5th Gr. Bk 1 p.106 National Afforestation Day table*	C3 (national symbol) → T: Decor → M: Low (comparison, procedures without connections) → E: Weak (E2 only). Diagnostic: Remove context → task unchanged. Stein et al. (2007), task degradation risk is low here because the task is already degraded at design level; teacher enactment cannot elevate it without fundamental redesign. CSP implication: task appropriates national environmental value without sustaining any mathematical knowledge about afforestation ecology.
D2 Contextual	*5th Gr. Bk 2 p.46 TÜİK traffic accident proportions + branched tree*	C2 (national data) → T: Directive → M: Moderate (logical conditional reasoning + comparison) → E: Partial (E3+E4: branching paths + peer comparison). Diagnostic: Remove context → task significantly simplified (generic branched tree without authentic proportions loses analytical meaning). Teacher enactment opportunity (Brown, 2009): an adaptive teacher can elevate to D3 by asking "What would need to change in driver behavior to make roads safer? How much?" — adding a modeling/validation layer. CSP implication: task uses national safety data authentically but does not invite critique of road infrastructure or policy.
D3 Transform.	*6th Gr. Bk 1 p.64 TÜİK elderly population + privacy rights*	C2+C4 (public data + value/ethical: elderly rights, personal privacy) → T: Modeling Agent (ethical framing determines what research questions are permissible, what constraints apply, what limitations must be acknowledged) → M: High (research question design + variable-type reasoning + methodological limitations) → E: Strong E4+E5 (learner agency in question formulation; power-identity: privacy as a right of research participants). Diagnostic: Remove ethical framing → task degrades to generic variable-type classification. Gutstein (2006), this task exemplifies "reading the world with mathematics" — students use statistical reasoning to examine the rights and vulnerabilities of elderly individuals. Rosa & Orey (2016), this task enacts the glocal ethnomodeling principle: national demographic data (etic) interpreted through the local lens of elderly community rights (emic). Teacher enactment note: protecting the E5 signal requires that the teacher not pre-answer the limitations discussion — leaving the constraint open is essential to D3 maintenance.

Alignment and Gap Map (RQ3)

Alignment A — Symbolic Crosswalk: OB5 tags and Virtue–Value–Action (Erdem–Değer–Eylem) icons operationalized in textbooks through labeling. Structural but insufficient—icon presence distributes across D1–D3. Consistent with Haggarty and Pepin's (2002) finding that textbooks carry implicit messages; the TYMM icon system sends the message that culture is present without ensuring it is mathematically active.

Gap 1 — Token Bridge: C (national/local data) + M low (procedure/comparison) + T Decor = D1. Most prevalent rupture (34.0%). Schmidt et al. (1997), this represents the "splintered" curriculum problem: cultural references are added as surface features without coherent integration into deeper mathematical learning.

OTL implication: students in these tasks are denied OTL for culturally grounded mathematical reasoning despite the cultural reference.

Gap 2 — Scaffold Without Agency: E2 near-universal (n=50) but E4 (n=28) and E5 (n=5) scarce. Brown (2009) offloading mode: step-by-step scaffolded tasks encourage teacher offloading (following the script), which locks the enacted curriculum at E2 and eliminates teacher-activated E4 opportunities. This creates a structural cycle: D1-oriented design → teacher offloading → enacted D1 → no evidence that D3 is possible.

Bridge Corridor: Statistical Investigation + C4 + Modeling Agent → D3 (6th grade Book 1, pp.63, 64, 74, 113). Rosa and Orey (2016), this corridor enacts glocal ethnomodeling: authentic national data (etic perspective) interpreted through local ethical constraints (emic perspective). Gutstein (2006), this corridor exemplifies CME practice: mathematics as a tool for reading social reality. Enactment implication (Stein et al., 2007): these tasks are high-demand and therefore most vulnerable to task degradation; teacher PD should specifically address how to maintain the ethical constraint and modeling cycle during enactment.

DISCUSSION: THEORETICAL AND PRACTICAL IMPLICATIONS

The Depth Gradient, T-Role, and the Written-to-Enacted OTL Gap

The overall depth profile (D1=34.0%, D2=43.4%, D3=22.6%) and the cross-grade gradient (Grade 5 D3=15.8% → Grade 6 D3=40.0%) reveal that cultural integration in TYMM materials is architecturally achievable—6th grade Book 1 demonstrates this—but uneven. The theoretical framework developed allows a more precise account of why. From the textbook use perspective (Remillard & Heck, 2014), the written curriculum depth profile represents a designed OTL ceiling: teachers enacting from these materials can achieve at most the depth designed into the task (Stein et al., 2007), and frequently achieve less through task degradation. The 22.6% D3 rate therefore represents the maximum available culturally sustaining and mathematically rich OTL in the current corpus—and the actual enacted OTL is likely lower if many teachers are operating in offloading mode (Brown, 2009).

The T-role is the proximate design mechanism: 10 of 12 D3 units carry Modeling Agent T-role, making it the single design feature most directly responsible for transformative outcomes. Tasks with Modeling Agent T-role require students to engage with the mathematical construction logic of cultural practices rather than merely label them—the configuration CSP associates with sustaining community knowledge and CME associates with "reading the world with mathematics" (Gutstein, 2006).

The Four Literature Strands: Convergent Implications for Design

The four literature strands converge on a coherent set of design implications for textbook cultural integration.

From teachers' textbook use: Textbook design depth is a ceiling, not a floor. D3 task design is necessary but not sufficient; teachers must be supported in protecting modeling cycles, maintaining ethical constraints, and resisting the task-degradation tendency. The toolkit in Section 7 addresses this through the remove-context diagnostic (operable during lesson planning) and the three upgrade patterns (enabling teacher-level D1→D3 transformation).

From culturally sustaining pedagogy and CME: The D1/D2 dominance in the corpus reflects a design default toward cultural responsiveness without cultural sustaining or critical consciousness. Expanding D3 requires both redesigning C3 tasks (cultural artifacts) to activate their construction logic (CSP dimension) and expanding C4 tasks (value/ethical) beyond the statistical investigation domain to geometry, algebra, and probability (CME dimension).

From OTL and textbook-task enactment: The "splintered curriculum" critique (Schmidt et al., 1997) applies to cultural integration depth: many tasks provide the appearance of cultural responsiveness (shallow distribution) without the depth that constitutes genuine OTL. Improving D3 rates is an OTL equity imperative because students who never encounter D3 tasks are denied the opportunity to develop culturally grounded mathematical reasoning and critical mathematical citizenship.

From ethnomodeling and task design: Rosa and Orey's (2015, 2016) trivium framework and glocal ethnomodeling approach provide the task design blueprint for D3: glocal tasks begin with authentic local/national data, require emic (culturally situated) interpretation, translate into etic (school-formal) mathematical representations, and validate the model against the cultural constraints. This four-step cycle (problem formulation → mathematization → validation → critique) directly operationalizes the Modeling Agent T-role and high-demand M that characterize D3.

Why Power-Identity Signaling (E5) Is Rare: A Contextual Analysis

Power-identity signaling (E5) is rare in the TYMM corpus (n=5, 9.4%) and concentrated exclusively in 6th grade Book 1 statistical investigation tasks—a pattern that signals not only an equity gap in the Hobbs (2018) sense but also a gap from both CSP and CME perspectives: the materials sustain cultural references predominantly at the level of national pride and heritage (D1–D2) while rarely inviting students to use mathematics to examine, critique, or transform social structures. At least four contextual factors explain this pattern.

First, TYMM's value framework is virtue-oriented, not critically oriented. The Virtue–Value–Action (Erdem–Değer–Eylem) framework centers on individual virtues (honesty, responsibility, respect)—philosophically distinct from CME's critical consciousness (Skovsmose, 1994) or Gutstein's (2006) "writing the world with mathematics." The five E5 instances involve institutionally recognized rights frameworks (KVKK data protection, accessibility law), suggesting E5 operates as legal-rights awareness within institutional legitimacy rather than as critique of power structures. Genuinely critical E5 tasks—examining whose data is collected and why, whose rights are protected by mathematical models—would require explicit policy space that the current TYMM framework does not create.

Second, statistical inquiry has a structural advantage for social data analysis. CME practice has historically gravitated toward statistics and data analysis because these domains require authentic social context (Frankenstein, 1983; Gutstein, 2006). The corpus confirms this structural pattern: all five E5 instances are in statistical investigation tasks. Extending E5 to geometry (spatial justice, accessibility of public spaces) and algebra (equitable distribution algorithms, fairness of mathematical models) would require deliberate design choices that work against the structural grain of these domains—possible but more effortful.

Third, E5 requires C4 + Modeling Agent T-role simultaneously. Of the 15 C4-coded units, only 5 (33%) reach E5—the remaining 10 C4 units involve value references in motivational framing without structuring the task. This confirms the C4 is necessary but not sufficient for E5: the T-role must assign the value/ethical dimension an active role in determining the mathematical inquiry constraints.

Fourth, systemic constraints inhibit E5 expansion. The implicit norm of mathematical neutrality in state-approved textbooks (Bishop, 1991; Gutiérrez, 2013), combined with TYMM's state-mandated character, makes it unlikely that textbook authors will design tasks that explicitly critique state institutions or power structures. Rosa and Orey's (2016) glocal ethnomodeling approach—which explicitly balances emic respect for local cultural perspectives with etic mathematical formalization—provides a framework for expanding E5 that is compatible with state-mandated educational models: glocal E5 tasks examine community rights and protections through mathematical analysis rather than positioning mathematics as a tool for political critique, which is more likely to be compatible with state educational governance.

Enactment Implications and Teacher Support

The textbook use literature (Brown, 2009; Remillard & Heck, 2014) establishes that even the best-designed D3 tasks can be degraded in enactment. Three degradation risks are particularly salient for culturally grounded D3 tasks in the TYMM context:

Ethical constraint stripping: A teacher planning the 6th grade Book 1 elderly population task (p. 64) may focus on the statistical objective and treat the privacy rights framing as motivational context, reducing the task from D3 to D2. Teachers need to understand that ethical constraints are mathematical constraints, not decorative packaging.

Research question pre-answering: D3 tasks require students to formulate their own research questions (E4). A teacher who pre-generates this question eliminates the open problem formulation stage—the most critical D3 design feature. Protecting this openness requires pedagogical restraint that is explicitly addressed in the toolkit (Section 7).

Peer discussion formalization: D3 tasks with E5 (power-identity) dimensions involve discussion of rights, fairness, and social implications of mathematical models. A teacher who channels this discussion into predetermined conclusions ("of course privacy is important, now back to the statistics") loses the critical consciousness opportunity. Maintaining genuine inquiry openness—being willing to hear students challenge assumptions and arrive at unexpected conclusions—is a pedagogical disposition that requires explicit PD support.

International Comparative Positioning

The TYMM corpus findings are positioned relative to the three comparative studies reviewed. Shield and Dole's (2013) "surface-level cultural contextualization" pattern is replicated in TYMM's D1=34.0% and Decor T-role prevalence. Xenofontos and Andrews's (2014) national differences in the structural role of cultural context are evidenced within TYMM itself: 6th grade Book 1 operates in the "epistemically active" model while 6th grade Book 2 operates in the "motivational" model—demonstrating that design choices within a national curriculum can reproduce the between-country variation documented comparatively. Pöhler and Prediger's (2015) "scaffolding without rigor" pattern is precisely the E2-without-E4 gap in TYMM materials. What distinguishes TYMM from these comparators is the existence—and concentration—of a genuine D3 corridor: the glocal ethnomodeling corridor in 6th grade Book 1 achieves a depth level that goes beyond what Shield and Dole (2013) documented as typical in first-generation culturally responsive materials. The design challenge is replication and extension.

SOLUTIONS AND RECOMMENDATIONS: A BRIDGING TOOLKIT

The Depth-of-Bridging Rubric (0–3)

Table 7 converts the C–T–M–E depth analysis into a practical four-point scoring instrument (0–3) that curriculum writers, textbook authors, and teachers can apply directly when designing or reviewing a task. For each score, the table specifies the expected cultural (C) role, mathematical (M) demand, and equity (E) signals, together with the diagnostic outcome of removing the cultural context and a note on what teacher enactment can or cannot do to raise the task's depth. This rubric operationalizes the depth levels defined conceptually in Section 2.10.5 and illustrated with corpus examples in Table 6, translating them into a tool usable outside the coding context.

Table 7. Depth-of-bridging rubric (for textbook design, teacher planning, and curriculum review)

Score	Level	C Role	M Demand	E Signals	Diagnostic + Enactment Note
0	No bridge	No cultural anchor	Routine only	None	Context removal: no loss. Teacher cannot elevate.
1	D1 Tokenistic	Incidental; Decor T-role	Low; procedures without connections	E2 only; no agency/identity	Remove → same task. Teacher must redesign (Pattern A or B) to elevate.
2	D2 Contextual	Meaningful; Directive T-role	Moderate; explanation/ comparison	E2+E3 or E2+E4; partial agency	Remove → task simplified. Adaptive teacher can elevate to D3 (Pattern C) by adding research question and validation step.
3	D3 Transformative	Epistemic; Modeling Agent T-role	High; modeling/generalization/ critique	E3+E4 min.; ideally E5	Remove → meaningless. Teacher must protect ethical constraint and open formulation stage from degradation (Stein et al., 2007).

Token-to-Transformative Transformation Playbook

The three patterns below are grounded in the corpus analysis, aligned with the four theoretical strands, and designed to be operable during teacher lesson planning without full textbook restructuring.

Pattern A: National/Local Data for Numeracy → Statistical Inquiry Cycle (D1→D3)

Theoretical basis: Rosa & Orey (2016) glocal ethnomodeling; Gutstein (2006) reading the world with mathematics; Stein et al. (2007) OTL enhancement through teacher adaptation. Starting point: Authentic national data table used only for place-value routines or simple comparison. Upgrade moves: (1) Add a research question the data was collected to answer — activating the problem formulation stage of the ethnomodeling cycle. (2) Require representation choice and justification — "Would a bar chart or line plot better communicate this trend? Defend your choice" — activating the etic representation stage. (3) Add interpretation with evidence — "What pattern do you see? Which data points support it?" — activating the validation stage. (4) Add validity/limitation checks — "What might be missing? What could make it misleading?" — activating the critique stage. (5) Add peer comparison — activating E3+E4 and, if the data concerns a social issue, potentially E5. Enactment note: the research question in step (1) should remain open — teacher does not pre-generate it. This requires pedagogical restraint that should be explicitly addressed in teacher PD.

Pattern B: Cultural Artifact as Visual Background → Ethnomodeling Cycle (D1→D3)

Theoretical basis: Rosa & Orey (2013, 2015) ethnomodeling; Paris (2012) culturally sustaining pedagogy; Fouze & Amit (2018) ethnomathematics-based task design. Starting point: Architectural ornament or cultural artifact used as visual background for shape identification. Upgrade moves: (1) "What mathematical constraints does this pattern impose?" — activating reverse-engineering of construction logic (sustaining community mathematical knowledge). (2) "If you were to recreate this with only compass and straightedge, what would your first step be? Why?" — activating mathematization from cultural constraint. (3) Compare two cultural construction traditions mathematically — activating glocal ethnomodeling (emic + etic perspectives). (4) "What class of patterns can be built using these construction rules?" — activating generalization. Enactment note: steps (1)–(2) require that the teacher genuinely does not know the answer and is willing to investigate with students — the task must be positioned as mathematical inquiry, not demonstration.

Pattern C: Fully Guided Task → Restore Agency (D2→D3)

Theoretical basis: Brown (2009) teacher-tool improvisation mode; Hobbs (2018) equity through agency design; Stein et al. (2000) high-demand mathematical work. Starting point: Real-life problem with step-by-step solution format. Upgrade moves: (1) Remove 2–3 critical prescribed steps; replace with "Choose a strategy and justify." (2) Ask for two distinct solution approaches. (3) Add peer review prompt: "Evaluate another group's reasoning — is it valid? More or less efficient than yours?" (4) Add model validation: "Does your solution make practical sense in the real-world context? What would you check?" Enactment note: this pattern transitions teacher role from script follower (offloading) to activity facilitator (improvising). Teacher PD should address how to manage the increase in classroom ambiguity and divergent solution paths that accompany this transition.

Recommendations by Stakeholder

For curriculum writers: (1) Provide D3 exemplars with explicit ethnomodeling cycle annotations in official guidance — making "transformative integration" concrete for textbook authors. (2) Develop assessment frameworks that require evidence of mathematical reasoning about cultural context, not only procedural answers. (3) Create design guidelines that operationalize CSP (sustaining community mathematical knowledge) and CME (mathematical critique of social reality) as curriculum-level standards, not only icon tags.

For textbook authors: (1) Ensure each thematic unit contains at least one D3 task with Modeling Agent T-role. (2) Diversify C-codes toward C1 and C4 in geometry and algebra domains. (3) In step-by-step formats, leave 2–3 critical decisions for students (E4 activation). (4) Expand E5 contexts: geometry (accessibility of public spaces), algebra (fairness of distribution algorithms), probability (equity of game outcomes). (5) Annotate D3 tasks with "enactment notes" that explicitly identify the open formulation stages requiring teacher restraint — borrowing the "educative curriculum materials" design approach advocated by Davis and Krajcik (2005).

For teachers: (1) Apply the remove-context test during lesson planning. (2) Use Pattern A to upgrade any national data comparison task by adding a research question and validation step. (3) Use Pattern C to restore agency in any over-scaffolded guided task by removing 2–3 prescribed steps. (4) In enacting D3 tasks, identify and protect the open formulation stage — resist the task-degradation tendency by not pre-answering the ethical or research question that the task is designed to keep open.

FUTURE RESEARCH DIRECTIONS

Longitudinal TYMM analysis: As Grades 7–8 textbooks are published, systematic replication of the C–T–M–E coding protocol will enable longitudinal analysis of whether depth improves as the reform matures. The teachers' textbook use framework (Remillard & Heck, 2014) suggests that teacher familiarity with the new framework should increase enacted depth over time even if the written curriculum remains constant.

Enacted curriculum study: Document analysis establishes designed OTL; enacted OTL depends on teacher interpretation. Design-based studies pairing C–T–M–E coding with classroom observation and teacher interviews would test whether D3 tasks produce richer participation in practice and would document the degradation patterns predicted by Stein et al. (2007) — providing the link between the written and enacted curriculum that this study cannot establish alone.

CSP-CME expansion: Future participatory design research could develop and validate D3 task templates embedding CSP (sustaining community mathematical knowledge) and CME (mathematical critique of social reality) in geometry, algebra, and probability — addressing the E5 expansion gap documented. Rosa and Orey's (2016) glocal ethnomodeling framework provides the design scaffold for creating tasks that are both locally grounded (emic) and school-formally rigorous (etic) without requiring state-sensitive political critique.

Comparative studies: The C–T–M–E protocol is transferable. Comparative application to textbooks from other national systems — particularly those with similar centralized governance structures (single publisher, national distribution) — would test whether the Decor T-role dominance and scaffold-without-agency patterns are universal design defaults or system-specific tendencies amplified by centralization.

Educative curriculum materials: Davis and Krajcik's (2005) framework for educative curriculum materials — materials that include annotations and support content for teachers, not only content for students — could be applied to TYMM D3 tasks. Future design research could develop and validate teacher support annotations for D3 tasks that specifically address how to protect Modeling Agent T-role and open formulation stages during enactment.

CONCLUSION

This chapter addressed a persistent practical problem: cultural references in mathematics curricula do not automatically produce meaningful mathematical work or equitable participation. The C–T–M–E framework developed here tackles this by integrating four literature strands—ethnomathematics/ethnomodeling, culturally sustaining pedagogy and CME, teachers' textbook use, and OTL/enactment research—into a single diagnostic model that operates at the level of individual textbook tasks.

These four strands converge in the C–T–M–E model (Figure 2), which operationalizes the bridge between ethnomathematics and CRME through four coupled components (Cultural resources → Task/representation features → Mathematical work → Equity/access signals) and three depth levels (D1 tokenistic / D2 contextual / D3 transformative). The model's central diagnostic—"Bağlam kalksa soru değişir mi?" (Does removing the context change the task?)—is simultaneously a theoretical claim about what cultural integration means and a practical tool for textbook authors, curriculum reviewers, and teachers.

Empirically, the concept-mapped analysis of four TYMM Grades 5–6 textbook volumes—grounded in systematic C–T–M–E coding of 53 cultural elements with inter-rater reliability κ=0.926, visualized through two concept maps (Figures 3–4) and a Sankey flow (Figure 5)—revealed a picture simultaneously encouraging and sobering. The encouraging finding is that transformative integration is achievable: 6th grade Book 1 produces D3=46.2%, driven by the glocal ethnomodeling corridor (Rosa & Orey, 2016) of statistical investigation tasks that combine TÜİK data with ethical constraints and assign cultural context a modeling agent T-role. These tasks exemplify "reading the world with mathematics" (Gutstein, 2006) in a state-mandated educational context. The sobering finding is that D3 represents only 22.6% of the corpus, the Decor T-role (34.0% of tasks) perpetuates a "splintered" cultural OTL (Schmidt et al., 1997), and power-identity signaling (E5) is rare (9.4%) for contextually specific reasons analyzed. The written curriculum provides at most 22.6% culturally sustaining and mathematically transformative OTL—and enacted OTL is likely lower given teacher task-degradation tendencies (Stein et al., 2007; Brown, 2009).

The practical toolkit—the depth rubric (Table 6), three upgrade patterns, and stakeholder recommendations—translates this diagnosis into actionable design and enactment criteria. The toolkit is grounded in what already works in the corpus (particularly 6th grade Book 1's glocal corridor), theoretically justified by the four-strand framework, and practically designed for the specific mediation structures of the TYMM system: centralized textbook production, single-publisher distribution, and the teacher enactment context that mediates between designed and experienced OTL. In this way, the chapter offers not a critique but a roadmap—one that takes seriously both the cultural equity aspirations of the TYMM reform and the design and enactment conditions necessary to realize those aspirations in students' actual mathematical learning opportunities.

REFERENCES

Aguirre, J. M., & Zavala, M. R. (2013). Making culturally responsive mathematics teaching explicit: A lesson analysis tool. *Pedagogies*, *8*(2), 163–190. DOI: 10.1080/1554480X.2013.768518

Bingölbali, E., Arslan, S., & Zembat, İ. Ö. (Eds.). (2016). *Matematik eğitiminde teoriler*. Pegem Akademi.

Bishop, A. J. (1991). *Mathematical enculturation: A cultural perspective on mathematics education*. Kluwer Academic Publishers.

Bishop, A. J. (1997, August). The relationship between mathematics education and culture. Opening address delivered at the Iranian Mathematics Education Conference in Kermanshah, Iran.

Bowen, G. A. (2009). Document analysis as a qualitative research method. *Qualitative Research Journal*, *9*(2), 27–40. DOI: 10.3316/QRJ0902027

Brown, M. W. (2009). The teacher-tool relationship: Theorizing the design and use of curriculum materials. In Remillard, J. T., Herbel-Eisenmann, B. A., & Lloyd, G. M. (Eds.), *Mathematics teachers at work: Connecting curriculum materials and classroom instruction* (pp. 17–36). Routledge.

Carroll, J. B. (1963). A model of school learning. *Teachers College Record*, *64*(8), 723–733. DOI: 10.1177/016146816306400801

Cohen, J. (1960). A coefficient of agreement for nominal scales. *Educational and Psychological Measurement*, *20*(1), 37–46. DOI: 10.1177/001316446002000104

D'Ambrosio, U. (1985). Ethnomathematics and its place in the history and pedagogy of mathematics. *For the Learning of Mathematics*, *5*(1), 44–48.

D'Ambrosio, U. (2001). *Ethnomathematics: Link between traditions and modernity*. Sense Publishers.

Davis, E. A., & Krajcik, J. S. (2005). Designing educative curriculum materials to promote teacher learning. *Educational Researcher*, *34*(3), 3–14. DOI: 10.3102/0013189X034003003

Fouze, A. Q., & Amit, M. (2018). Development of mathematical thinking through integration of ethnomathematic folklore game in math instruction. *Eurasia Journal of Mathematics, Science and Technology Education*, *14*(2), 617–630. DOI: 10.12973/ejmste/80626

Frankenstein, M. (1983). Critical mathematics education: An application of Paulo Freire's epistemology. *Journal of Education, 165*(4), 315–339. DOI: 10.1177/002205748316500403

Freire, P. (2014). *Pedagogy of the oppressed* (30th anniversary ed.). Bloomsbury Academic.

Gay, G. (2018). *Culturally responsive teaching: Theory, research, and practice* (3rd ed.). Teachers College Press.

Gutiérrez, R. (2008). A 'gap-gazing' fetish in mathematics education? Problematizing research on the achievement gap. *Journal for Research in Mathematics Education, 39*(4), 357–364. DOI: 10.5951/jresematheduc.39.4.0357

Gutiérrez, R. (2013). The sociopolitical turn in mathematics education. *Journal for Research in Mathematics Education, 44*(1), 37–68. DOI: 10.5951/jresematheduc.44.1.0037

Gutstein, E. (2006). *Reading and writing the world with mathematics: Toward a pedagogy for social justice*. Routledge.

Gutstein, E., & Peterson, B. (2005). Rethinking mathematics: Teaching social justice by the numbers. *Rethinking Schools.*

Haggarty, L., & Pepin, B. (2002). An investigation of mathematics textbooks and their use in English, French and German classrooms: Who gets an opportunity to learn what? *British Educational Research Journal, 28*(4), 567–590. DOI: 10.1080/0141192022000005832

Hobbs, L. (2018). *Developing culturally responsive mathematics teachers: Understanding equity and access in math education* [Doctoral thesis, University of Houston]. University of Houston Institutional Repository.

Husén, T. (Ed.). (1967). *International study of achievement in mathematics: A comparison of twelve countries* (Vols. 1–2). Almqvist & Wiksell; John Wiley.

Işık, C., & Kar, T. (2012). Sınıf öğretmeni adaylarının sözel problemlere yönelik kurdukları problemlerin analizi. *Kuram ve Uygulamada Eğitim Bilimleri, 12*(1), 201–221.

Ladson-Billings, G. (1995). Toward a theory of culturally relevant pedagogy. *American Educational Research Journal, 32*(3), 465–491. DOI: 10.3102/00028312032003465

Landis, J. R., & Koch, G. G. (1977). The measurement of observer agreement for categorical data. *Biometrics, 33*(1), 159–174. DOI: 10.2307/2529310 PMID: 843571

Ministry of National Education. (2024a). *Türkiye Yüzyılı Maarif Modeli: Ortak metinler*. MEB.

Ministry of National Education. (2024b). *Matematik dersi öğretim programı (1–8. sınıflar)*. MEB.

Ministry of National Education. (2024c). *Matematik ders kitabı 5. sınıf, 1. kitap*. MEB Yayınları.

Ministry of National Education. (2024d). *Matematik ders kitabı 5. sınıf, 2. kitap*. MEB Yayınları.

Ministry of National Education. (2025a). *Matematik ders kitabı 6. sınıf, 1. kitap*. MEB Yayınları.

Ministry of National Education. (2025b). *Matematik ders kitabı 6. sınıf, 2. kitap*. MEB Yayınları.

Ministry of National Education. (2025c). *Ders kitabı dağıtım bilgileri*. MEB TYMM Portalı. Retrieved March 2025, from https://tymm.meb.gov.tr

National Council of Teachers of Mathematics. (2000). *Principles and standards for school mathematics*. NCTM.

National Research Council. (2001). *Adding it up: Helping children learn mathematics*. National Academies Press.

Novak, J. D., & Cañas, A. J. (2008). *The theory underlying concept maps and how to construct and use them*. Technical Report IHMC CmapTools 2006-01 Rev 01-2008, Florida Institute for Human and Machine Cognition, Pensacola. https://cmap.ihmc.us/Publications/ResearchPapers/TheoryUnderlyingConceptMaps.pdf

Paris, D. (2012). Culturally sustaining pedagogy: A needed change in stance, terminology, and practice. *Educational Researcher, 41*(3), 93–97. DOI: 10.3102/0013189X12441244

Paris, D., & Alim, H. S. (2017). *Culturally sustaining pedagogies: Teaching and learning for justice in a changing world*. Teachers College Press.

Pepin, B., & Haggarty, L. (2001). Mathematics textbooks and their use in English, French and German classrooms: A way to understand teaching and learning cultures. *ZDM Mathematics Education, 33*(5), 158–175. DOI: 10.1007/BF02656616

Pöhler, B., & Prediger, S. (2015). Intertwining lexical and conceptual learning trajectories. *Eurasia Journal of Mathematics, Science and Technology Education, 11*(6), 1697–1716. DOI: 10.12973/eurasia.2015.1497a

Remillard, J. T. (2005). Examining key concepts in research on teachers' use of mathematics curricula. *Review of Educational Research, 75*(2), 211–246. DOI: 10.3102/00346543075002211

Remillard, J. T., & Heck, D. J. (2014). Conceptualizing the curriculum enactment process in mathematics education. *ZDM Mathematics Education, 46*(5), 705–718. DOI: 10.1007/s11858-014-0600-4

Rosa, M., & Orey, D. C. (2013). Ethnomodeling as a research theoretical framework on ethnomathematics and mathematical modeling. *Journal of Urban Mathematics Education, 6*(2), 62–80. DOI: 10.21423/jume-v6i2a195

Rosa, M., & Orey, D. C. (2015). A trivium curriculum for mathematics based on literacy, matheracy, and technoracy: An ethnomathematics perspective. *ZDM Mathematics Education, 47*(4), 587–598. DOI: 10.1007/s11858-015-0688-1

Rosa, M., & Orey, D. C. (2016). State of the art in ethnomodelling: Research on local mathematical knowledge and perspectives. In Rosa, M., D'Ambrosio, U., Orey, D. C., Shirley, L., Alangui, W. V., Palhares, P., & Gavarrete, M. E. (Eds.), *Current and future perspectives of ethnomathematics as a program* (pp. 71–80). Springer. DOI: 10.1007/978-3-319-30120-4

Schmidt, W. H., McKnight, C. C., & Raizen, S. A. (1997). *A splintered vision: An investigation of U.S. science and mathematics education.* Kluwer Academic Publishers.

Shield, M., & Dole, S. (2013). Assessing the potential of mathematics textbooks to promote deep learning. *Educational Studies in Mathematics, 82*(2), 183–199. DOI: 10.1007/s10649-012-9415-9

Skovsmose, O. (1994). *Towards a philosophy of critical mathematics education.* Kluwer Academic Publishers. DOI: 10.1007/978-94-017-3556-8

Stein, M. K., Remillard, J. T., & Smith, M. S. (2007). How curriculum influences student learning. In Lester, F. K. (Ed.), *Second handbook of research on mathematics teaching and learning* (pp. 319–370). Information Age Publishing. DOI: 10.1108/978-1-28141-200-320251012

Stein, M. K., Smith, M. S., Henningsen, M. A., & Silver, E. A. (2000). *Implementing standards-based mathematics instruction: A casebook for professional development.* Teachers College Press.

Tarr, J. E., Reys, R. E., Reys, B. J., Chávez, Ó., Shih, J., & Osterlind, S. J. (2008). The impact of middle-grades mathematics curricula and the classroom learning environment on student achievement. *Journal for Research in Mathematics Education, 39*(3), 247–280. https://www.jstor.org/stable/30034970

Valverde, G. A., Bianchi, L. J., Wolfe, R. G., Schmidt, W. H., & Houang, R. T. (2002). *According to the book: Using TIMSS to investigate the translation of policy into practice through the world of textbooks*. Springer. DOI: 10.1007/978-94-007-0844-0

Xenofontos, C., & Andrews, P. (2014). Defining mathematics teaching: A comparative study of Cypriot and English primary teachers' beliefs. *Teaching and Teacher Education, 37*, 1–10. DOI: 10.1016/j.tate.2013.08.006

KEY TERMS AND DEFINITIONS

C–T–M–E Bridging Model: A four-component analytic model tracing how Cultural resources (C) are embedded through Task/representation features (T) into Mathematical work (M) with Equity/access/participation signals (E). See Section 2.10 and Figure 2 for the comprehensive overview.

Critical Mathematics Education (CME): A framework, rooted in Freire's (2014) critical pedagogy, that positions mathematics as a tool for reading and writing the world—using mathematical reasoning to understand, critique, and transform social structures (Skovsmose, 1994; Frankenstein, 1983; Gutstein, 2006).

Culturally Sustaining Pedagogy (CSP): An extension of culturally relevant pedagogy that aspires to sustain and revitalize the cultural practices, languages, and ways of knowing of nondominant communities rather than merely leveraging them as pedagogical bridges (Paris, 2012; Paris & Alim, 2017).

Depth of Integration: D1 (tokenistic—context incidental/Decor; remove → unchanged); D2 (contextual—context shapes data/representation/Directive; remove → simplified); D3 (transformative—context is epistemic driver/Modeling Agent; remove → meaningless).

Enacted Curriculum: What students actually experience in classrooms, as distinct from the written curriculum (designed textbook) and the intended curriculum (teacher's interpretation). Teacher enactment mediates between designed and experienced OTL (Remillard & Heck, 2014; Stein et al., 2007).

Ethnomathematics: The study and educational use of mathematical ideas and practices developed within cultural groups, emphasizing mathematizing of cultural practice as epistemic resource (D'Ambrosio, 1985, 2001).

Ethnomodeling (Glocal): Translating culturally situated practices (emic perspective) into formal models, representations, and generalizations (etic perspective), respecting the integrity of both and creating a glocal bridge between community knowledge and school mathematics (Rosa & Orey, 2013, 2016).

Modeling Agent (T-role): The task-representation role in which cultural context determines modeling constraints, assumptions, and validation criteria—the proximate design mechanism of D3 depth.

Opportunity to Learn (OTL): The learning opportunities made available to students through curriculum, instruction, and assessment. Textbooks are a major OTL mediator; written OTL depth is a ceiling for enacted OTL (Schmidt et al., 1997; Stein et al., 2007).

Power-Identity Signaling (E5): Tasks positioning students as citizens who can use mathematics to examine justice, privacy, fairness, or power relations (Gutiérrez, 2008; Gutstein, 2006).

Task Degradation: The tendency for high-demand tasks to be reduced in cognitive complexity during teacher enactment—through excessive scaffolding, pre-answering open questions, or converting modeling cycles into procedural sequences (Stein et al., 2007).

Tokenistic Contextualization (D1): Cultural references present but task unchanged if context removed. Diagnostic test: "Bağlam kalksa soru değişir mi?" → No change = D1.

Chapter 5
Ethnomathematics for Teaching Geometry and Measurement: A Design-Based Research Approach

Mustafa Serkan Pelen
http://orcid.org/0000-0002-2113-6383
Ministry of National Education, Turkey

ABSTRACT

Mathematics education is often perceived as culturally neutral, limiting students' connections to meaningful contexts. Ethnomathematics addresses this challenge by recognizing mathematical practices embedded in cultural traditions. This chapter presents a design-based research (DBR)-informed instructional design using Anatolian motifs to teach geometry and measurement. Through a four-phase DBR process (analysis, design, exploratory piloting, and refinement), cultural artifacts were transformed into curriculum-aligned learning experiences. Embodying polygons, symmetry, and geometric transformations, Anatolian motifs are positioned as pedagogical tools rather than folkloric decorations. Four ethnomathematical design principles are proposed: cultural authenticity, mathematical transparency, exploratory construction, and structural decomposition. The chapter offers a classroom-ready instructional model for integrating culturally responsive tasks into geometry and measurement curricula, with implications for teachers, curriculum developers, and mathematics education researchers.

DOI: 10.4018/979-8-2600-3485-9.ch005

INTRODUCTION

Problem Statement and Gap

Mathematics education is generally perceived as a culture-free field consisting solely of numerical calculations, rigid equations, and objective technical procedures (Rosa & Orey, 2011). This conventional detachment between students' out-of-school cultural experiences and the formal academic curriculum contributes significantly to low mathematical achievement, alienating students from diverse backgrounds and limiting opportunities for meaningful understandings (Asare, 2026). When mathematics is presented as an isolated, Eurocentric domain disconnected from lived experience, students frequently struggle to perceive its relevance (Aktuna, 2013).

Ethnomathematics addresses this issue by recognizing the mathematical practices embedded within distinct cultural traditions (D'Ambrosio, 1985). As Bishop (1988) identified, mathematical development across all cultures is driven by six universal activities: counting, locating, measuring, designing, playing, and explaining. Applying ethnomathematics enriches learning by allowing students to investigate the mathematical structures within cultural products. Importantly, ethnomathematics is not positioned as an alternative or a replacement for formal academic mathematics; rather, it serves as a complementary approach that enriches the curriculum. By relating with cultural practices, mathematics can be perceived as a cohesive whole rather than a set of isolated, disconnected rules. Ultimately, demonstrating that mathematics is a human endeavor fosters deeper mathematical thinking, enhances interdisciplinary understanding, and strengthens students' cultural identity (Bahadır, 2021).

Despite a growing theoretical consensus on the value of culturally responsive pedagogy, a significant gap remains in its practical classroom implementation. Research indicates that highly centralized and standardized curricula often limit the flexible inclusion of local cultural contexts in mathematics teaching (Günay, 2023; Ketenci Tona & Akman, 2026). Although teachers acknowledge the value of culturally relevant tasks, they frequently lack the structured pedagogical guides, resources, and specific content knowledge required to blend cultural artifacts with strict mathematical learning outcomes (Fouze & Amit, 2023). Furthermore, an examination of current middle school mathematics textbooks reveals that cultural elements are frequently relegated to superficial visual aids or end-of-chapter reading passages rather than being utilized as active, structural pedagogical tools within the evaluation and learning processes (Demir & Seçkin Kapucu, 2023). While recent studies have documented the ethnomathematical resources available, researchers consistently emphasize the urgent need for robust design-based studies to translate these cultural artifacts into classroom-ready materials and to measure their actual impact on student outcomes (Anderson & Shattuck, 2012; Ergene et al., 2020).

Purpose and Specific Objectives

The primary purpose of this chapter is to develop and present an ethnomathematics-based instructional design grounded in Anatolian motifs for teaching geometry and measurement concepts in middle school mathematics. Moving beyond the use of cultural elements as superficial folkloric illustrations, this study employs them as structured pedagogical tools to cultivate mathematical reasoning and spatial understanding.

The specific objectives of this chapter are to:

- Develop an ethnomathematics-based instructional model utilizing Anatolian motifs as authentic cultural mathematical artifacts.
- Demonstrate how the mathematical practices embedded in these motifs can support students' comprehension of geometric properties, transformations (reflection, rotation, translation), and measurement (area and perimeter).
- Design and present detailed, classroom-ready ethnomathematical learning activities aligned with middle school mathematics curricula.
- Derive and articulate transferable design principles for integrating ethnomathematics into geometry and measurement teaching through the DBR framework.
- Offer exploratory pilot-informed observations on the instructional potential and classroom feasibility of this culturally responsive approach.

Significance and Overview of Chapter Structure

This chapter contributes to the edited volume by demonstrating how culturally grounded mathematical knowledge can be systematically transformed into replicable, classroom-ready instructional designs. By integrating ethnomathematics with Design-Based Research, the chapter provides both a robust theoretical perspective and practical tools for curriculum developers and educators. This approach answers the call in the literature to move beyond purely descriptive ethnographic accounts and offer actionable pedagogical frameworks (Bakker, 2018).

The Turkish educational context provides a particularly relevant setting for this work. The 2024 MoNE curriculum, framed under the Turkish Century Education Model (Türkiye Yüzyılı Maarif Modeli), explicitly adopts a holistic approach to education and encourages teachers to utilize traditional Turkish arts - including tile making, ceramics, and weaving - to contextualize geometry and measurement learning (MoNE, 2024). Anatolian motifs, structurally governed by geometric transformations, symmetry axes, and polygonal compositions, align naturally with these curricular objectives (Dumlu & Alkaş Ulusoy, 2025; Günay & Takunyacı, 2023).

The chapter is structured as follows. Background establishes the theoretical foundation linking ethnomathematics, geometry learning, and DBR. Methodological Approach outlines the four-phase DBR framework. Instructional Design describes the cultural, mathematical, and pedagogical components of the model. Classroom Activities presents four practical, curriculum-aligned activities. Design Rationale and Principles details how iterative design reasoning informed the four core principles. Finally, Discussion and Future Research synthesizes the contributions and implications for teachers, curriculum developers, and researchers.

BACKGROUND

Ethnomathematics as a Theoretical Lens

The field of ethnomathematics, formally introduced by Brazilian mathematician Ubiratan D'Ambrosio, challenges the Eurocentric paradigm by treating mathematics as a culturally embedded human endeavor (Akdemir & Şengül Akdemir, 2024; Rosa et al., 2016). D'Ambrosio conceptualized the term through the synthesis of three roots: "ethno" (representing the natural, social, cultural, and imaginary environment), "mathema" (meaning explaining, learning, knowing, and coping with reality), and "tics" (derived from techné, referring to modes, styles, arts, and techniques) (D'Ambrosio, 2006). Viewed through this lens, mathematics is not an isolated academic domain but a compendium of practices developed by diverse cultural groups to navigate their environments (Rosa & Orey, 2011).

Bishop (1988) identified six universal pre-mathematical activities present in every culture: counting, locating, measuring, designing, playing, and explaining. These universal activities illustrate that while the capacity for mathematical thinking is pan-human, the resulting symbolic technologies, representations, and educational expressions differ substantially from one cultural group to another (Bishop, 1988; Ketenci Tona & Akman, 2026). Ethnomathematics thus aims to bridge the gap between formal academic mathematics and students' daily lives by validating the mathematical ideas that already exist within their cultural identities (Arı & Demir, 2022; Rosa et al., 2016).

Cultural Artifacts in Mathematics Education

Integrating cultural artifacts into the mathematics classroom serves to contextualize abstract concepts and provides a tangible medium for geometric and spatial reasoning (Setiaputra et al., 2025). Cultural artifacts and daily life practices are natural learning spaces where abstract mathematical concepts are concretized, showing students that mathematical thinking is an inherent component of their heritage (Rosa & Orey, 2011).

Globally, researchers have documented the rich mathematical knowledge embedded in indigenous crafts, games, and architecture. For example, studies in the African context reveal that the symmetrical designs of Ghanaian Adinkra motifs embody concepts of transformation and reflection, while the intricate patterns of woven Kente cloths demonstrate geometric tessellations and proportional reasoning (Asare, 2026). Traditional games, such as Oware, Mancala, Morabaraba, and the South African rope-skipping game Kgati, have been shown to foster strategic planning, algebraic reasoning, and spatial analysis (Batiibwe, 2024; Kusi & Bonyah, 2025; Meeran & Niranjan, 2026). Similarly, the Sona sand drawing tradition of Angola showcases advanced concepts of Eulerian graphs and geometry (Batiibwe, 2024; Verner et al., 2019; Yazıcı, 2021), and Bedouin embroidery utilizes precise counting methods to create right triangles and symmetries (Fouze & Amit, 2018; Fouze & Amit, 2023). These artifacts are not merely decorative elements or folkloric curiosities; they are structurally robust mathematical resources that require sophisticated spatial reasoning, problem-solving, and logical deduction (Asare, 2026; Meeran & Niranjan, 2026). When utilized as pedagogical tools, they allow students to actively explore mathematics through familiar materials, deepening conceptual understanding while increasing motivation and engagement (Fouze & Amit, 2018; Rosa et al., 2016).

Cultural artifacts become pedagogically powerful when curriculum bridges students' cultural knowledge and formal mathematics. Adam (2004) structures this connection through the framework of thinking mathematical - the processes linking cultural practices to school mathematics (see Figure 1). This framework positions mathematical thinking as the active bridge between students' cultural activities and conventional mathematics, providing the theoretical foundation for the ethnomathematics-based instructional design developed in this chapter.

Figure 1. Framework for ethnomathematical instructional design (Adapted from Adam, 2004)

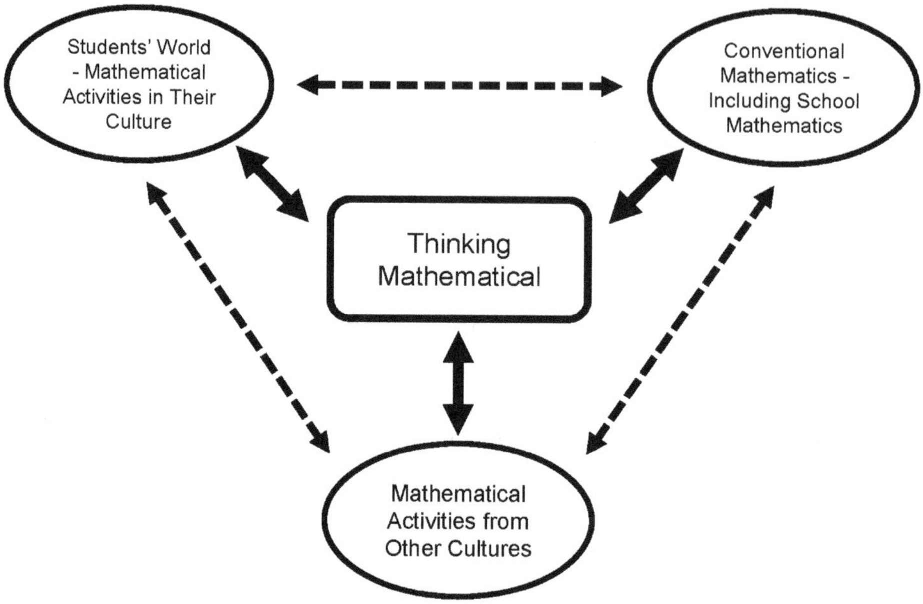

Ethnomathematics in Türkiye: Prior Work, Findings, and Limitations

Türkiye, with its deep historical roots and diverse cultural mosaic, offers a rich landscape for ethnomathematical inquiry, though research in this area remains in a developmental stage compared to international literature (Ketenci Tona & Akman, 2026). Recent studies in the Turkish context have demonstrated positive cognitive and affective outcomes. Ergene et al. (2020) found that engaging pre-service mathematics teachers in designing ethnomathematical activities significantly increased their cultural awareness and expanded their epistemological perspectives on the nature of mathematics.

In terms of student outcomes, Günay and Takunyacı (2023) provided evidence that instruction utilizing ethnomathematical elements (such as traditional tiles and rug patterns) enhanced students' problem-solving skills in transformation geometry, with marked improvements in their ability to understand problems, devise plans, and apply mathematical knowledge to everyday situations. Furthermore, Dumlu and Alkaş Ulusoy (2025) observed that integrating Turkish rug motifs into geometry lessons helped students relate abstract concepts like reflection, rotation, and axes of symmetry to real-life applications, thereby increasing their motivation and questioning skills.

Despite these affordances, systemic limitations hinder widespread implementation. The highly centralized and standardized structure of the Turkish mathematics curriculum often restricts the flexible inclusion of local cultural contexts in daily teaching processes, leaving teachers without the structured pedagogical guides and ready-to-use materials needed to blend cultural artifacts with mandatory mathematical learning outcomes (Çenberci & Horzum, 2023; Ketenci Tona & Akman, 2026). This chapter builds directly on these foundational Turkish studies while addressing their shared limitation: the absence of a systematic, DBR-informed instructional model ready for classroom use.

Geometry and Measurement Learning in Middle School

Geometry and measurement are critical domains within the middle school mathematics curriculum, essential for developing spatial relationships, logical reasoning, and quantitative comprehension. In Türkiye, the Ministry of National Education (MoNE) Middle School Mathematics Curriculum, recently updated under the framework of the Turkish Century Education Model, strongly emphasizes themes such as "Geometric Shapes," "Geometric Quantities," and "Transformation" (MoNE, 2024). Specifically, the curriculum requires students to infer geometric properties and construct the images of geometric shapes under reflection, translation, rotation, and symmetry axis transformations (Dumlu & Alkaş Ulusoy, 2025).

However, geometry and measurement frequently present conceptual challenges for students when taught through rigid, abstract, and purely symbolic methods (Demir & Seçkin Kapucu, 2023). Students often struggle to grasp dynamic transformations or area-perimeter relationships because formal schooling typically prioritizes procedural memorization over contextual understanding (Bahadır, 2021; Prahmana & D'Ambrosio, 2020). Culturally grounded tasks mitigate this abstraction by providing a concrete, visual, and tactile medium through which geometric properties can be actively reconstructed (Dumlu & Alkaş Ulusoy, 2025; Prahmana & D'Ambrosio, 2020). The Turkish curriculum explicitly supports this approach, advising teachers to utilize traditional arts - such as tile making, ceramics, and weaving - to help students contextualize and discover these geometric relationships (Demir & Seçkin Kapucu, 2023; MoNE, 2024). Anatolian motifs are particularly well suited to this purpose, as their structural composition directly instantiates the transformation geometry and measurement concepts required by the curriculum.

Anatolian Motifs: Mathematical Structures, Geometric Properties, and Cultural Significance

Anatolia's traditional carpets, rugs, and handicrafts have served as cultural mediums for centuries, conveying the society's beliefs, histories, sorrows, and expectations (Küçük, 2014). Beyond their artistic and anthropological value, these motifs are structurally governed by advanced mathematical axioms and geometric rules (Dumlu & Alkaş Ulusoy, 2025).

Traditional Anatolian motifs - such as the ram's horn (koç boynuzu), hands on hips (eli belinde), tree of life (hayat ağacı), and waterway (su yolu) - systematically integrate complex polygons including triangles, squares, rhombuses, and hexagons into their designs (Dumlu & Alkaş Ulusoy, 2025). Their formation requires the precise execution of geometric transformations (Küçük, 2014). Artisans and weavers apply reflection, translation, rotation around specific angles, and defined axes of symmetry to maintain aesthetic balance and proportionality (Dumlu & Alkaş Ulusoy, 2025). Representative examples of these motifs and their underlying geometric structures are illustrated in Figure 2.

Figure 2. Examples of traditional Anatolian motifs and their geometric structures

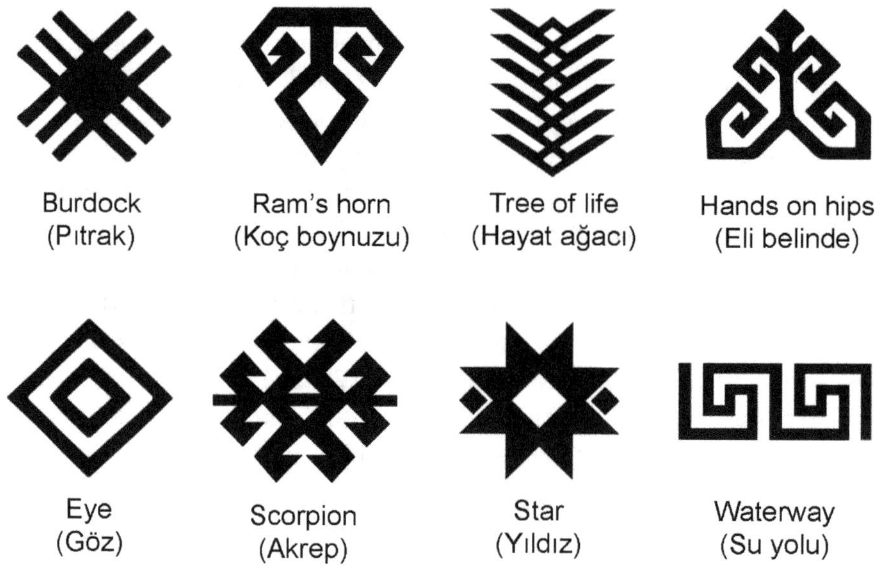

While the weavers may not utilize formal academic nomenclature, they intuitively and practically "mathematize" spatial structures (Dumlu & Alkaş Ulusoy, 2025). They engage in complex counting and geometric planning to ensure that patterns repeat consistently across the textile (Kırıkçılar, 2023). Because Anatolian motifs are inherently composed of geometric transformations and symmetries, they serve as powerful pedagogical instruments for concretizing abstract transformation geometry (Dumlu & Alkaş Ulusoy, 2025). They transform a potentially abstract curricular requirement into an engaging, culturally resonant cognitive experience that validates the students' heritage (Ketenci Tona & Akman, 2026).

Design-Based Research in Mathematics Education

Design-Based Research (DBR) is an iterative, interventionist methodology specifically aimed at bridging the gap between educational theory and classroom practice. Unlike traditional experimental designs that isolate variables to test "what works" in controlled environments, DBR investigates "how, when, and why" an intervention functions within the complex, authentic ecology of a real classroom (Karademir & Girgin, 2024; McKenney & Reeves, 2012).

The DBR methodology typically progresses through cyclical phases: analysis and exploration, design and construction, evaluation and reflection, and implementation and refinement. In the analysis phase, researchers work with practitioners to identify practical problems and review literature. During design and construction, instructional prototypes are developed. The evaluation phase involves pilot testing these interventions in real classrooms, observing student interactions, and refining the design based on collected insights (McKenney & Reeves, 2012).

DBR is particularly well suited to ethnomathematics integration in formal schooling (Kırıkçılar, 2023). Simply bringing a cultural artifact into a classroom as a visual aid is insufficient for deep learning (Batiibwe, 2025). The pedagogical processing, curricular alignment, and structural decomposition of the artifact must be systematically engineered (Rosa et al., 2016). Central to this process is ethnomodeling - a pedagogical action that applies mathematical modeling techniques to translate local mathematical practices into the academic curriculum (Arı & Demir, 2022; Rosa et al., 2016). Ethnomodeling operates through a dialogue between two analytical perspectives: the emic approach, which examines mathematical practices from the insider's perspective of the cultural community itself, and the etic approach, which applies standardized academic frameworks to observe and translate those practices into formal mathematical knowledge (Rosa & Orey, 2021). By employing DBR, researchers ensure that local cultural knowledge is translated into viable, classroom-ready instructional models, yielding both practical educational solutions and transferable theoretical design principles (Bakker, 2018; McKenney & Reeves, 2012).

Universal Challenges of Ethnomathematics Integration

While the ethnomathematics holds great promise for inclusive mathematics education, its global implementation encounters several recurring challenges. The primary barrier is teacher preparedness and knowledge (Batiibwe, 2025; Ketenci Tona & Akman, 2026). Teachers frequently lack the pedagogical content knowledge to recognize the mathematics embedded in diverse cultural practices, and they struggle to design lessons that integrate these cultural elements with mandatory curriculum standards (Ergene et al., 2020; Çenberci & Horzum, 2023).

Another significant pedagogical danger is cultural appropriation or trivialization (Batiibwe, 2025). There is a risk that cultural artifacts are reduced to superficial "folkloric" presentations - such as merely showing pictures of rugs or traditional foods - without engaging students in genuine, rigorous mathematical inquiry (Ketenci Tona & Akman, 2026). Ethnomathematical instruction must carefully balance the preservation of cultural authenticity with the demand for mathematical transparency, ensuring that the mathematics remains the focal point of the lesson (Rosa et al., 2016).

Furthermore, the field suffers from assessment constraints. Traditional, standardized testing models heavily emphasize decontextualized academic mathematics, which structurally constrains teachers from adopting culturally based teaching practices (Ketenci Tona & Akman, 2026; Rosa et al., 2016). Finally, there is a scarcity of robust, empirically tested instructional materials, leaving teachers without clear guidelines on how to implement this approach effectively in their daily routines (Batiibwe, 2025; Ketenci Tona & Akman, 2026). These challenges collectively underscore the need for structured, replicable instructional designs of the kind this chapter develops.

Gap in the Literature and Rationale for This Study

A critical examination of the existing ethnomathematics literature reveals that the majority of research consists of qualitative and ethnographic descriptions documenting the mathematics inherent in various cultural practices (Ketenci Tona & Akman, 2026). While foundational for the field, mathematics education researchers emphasize the urgent need to transition from these descriptive anthropological accounts to design-based studies that translate cultural artifacts into formal educational settings (Batiibwe, 2025; Ketenci Tona & Akman, 2026). Specifically: (a) descriptive ethnomathematics studies establish what mathematical knowledge exists within cultural practices; (b) recent Turkish DBR-adjacent studies demonstrate the positive impact of cultural artifacts on student motivation and geometric reasoning; but (c) a residual gap remains in systematic, curriculum-aligned instructional designs that teachers can reliably implement across middle school geometry and measurement topics.

To bridge this divide, scholars advocate for ethnomodeling - applying mathematical modeling techniques to translate local mathematical practices into the academic curriculum through the emic-etic dialogue introduced above (Arı & Demir, 2022; Rosa et al., 2016), as illustrated in Figure 3.

Figure 3. Ethnomodeling as the intersection of mathematical modeling, cultural anthropology, and ethnomathematics (based on Orey & Rosa, 2010)

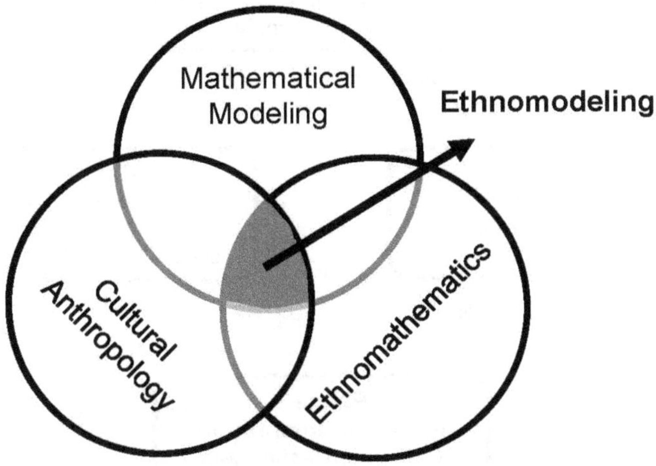

This chapter addresses this identified gap by utilizing the DBR framework to systematically transform Anatolian motifs from passive cultural illustrations into dynamic, ethnomodeling-based instructional tools, offering a replicable model for middle school geometry and measurement education alongside transferable design principles for culturally responsive mathematics instruction.

METHODOLOGICAL APPROACH

Building on the DBR framework introduced in the previous section, this chapter adopts a four-phase macro-cycle - analysis and exploration, instructional design, exploratory piloting, and refinement - to systematically transform Anatolian motifs into structured, curriculum-aligned pedagogical tools for middle school geometry and measurement (McKenney & Reeves, 2012). Rather than constituting a large-scale empirical investigation, this process represents a design-oriented inquiry in which iterative development, classroom experience, and theoretical reflection collectively informed the instructional model and design principles presented in this chapter. The four phases and their iterative relationships are illustrated in Figure 4.

Figure 4. Four-phase DBR cycle used in this chapter (Adapted from McKenney & Reeves, 2012)

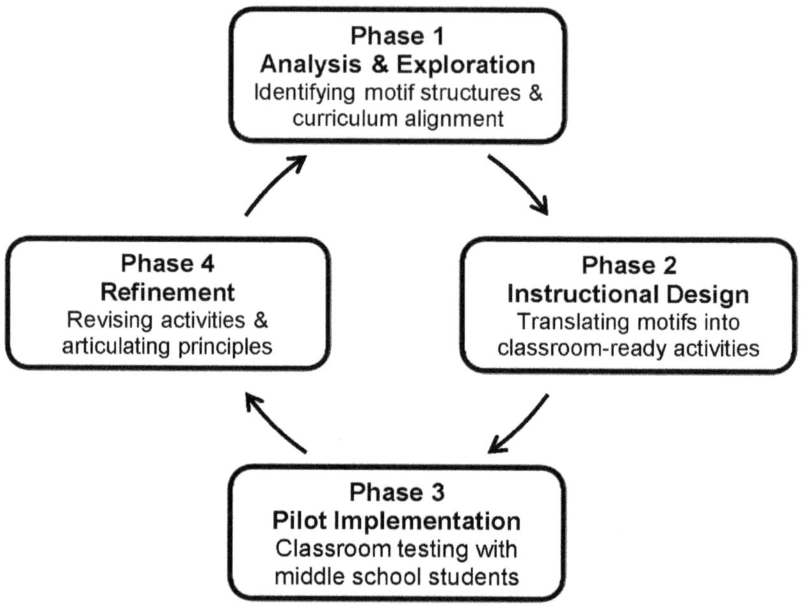

Table 1. Overview of the DBR design process

Phase	Focus	Activities Undertaken	Sources of Design Feedback
1. Analysis and Exploration	Needs and context analysis	Motif analysis; curriculum mapping	Literature; curriculum documents
2. Instructional Design	Design and construction	Activity development; task sequencing	Design principles; practitioner knowledge
3. Exploratory Piloting	Classroom enactment	Limited-scale classroom implementation	Student work; observations; teacher reflections
4. Refinement	Evaluation and revision	Activity revision; principle articulation	Cross-phase design insights

Phase 1 – Analysis and Exploration

The first phase of the DBR process involves the analysis of practical problems through the collaboration between researchers and practitioners, paired with a systematic literature review (Štemberger & Cencič, 2016). This phase aims to gain theoretical inputs that shape the understanding of the problem, the context, and the needs of the stakeholders. It is an exploratory stage where the boundaries of feasible changes and partial design requirements are determined (McKenney & Reeves, 2012).

In this chapter, this phase focused on an in-depth ethnomathematical examination of Anatolian cultural artifacts. Traditional motifs were systematically analyzed to identify their embedded mathematical structures, such as symmetries, geometric transformations, and polygonal compositions. Concurrently, an analysis of the Ministry of National Education (MoNE) middle school mathematics curriculum was conducted to map these cultural structures directly onto formal learning outcomes for geometry and measurement. This needs and context analysis ensured that the subsequent design would function as a mathematically rigorous tool aligned with the educational context and curricular scope rather than a superficial folkloric addition (McKenney & Reeves, 2012).

Phase 2 – Instructional Design

The second phase, design and construction, represents a deliberative-generative cycle where theoretical insights and contextual boundaries identified in Phase 1 are translated into a tangible educational solution. Researchers develop innovative learning materials informed by existing design principles and practitioner craft wisdom, progressing from a skeleton design to detailed instructional specifications (McKenney & Reeves, 2012).

In this chapter, the instructional design phase involved translating the mathematical properties of Anatolian motifs into concrete, classroom-ready activities. The instructional tasks were specifically designed to support student exploration, reasoning, and classroom discussion. The pedagogical component was structured to move students from visual observation of the motifs to active geometric composition, decomposition, and transformation. This phase required careful attention to maintaining both cultural authenticity and mathematical transparency, ensuring that the embedded geometry was explicitly accessible to middle school learners without losing its cultural resonance.

Phase 3 – Exploratory Piloting

The third phase involves enacting the designed intervention within an authentic learning environment to evaluate its practical viability. DBR is inherently interventionist, investigating new forms of learning by actively bringing them about in real classroom settings (Cobb et al., 2003; Design-Based Research Collective, 2003; McKenney & Reeves, 2012).

In this chapter, the designed ethnomathematical activities were explored on a limited, exploratory scale with middle school students to examine their pedagogical feasibility and instructional potential. This exploratory piloting allowed the authors to observe how the Anatolian motifs functioned as active learning tools in classroom settings - how students engaged with transformation geometry and area-perimeter relationships when the concepts were anchored in familiar cultural artifacts, and what cognitive or logistical challenges that emerged during instruction. Consistent with the exploratory nature of this phase, the insights gathered are presented as design-informing observations rather than generalizable empirical findings.

Phase 4 – Refinement

The final phase entails evaluation, reflection, and refinement. Design experiments are conjecture-driven tests; if initial conjectures about how a design supports learning are only partially realized, new conjectures are generated and the design is iteratively revised. Through retrospective analysis, researchers synthesize the empirical findings to produce both practical improvements to the intervention and transferable theoretical understanding, such as design principles (Cobb et al., 2003; McKenney & Reeves, 2012).

Following the exploratory piloting, the ethnomathematical activities in this chapter were carefully revised based on design insights gathered from the classroom experience. This reflection process evaluated the alignment between the intended mathematical outcomes and the student reasoning observed. The iterative refinement resulted in an improved set of learning activities for geometry and measurement, and facilitated the articulation of four core ethnomathematical design principles (cultural authenticity, mathematical transparency, exploratory construction, and structural decomposition) that can guide future research and curriculum development.

Sources of Design Feedback

To inform the iterative refinement of the instructional design, feedback was gathered through three complementary sources that together provided a rounded picture of how the activities functioned in classroom settings (Cobb et al., 2003; McKenney & Reeves, 2012).

Student work: Worksheets and task outputs were examined to gain insights into students' mathematical reasoning and engagement with motif-based tasks. Analyzing these artifacts allowed the design team to observe how students approached decomposition strategies, identified symmetry, and applied geometric transformations when reconstructing the Anatolian motifs (Aktuna, 2013; Dumlu & Alkaş Ulusoy, 2025).

Classroom observations: Observation of classroom sessions, including researcher field notes, were used to capture the dynamic processes of enactment - classroom discourse, student interactions, and the ways in which learners engaged visually and physically with the cultural artifacts (Cobb et al., 2003; Dumlu & Alkaş Ulusoy, 2025).

Teacher reflections: The implementing teacher maintained a reflection diary recording impressions of student motivation, the effectiveness of ethnomathematical connections, pedagogical challenges encountered, and notable classroom moments. This instrument captured practitioner perspectives on the practical viability of the instructional design and highlighted areas requiring refinement (Dumlu & Alkaş Ulusoy, 2025; McKenney & Reeves, 2012).

Together, these sources of design feedback informed the iterative refinement of the activities and the articulation of the four design principles presented later in the chapter. They are offered not as a formal data corpus but as the design evidence base that shaped the instructional model - consistent with the exploratory, design-oriented nature of this work.

INSTRUCTIONAL DESIGN

The instructional design developed in this chapter serves as a systematic bridge between the theoretical principles of ethnomathematics and practical classroom application. By utilizing ethnomodeling - a pedagogical action that merges cultural anthropology, mathematics, and mathematical modeling - the design transforms students' out-of-school cultural realities into structured academic knowledge (Arı & Demir, 2022). Through the emic-etic dialogue introduced in the Background, the implicit geometric logic embedded in Anatolian motifs is made explicitly accessible within the formal curriculum. The learning environment is organized around three interconnected components: cultural, mathematical, and pedagogical. These components are deliberately synchronized to meet formal curriculum standards, ensuring that ethnomathematical integration enriches rather than displaces rigorous geometric and measurement learning.

Cultural Component: Anatolian Motifs as Mathematical Artifacts

In this instructional design, traditional Anatolian motifs woven into carpets and rugs are positioned as the primary cultural anchors. Historically, these textiles have functioned as a significant medium for social communication in Turkey, where distinct motifs such as ram's horn, hands on hips, tree of life, abundance, bird, and waterway convey beliefs, expectations, migrations, and emotions (Dumlu & Alkaş Ulusoy, 2025). Within this educational framework, these artifacts are not treated merely as historical curiosities; rather, they are elevated to the status of authentic mathematical artifacts (Barton, 1996; Demir & Seçkin Kapucu, 2023).

The design recognizes that the weavers and artisans who created these patterns employed an intuitive yet mathematically sophisticated logic to achieve spatial harmony, aesthetic balance, and proportionality (Küçük, 2014; Prahmana & D'Ambrosio, 2020). By bringing these familiar cultural artifacts into the classroom, the instructional design provides a culturally resonant foundation that validates students' local heritage and transforms abstract geometric concepts into concrete, tangible experiences (Dumlu & Alkaş Ulusoy, 2025; Payadnya et al., 2024).

Mathematical Component: Geometry (Polygons, Symmetry, Transformations) and Measurement (Area, Perimeter)

The mathematical core of the design focuses on advancing students' spatial reasoning through transformation geometry and measurement. The structural analysis of Anatolian motifs reveals that their composition relies on the systematic arrangement of polygons, including triangles, squares, rhombuses, and hexagons. The instructional tasks are designed to require students to deconstruct and reconstruct these motifs using reflection, translation, rotation, and axes of symmetry. For example, specific activities prompt students to reflect the shackle motif across the y-axis or translate the hands on hips motif along the coordinate plane (Dumlu & Alkaş Ulusoy, 2025).

The design explicitly incorporates measurement concepts by having students investigate the area and perimeter of these culturally embedded shapes. Through structural decomposition, students break down complex, composite motifs into basic geometric units (such as right triangles and rectangles) to calculate their area and perimeter accurately. This allows students to explore how geometric transformations preserve lengths and areas, linking spatial visualization directly to quantitative measurement (Ascher, 1991; Bishop, 1988).

Pedagogical Component: Exploration, Discussion, and Written Justification Tasks

To maximize cognitive engagement, the pedagogical component shifts the classroom dynamic from traditional, teacher-centered transmission to a student-centered, constructivist environment (Aktuna, 2013; Kusi & Bonyah, 2025). The learning process is designed to unfold through active exploration, collaborative discussion, and rigorous written justification.

The intended instructional sequence begins with exploratory classroom discussions in which the teacher presents authentic rug visuals and guides students to identify geometric shapes, symmetries, and potential transformation rules governing the patterns. Following this orientation, students are designed to engage in both hands-on and digital exploration. Given that the technology integration is strongly recommended for transformation geometry learning, the activities are structured to allow students to utilize dynamic geometry software to test point motions, verify transformation rules, and create their own ethnomathematical designs (Dumlu & Alkaş Ulusoy, 2025; Nasrum et al., 2025).

A critical element of this pedagogical design is the requirement for written mathematical justification. Student worksheets explicitly prompt learners to explain their geometric reasoning after performing operations – a design decision intended to prevent reliance on rote memorization and support deep conceptual understanding (Supriyadi et al., 2024). Additionally, the design incorporates reflective thinking forms at the end of each lesson to evaluate their understanding of the relationship between geometry and cultural artifacts (Dumlu & Alkaş Ulusoy, 2025).

Curriculum Alignment: MoNE Middle School Mathematics

For an ethnomathematical intervention to be viable in formal education, it must align with mandatory academic standards. This instructional design is mapped primarily to the Ministry of National Education (MoNE) Middle School Mathematics Curriculum in Türkiye (MoNE, 2024). The MoNE curriculum explicitly mandates that teachers utilize historical and cultural artifacts - specifically mentioning traditional arts, tile, ceramics, carpet decorations, and weaving - to contextualize geometry and measurement. The designed activities directly fulfill transformation geometry learning outcomes requiring students to draw and construct the images of polygons resulting from translation, reflection, and rotation. The measurement tasks align similarly with learning outcomes concerning area and perimeter estimation and calculation using structural patterns (Demir & Seçkin Kapucu, 2023; MoNE, 2024).

While the design's primary curricular anchor is the Turkish national curriculum, its underlying principles resonate with broader international frameworks. NCTM standards emphasize the use of multiple representations to organize and communicate mathematical ideas, and the modeling of physical and cultural phenomena – both of which are central to this design (NCTM, 2000). This alignment suggests that the instructional model, while developed within the Turkish context, carries potential relevance for mathematics educators working within other culturally diverse educational systems.

Together, the three components of this instructional design - cultural, mathematical, and pedagogical - operationalize the four design principles articulated later in the chapter. Cultural authenticity is embodied primarily through the Cultural Component, which ensures motifs are used as genuine mathematical artifacts rather than decorative props. Mathematical transparency is realized through the Mathematical Component, which makes the geometric structures of motifs explicitly visible and accessible. Exploratory construction and structural decomposition are enacted through the Pedagogical Component, which requires students to actively build, decompose, and justify their mathematical reasoning. A summary mapping of each classroom activity to these components, the corresponding design principles, and the relevant MoNE learning outcomes is provided in Table 2 at the opening of the Classroom Activities section.

CLASSROOM ACTIVITIES

The operationalization of ethnomathematics within the formal classroom environment requires the translation of cultural artifacts into rigorous, curriculum-aligned mathematical tasks. This section details four classroom-ready activities developed through the ethnomodeling process, which bridges students' cultural realities with academic mathematics (Rosa et al., 2016). These activities are specifically aligned with the MoNE middle school mathematics curriculum, particularly targeting the geometry and measurement learning areas (MoNE, 2024). A summary mapping of each activity to the design components, design principles, and relevant MoNE learning outcomes is provided in Table 2. Each activity utilizes traditional Anatolian motifs to develop students' spatial reasoning, measurement skills, and conceptual understanding of geometry concepts.

Table 2. Mapping of classroom activities to design components, design principles, and MoNE learning outcomes

	Activity 1 Analyzing Symmetry in Anatolian Motifs	**Activity 2 Reconstructing Motifs: Composition and Decomposition**	**Activity 3 Area and Perimeter Relationships Within Motif Structures**	**Activity 4 Identifying and Applying Transformations**
Primary Design Component	**Cultural + Mathematical**	**Mathematical + Pedagogical**	**Mathematical + Pedagogical**	**Cultural + Mathematical + Pedagogical**
Design Principles Instantiated	Cultural Authenticity, Mathematical Transparency	Mathematical Transparency, Structural Decomposition	Structural Decomposition, Exploratory Construction	Cultural Authenticity, Exploratory Construction, Mathematical Transparency
Key Mathematical Concepts	Reflectional symmetry, lines of symmetry, rotational symmetry	Polygon identification, geometric composition and decomposition, proportional reasoning	Area, perimeter, non-standard and standard units of measurement, area-perimeter relationship	Translation, reflection, transformation properties, coordinate plane
Materials and Tools	Rug images, symmetry mirrors, paper folding	Squared grid paper, dot paper, GeoGebra/ Polypad (optional)	Tangram pieces, Polypad grid, squared grid paper	GeoGebra, grid paper, tracing paper
Approximate Grade Level	5th–6th grade	6th grade	5th–6th grade	7th–8th grade
Relevant MoNE Learning Outcomes	MAT.5.3.5, MAT.5.3.6: Interpret and analyze properties of polygons; MAT.7.3.1(d): Determine whether two congruent shapes are symmetric with respect to a line	MAT.5.3.5, MAT.5.3.6: Construct and analyze polygons; MAT.6.4.2: Transfer experiences with rectangle area to parallelogram and triangle area formulas	MAT.5.4.2: Evaluate rectangle area starting from unit squares; MAT.5.4.3: Interpret the relationship between area and perimeter; MAT.6.4.1: Reason analogically about relationships between length and area measurement units	MAT.7.3.1: Make inferences about images of shapes under reflection transformation; MAT.8.5.1: Analyze translation transformation using mathematical tools and technology; MAT.8.5.2: Make inferences about changes in coordinates under translation and reflection in the coordinate system; MAT.8.5.3: Solve problems involving translation and reflection transformations

Activity 1: Analyzing Symmetry in Anatolian Motifs

Task description: This introductory activity grounds students in the concept of symmetry by analyzing authentic Anatolian rugs (kilim) and cross-stitch (etamin) patterns (Dumlu & Alkaş Ulusoy, 2025; Kırıkçılar, 2023). Students are provided with high-quality images of traditional textiles and are tasked with identifying motifs that exhibit reflectional symmetry. Using symmetry mirrors and paper-folding techniques, students locate and draw all possible vertical, horizontal, and diagonal lines of symmetry on various motifs. Students are also asked to compare different motifs and categorize them based on the number of symmetry axes they possess (MoNE, 2024).

Expected student reasoning: Students will initially rely on visual intuition to identify symmetry, easily spotting vertical and horizontal axes in rectilinear patterns. When examining the motifs, students are expected to realize that while some shapes (like squares or certain cross-stitch patterns) have multiple lines of symmetry, others exhibit strictly bilateral symmetry. Through guided inquiry, students are expected to reason about the relationship between a polygon's regularity and its number of symmetry axes — recognizing, for example, that a square has four lines of symmetry while a non-regular rhombus has only two, and connecting this to the visual balance they observe in the rug patterns (Dumlu & Alkaş Ulusoy, 2025; MoNE, 2024).

Teacher guidance notes: The teacher should facilitate exploratory discussions rather than directly transmitting knowledge. The formal term "line of symmetry" should be introduced only after students have informally described the "fold lines" or "mirror lines". Teachers should encourage students to find lines of symmetry using rulers and protractors. If students struggle to identify symmetry, the teacher may prompt them to use a symmetry mirror on the exact center of the motif to verify whether the reflected half completes the image (MoNE, 2024).

Adaptation tips: For advanced learners, the teacher can introduce the rotational symmetry, asking students to determine the angle of rotation required for a motif to map onto itself. For students needing additional support, pre-cut paper models of the motifs allow physically folding to find lines of symmetry (Fouze & Amit, 2023; MoNE, 2024).

Activity 2: Reconstructing Motifs - Geometric Composition and Decomposition

Task description: In this activity, students transition from observing to actively constructing. Students are provided with complex Anatolian motifs (such as the tree of life or abundance patterns) and are asked to decompose these composite shapes into constituent geometric polygons (such as triangles, squares, rectangles, and rhombuses). Using squared grid paper or dot paper, which simulates the woven grid of a rug or a cross-stitch canvas, students must mathematically reconstruct the motifs by plotting the component polygons accurately (Dumlu & Alkaş Ulusoy, 2025; Kırıkçılar, 2023; MoNE, 2024).

It is important to note that the specific polygons students identify will depend on the regional variant of the motif selected. For example, the tree of life appears in numerous Anatolian variants: some versions are primarily composed of stacked isosceles triangles and chevrons arranged in translational symmetry along a vertical axis of reflectional symmetry, while others incorporate squares, rectangles, and rhombuses within the trunk and branch structures. This regional variation is itself pedagogically valuable – teachers can present two or more variants of the same motif side by side, prompting students to compare which polygons appear across versions and which are specific to a particular regional tradition. Regardless of the variant selected, teachers should ensure that the polygon set they ask students to identify is genuinely present in the specific motif image being used, verifying geometric accuracy before the lesson.

Expected student reasoning: When deconstructing the motifs, students are expected to recognize that complex cultural symbols are actually composite structures built from repeating geometric units. During reconstruction on grid paper, students will likely use counting strategies (e.g., counting unit squares to determine the dimensions of a triangle within the motif) to ensure proportional accuracy. Students are expected to reason that preserving the structural integrity of the motif requires maintaining the side lengths and internal angles of the constituent polygons (Aktuna, 2013; Dumlu & Alkaş Ulusoy, 2025; MoNE, 2024).

Teacher guidance notes: Teachers should emphasize that the weavers of these rugs utilized spatial composition and geometric order to create aesthetic balance. Guide students to explicitly name the polygons they identify within the motif (e.g., recognizing that a diamond-shaped motif is a rhombus composed of two isosceles triangles). The teacher should circulate and pose probing questions, such as: "How do you know this part of the motif is a square and not a rectangle?" prompting students to verify properties using the grid (Dumlu & Alkaş Ulusoy, 2025; MoNE, 2024).

Adaptation tips: This activity can be adapted into a technology-integrated task by having students reconstruct the motifs using dynamic geometry software (like GeoGebra or Polypad). Alternatively, students can use physical manipulatives like geoboards or pattern blocks to build the motifs on their desks, providing a tactile medium for spatial decomposition (Dumlu & Alkaş Ulusoy, 2025; Kırıkçılar, 2023; MoNE, 2024).

Activity 3: Area and Perimeter Relationships Within Motif Structures

Task description: This activity uses the bird motif and introduces tangram pieces as a pedagogical tool for analyzing its geometric structure, developing students' understanding of area and perimeter through both non-standard and standard units of measurement. Students are first presented with numerous bird motifs constructed from the complete seven-piece tangram set, observing how the same pieces can produce different bird forms. They are then given silhouettes and asked to reconstruct them using all seven pieces - two large right triangles, one medium right triangle, two small right triangles, one square, and one parallelogram. Because each bird silhouette is assembled from the complete seven-piece tangram set, all bird forms share the same area of 16 small-triangle units, even though their outlines and perimeter vary. Students establish the value of each piece using the small triangle as the non-standard reference unit, confirming this shared area across different bird forms.

Different groups are then asked to measure the same bird motif using different tangram pieces as their unit - one group uses the small triangle, another uses the square, another uses the parallelogram. Groups compare their results and encounter the key question: "If one weaver measures the bird motif using small triangles and another uses squares, why do they get different numbers for the same motif?" Students work through this conflict to recognize that different non-standard units produce different numerical values for the same area, generating the need for a shared standard unit, and transition to unit squares on the Polypad grid (Dumlu & Alkaş Ulusoy, 2025; Kırıkçılar, 2023; MoNE, 2024).

Expected student reasoning: Students may initially predict that birds with wider wingspans have larger areas than compact forms. Through reconstruction, using all seven pieces, they are expected to confirm that all bird silhouettes share the same area regardless of their visual appearance – recognizing that area is determined by the pieces used, not the shape's outline. The critical reasoning moment arises when groups using different reference pieces report different numbers for the same motif. Students may initially be puzzled by this inconsistency, and through discussion are expected to deduce that the numerical value of an area measurement depends on the chosen unit, making a universally agreed standard unit necessary for measurement to be communicable and comparable. This mirrors the historically embedded nature of measurement practices across cultures (Aktuna, 2013; MoNE, 2024).

Teacher guidance notes: Teachers should not pre-explain the piece-to-piece conversion relationships, instead allowing students to discover them by combining pieces on Polypad or by physically overlapping tangram shapes. When assigning groups, teachers should deliberately give each group a different tangram piece as their measurement unit before results are compared, ensuring the cognitive conflict arises naturally from the task structure. Teachers should pose the reflective question: "Can two different motifs have the same perimeter but different areas?" allowing students to test this hypothesis on their grid (Aktuna, 2013; Dumlu & Alkaş Ulusoy, 2025; MoNE, 2024).

Adaptation tips: For classrooms without computer access, physical tangram sets and squared grid paper provide a fully equivalent experience. For students needing additional support, provide a partially completed area relationship table so they can focus on the conversion and measurement tasks rather than deriving all piece relationships independently. For enrichment, introduce Pick's Theorem ($A = I + B/2 - 1$, where I is the number of interior lattice points and B is the number of boundary lattice points) as an alternative area strategy using the dot paper version of the Polypad grid. Note that Pick's Theorem applies specifically to simple polygons whose vertices fall on integer lattice points – students should be guided to verify this condition before applying the theorem, and to understand that it cannot be applied directly to curved or non-polygonal figures (Kırıkçılar, 2023; MoNE, 2024).

Activity 4: Identifying Transformations

Task description: This activity directly aligns with transformation geometry learning outcomes, requiring students to construct the images of polygons resulting from translation, reflection, and rotation. Students examine selected Anatolian motif examples and explore how shapes are repeated and organized within the pattern, focusing on how geometric transformations contribute to the formation of these structures. Students then design their own waterway motif using dynamic geometry software such as GeoGebra or on grid paper. In this design process, students define a base unit and generate a pattern by systematically applying geometric transformations. Their designs are expected to incorporate reflection and translation in a coherent and mathematically justified way, maintaining the repeating structure characteristic of waterway motifs. Rotation may be incorporated as an extension for students working at the 8th grade level or above. Students are also asked to explain their construction process in writing (Dumlu & Alkaş Ulusoy, 2025; Günay & Takunyacı, 2023).

Expected student reasoning: In an ethnomathematics-based environment, students are likely to initially describe transformations using informal, everyday language – referring to a translation as "shifting" or "repeating the shape," and a reflection as "turning it upside down" or "reversing it". As they analyze motif structures, students are expected to recognize that translation preserves the orientation of the motif, whereas a reflection alters it. Some students may exhibit misconception of identifying a translation when the pattern actually requires a reflection, particularly where the motif is highly symmetric. Through exploration and comparison of different patterns, students are expected to refine and correct these interpretations. During the design of their own waterway motifs, students are expected to apply transformations more intentionally, making decisions about how to extend and organize patterns, and recognizing that different combinations of transformations lead to different structural and aesthetic outcomes. Their written explanations provide insight into how they coordinate transformations within their designs.

Teacher guidance notes: Accept and validate students' informal cultural terminology ("shifting," "reversing") before explicitly connecting these actions to the formal academic vocabulary (translation, reflection). During the analysis phase, guide students with questions that focus on how patterns are generated, rather than simply identifying transformation types. During the design phase, require students to explicitly write down the coordinates of their original motif and its transformed image (Dumlu & Alkaş Ulusoy, 2025; Günay & Takunyacı, 2023; MoNE, 2024). If dynamic geometry tools such as GeoGebra are used, guide students to construct transformations step by step, encouraging them to test, revise, and justify their designs. Teachers may also emphasize the structural characteristics of waterway motifs, supporting students in maintaining continuity and coherence in their repeating patterns.

Adaptation tips: To support visual learners, provide transparent tracing paper. Students can trace the motif, physically slide it (translate) or flip it over (reflect) to predict the outcome before constructing their own patterns. For a cross-disciplinary approach, students can recreate their mathematically transformed designs using physical weaving looms or digital coding programs (Kırıkçılar, 2023; MoNE, 2024).

DESIGN RATIONALE AND PRINCIPLES

A primary objective of Design-Based Research is to generate both practical educational interventions and transferable theoretical understanding through iterative design and classroom experience (McKenney & Reeves, 2012). In this chapter, the exploratory piloting phase and subsequent refinement process provided design-informing observations that shaped the development of the instructional model and the articulation of four core ethnomathematical design principles. This section presents the design rationale underlying each principle, drawing on theoretical grounding, iterative design reasoning, and exploratory classroom observations. Consistent with the design-oriented focus of this chapter, these observations are offered as preliminary insights that informed the refinement of the activities rather than as generalizable empirical findings.

Exploratory Classroom Observations

During the exploratory piloting phase, classroom experience provided several design-informing observations. When engaging with motif analysis tasks, students appeared to move progressively from treating Anatolian motifs as purely decorative toward engaging with their geometric structures – identifying symmetry axes, applying decomposition strategies, and connecting transformation rules to the visual patterns before them. In area and perimeter tasks, students appeared to engage productively with structural decomposition – breaking composite motif shapes into constituent triangles and rectangles to approach area calculation. These observations suggested that embedding geometric concepts in culturally familiar artifacts may support students in connecting formal mathematical operations to spatial and visual reasoning (Kusi & Bonyah, 2025; Payadnya et al., 2024).

Classroom observations also indicated a notable shift in participation dynamics. Students who were less engaged in more abstract, formula-driven lessons appeared to respond more actively to the motif-based tasks – volunteering observations, initiating peer discussions, and engaging with the cultural context of the artifacts (Ergene et al., 2020). Teacher reflections noted that the ethnomathematical activities brought a humanizing dimension to mathematics lessons, helping to position mathematics as a meaningful cultural endeavor rather than a set of isolated procedures (Ergene et al., 2020). At the same time, teacher reflections highlighted an important design tension: while students were drawn to the cultural narratives of the motifs, explicit pedagogical guidance was needed to maintain focus on rigorous mathematical abstraction rather than allowing the cultural story to overshadow the geometric content (Cadena et al., 2024). This tension directly informed the articulation of the Mathematical Transparency principle discussed below.

Four Design Principles: Cultural Authenticity, Mathematical Transparency, Exploratory Construction, Structural Decomposition

A primary objective of DBR is to produce generalized design practices, or design principles, that can guide future research and be enacted by other educators and curriculum developers (Bakker, 2018). Based on the theoretical framework established in the Background, the iterative design process, and the exploratory classroom observations described above, four core ethnomathematical design principles were formulated:

Cultural Authenticity: Instructional materials must utilize genuine cultural artifacts without distorting their original socio-cultural meaning or context. The cultural element should serve as a meaningful foundation for learning, honoring the heritage of the community rather than reducing it to a superficial or purely decorative hook (Janiola, 2024; Medina et al., 2024).

Mathematical Transparency: The mathematical concepts embedded within the cultural artifact must be explicitly visible and accessible to the learner. While artifacts possess deep cultural value, their geometric properties - such as symmetry, transformation, and proportional reasoning -must be brought to the forefront so that the mathematics is not overshadowed by the cultural narrative (Dumlu & Alkaş Ulusoy, 2025).

Exploratory Construction: Students should learn through active doing rather than passive observation. By physically or digitally reconstructing motifs, students engage in an exploratory process that bridges their intuitive, informal knowledge with formal academic mathematics (Hajazizi & Gooya, 2024; Verner et al., 2019).

Structural Decomposition: Complex cultural patterns should be approached by breaking them down into manageable, fundamental geometric units. This structural decomposition helps manage cognitive load, allowing learners to analyze area, perimeter, and transformations by isolating specific polygons within a larger composite design (Cherinda, 2024).

How Design Reasoning Informed Each Principle

In DBR, theoretical conjectures are continually tested and refined through design experience gathered in authentic settings (Bakker, 2018; McKenney & Reeves, 2012). The articulation of the four design principles was informed by the intersection of theoretical reasoning and the design tensions observed during exploratory piloting.

The Mathematical Transparency principle emerged from a recurring design challenge: while students appeared engaged by the cultural context of the motifs, classroom observations suggested that the cultural narrative could at times draw attention away from the geometric content. This indicated a need for more explicit visual scaffolding – such as dynamic geometry software to clearly highlight point motions and transformation properties – ensuring that the mathematical structures within the motifs remained the pedagogical focus (Dumlu & Alkaş Ulusoy, 2025; Cadena et al., 2024).

The Structural Decomposition principle was informed by a design insight regarding task complexity. When activities initially required students to engage with an entire composite motif at once, the visual complexity of the design appeared to present a significant cognitive challenge. Redesigning tasks to prompt students to isolate and work with smaller constituent polygons (individual triangles, squares, or rhombuses) before addressing the whole motif appeared to make the geometric relationships more accessible and supported more confident engagement with area and perimeter calculations (Cherinda, 2024).

The Cultural Authenticity principle was reinforced by observing the motivational dimension of culturally grounded tasks. When students encountered their own cultural heritage represented as a site of valid mathematical knowledge, their engagement with the tasks appeared to increase – suggesting that authentic cultural framing may serve not only as a pedagogical hook but as a sustained motivational resource (Verner et al., 2019). This observation underscored the importance of ensuring that cultural artifacts are used with genuine fidelity rather than as superficial illustrations.

Finally, the Exploratory Construction principle was supported by a design observation regarding the role of active making. Students appeared to develop a more secure understanding of transformation properties – particularly the preservation of lengths and angles – when they actively drew, plotted, and rebuilt motifs on grid paper, compared to when transformations were demonstrated visually by the teacher. This suggested that hands-on and digital construction tasks are not merely engaging supplements but may be structurally important for conceptual understanding in transformation geometry (Dumlu & Alkaş Ulusoy, 2025).

Together, these four principles constitute the transferable theoretical contribution of this chapter – a design framework that can guide future researchers, curriculum developers, and teachers seeking to integrate ethnomathematics into formal geometry and measurement instruction.

DISCUSSION

Theoretical Contributions: Connecting Ethnomathematics with DBR

A pervasive challenge within the field of ethnomathematics has been the predominance of qualitative, ethnographic, and descriptive research that documents mathematical practices in cultural contexts but stops short of classroom implementation (Rosa et al., 2016). Scholars have consistently emphasized the need for design-based studies that translate cultural artifacts into formal educational settings (Batiibwe, 2025; Ketenci Tona & Akman, 2026). This chapter makes a theoretical contribution by systematically bridging ethnomathematics with Design-Based Research, shifting the paradigm from anthropological observation to actionable educational intervention (Anderson & Shattuck, 2012; Bakker, 2018).

By employing the DBR framework, this chapter operationalizes ethnomodeling - the process of translating local, cultural (emic) practices into global, academic (etic) mathematical knowledge through a dialogic approach (Rosa & Orey, 2021; Orey & Rosa, 2024). The iterative design process demonstrates how the implicit geometric rules governing Anatolian motifs can be made pedagogically accessible and re-engineered into structured learning trajectories. In doing so, the chapter offers a theoretically grounded design framework – encapsulated in the four principles of cultural authenticity, mathematical transparency, exploratory construction, and structural decomposition – that extends beyond specific Turkish context to inform culturally responsive mathematics instruction more broadly (Anderson & Shattuck, 2012; Bakker, 2018).

Importantly, this chapter positions itself as theory-generating rather than theory-testing. The four design principles are offered not as empirically validated laws but as theoretically grounded and design-informed propositions that future researchers can adopt, adapt, and subject to more extensive empirical scrutiny across diverse cultural and educational contexts.

Practical Contributions: Replicable Model for Culturally Responsive Tasks

Despite the theoretical consensus on the benefits of culturally responsive mathematics education, a major barrier to its global integration remains the lack of structured, curriculum-aligned pedagogical materials. Teachers frequently cite a lack of practical task examples and specialized training as reasons why ethnomathematics remains at the level of theoretical aspiration rather than classroom reality (Asare, 2026; Batiibwe, 2025).

This chapter offers a practical contribution by providing a replicable, classroom-ready instructional model that is explicitly aligned with middle school geometry and measurement curricula (Hajazizi & Gooya, 2024). By detailing the full pedagogical progression - from analyzing symmetries in traditional rugs to performing coordinate transformations and area decompositions - the chapter equips educators with a concrete and adaptable blueprint. The four classroom activities are designed to function as standalone lessons or as a coherent instructional sequence, providing flexibility for teachers working within different scheduling and resource constraints.

Furthermore, the chapter demonstrates that when cultural artifacts are utilized as active cognitive tools rather than decorative illustrations, they can serve as productive bridges between abstract mathematical concepts and students' lived realities (Kusi & Bonyah, 2025). The design principles accompanying the activities provide curriculum developers with a transferable framework – one that can be adapted to other cultural contexts, artifact types, and mathematical domains beyond the geometry and measurement focus of this chapter.

Limitations and Boundary Conditions

While this chapter makes both theoretical and practical contributions, several limitations and boundary conditions should be acknowledged to guide its appropriate use and interpretation.

First, the instructional design presented here is developed within and primarily aligned with the Turkish national curriculum and the specific cultural context of Anatolian motifs. While the four design principles are intended to be transferable, the specific activities and motif selections reflect a particular cultural and curricular setting. Educators working in different national contexts will need to adapt both the cultural artifacts and the curriculum alignment to their own educational systems – the design principles provide a framework for this adaptation, but the specific activities are not directly portable without contextual modification.

Second, this chapter focuses on geometry and measurement – domains that align naturally with the visual and spatial structures of Anatolian motifs. The applicability of this specific instructional model to other mathematical domains, such as algebra, probability, or number theory, remains an open question requiring further design work. The design principles themselves may transfer more readily across domains than the specific activity structures.

Third, the successful implementation of these activities carries significant teacher preparation requirements. The integration of ethnomathematical tasks demands both mathematical proficiency and cultural sensitivity – teachers need to be comfortable facilitating open-ended geometric inquiry, managing the cognitive conflict that arises in tasks like Activity 3, and maintaining the balance between cultural authenticity and mathematical transparency (Batiibwe, 2025; Çenberci & Horzum, 2023). Without adequate professional development, there is a genuine risk of cultural trivialization or mathematical dilution. This boundary condition suggests that the instructional model is most effective when accompanied by structured teacher support materials – a direction that future work should address.

Finally, the design principles articulated in this chapter represent theoretically grounded propositions derived from design reasoning and exploratory classroom experience. Their robustness across different student populations, school contexts, and cultural settings remains to be established through future empirical research. The chapter is therefore best understood as a design contribution that opens a research agenda rather than closes one.

Taken together, the theoretical framework, instructional design, and design principles presented in this chapter suggest that the integration of ethnomathematics and DBR represents a productive and underexplored direction for mathematics education research and curriculum development. Anatolian motifs have demonstrated their potential not merely as cultural illustrations but as structurally rich mathematical artifacts capable of anchoring rigorous geometry and measurement learning. The four design principles – cultural authenticity, mathematical transparency, exploratory construction, and structural decomposition – offer a reusable conceptual vocabulary for researchers and practitioners working to bridge cultural knowledge and formal mathematics across diverse educational contexts. The limitations acknowledged above do not diminish this contribution; rather, they map the boundaries within which the contribution operates and point toward the empirical and design work that should follow.

FUTURE RESEARCH DIRECTIONS

While the design-based research framework and instructional model presented in this chapter offer a promising foundation, they represent an early step in a broader research agenda. Several directions merit priority attention in future work.

First, future research should scale these design principles to broader educational contexts through longitudinal and quasi-experimental studies. Such studies would provide the quantitative and longitudinal evidence needed to assess the long-term impact of ethnomathematics-based instruction on students' mathematical achievement, conceptual retention, and affective attitudes toward mathematics (Aktuna, 2013; Asare, 2026). Scaling across multiple schools and demographically diverse populations would help validate and refine the four design principles, moving them from locally informed propositions toward more robustly tested educational frameworks (Anderson & Shattuck, 2012).

Second, while this chapter focused on transformation geometry and measurement – domains that align naturally with the visual and spatial structures of Anatolian motifs (Batiibwe, 2024) – future work should explore how ethnomathematical contexts can scaffold learning in more abstract domains such as algebra, probability, data analysis, and numerical reasoning (Batiibwe, 2025; Kusi & Bonyah, 2025). Developing and empirically testing culturally grounded tasks for these domains would demonstrate the broader applicability of the ethnomathematics program across grade levels and mathematical content areas (Aktuna, 2013).

Third, future studies should explore a wider array of cultural traditions and artifact types. Although Anatolian motifs provided a structurally rich foundation for this chapter, the ethnomodeling approach developed here can be adapted to other local cultural elements – regional architecture, traditional games, indigenous weaving practices, and daily life routines – providing fresh pedagogical resources for diverse student populations (Batiibwe, 2024; Fouze & Amit, 2018). Cross-cultural comparative research could additionally shed light on universal versus culture-specific strategies for embedding ethnomathematics into formal curricula, enriching our understanding of global mathematical heritage (Asare, 2026; Orey & Rosa, 2024).

Finally, the teacher preparation dimension identified as a boundary condition in this chapter warrants dedicated research attention. Future studies should investigate how pre-service and in-service professional development programs can effectively equip educators to recognize the mathematics embedded in their communities' cultural practices and translate it into rigorous academic tasks (Batiibwe, 2025; Ergene et al., 2020). Collaborative efforts between researchers, curriculum developers, and practicing teachers will be essential for producing the comprehensive pedagogical guides and support materials needed to sustain ethnomathematics integration at scale (Asare, 2026; Cherinda, 2024; Naresh & Kasmer, 2018).

CONCLUSION

This chapter has demonstrated the educational potential of integrating ethnomathematics with Design-Based Research to teach middle school geometry and measurement. By systematically transforming traditional Anatolian motifs into structured, curriculum-aligned learning experiences, the chapter moves the field of ethnomathematics beyond purely descriptive or ethnographic accounts toward actionable instructional design (Design-Based Research Collective, 2003; Dumlu & Alkaş Ulusoy, 2025). The research illustrates that cultural artifacts are not merely folkloric decorations, but structurally rich cognitive tools that can support students' spatial reasoning, problem-solving skills, and conceptual understanding of geometric relationships (Günay & Takunyacı, 2023; Kusi & Bonyah, 2025). Through iterative cycles of analysis, design, exploratory piloting, and refinement, the chapter generated four core design principles – cultural authenticity, mathematical transparency, exploratory construction, and structural decomposition – offered as theoretically grounded and design-informed propositions for culturally responsive mathematics instruction (Design-Based Research Collective, 2003; McKenney & Reeves, 2012).

Implications for Teachers

For classroom practitioners, this chapter offers a practical and replicable instructional model for teaching transformation geometry, area, and perimeter through culturally grounded tasks (Günay & Takunyacı, 2023; Verner et al., 2019). The findings suggest that when teachers utilize ethnomathematics as an active pedagogical tool, they can bridge the gap between formal academic mathematics and students' lived realities, potentially reducing mathematical anxiety and fostering a more inclusive classroom environment where students feel their cultural heritage is valued (Fouze

& Amit, 2023; Sharma & Orey, 2017). To achieve this, teachers must act as facilitators and "ethnomodelers" – guiding students to uncover the mathematical structures embedded in their own cultural contexts while maintaining rigorous mathematical abstraction (Bishop, 1988; Rosa & Orey, 2021). The four design principles provide teachers with a practical decision-making framework for selecting, adapting, and sequencing culturally grounded tasks within their existing curriculum.

Implications for Curriculum Developers

The findings underscore a critical need for curriculum developers to move beyond the superficial inclusion of cultural elements - such as adding historical pictures to textbook margins - and instead embed culture structurally into mathematical tasks (Demir & Seçkin Kapucu, 2023). Curriculum frameworks, including the MoNE curriculum, should provide structured pedagogical guides, textbooks, and digital learning environments that explicitly integrate local ethnomathematics into core learning outcomes (Rosa et al., 2016). The four design principles developed in this chapter offer curriculum developers a concrete starting point for this work – providing criteria against which culturally grounded task designs can be evaluated for both cultural fidelity and mathematical rigor. Furthermore, developers should consider designing assessment strategies that evaluate students' contextualized mathematical reasoning alongside more conventional measures, moving toward a more balanced approach to mathematical competence that recognizes culturally grounded thinking as a legitimate form of mathematical knowledge (Batiibwe, 2025).

Implications for Researchers

For the broader research community, this chapter highlights the value of a methodological orientation that moves from observational anthropology toward interventionist Design-Based Research (Design-Based Research Collective, 2003). While this chapter applied the DBR framework to geometry and measurement using Anatolian motifs, future research must extend these design principles to other abstract mathematical domains and cultural contexts (Anderson & Shattuck, 2012). The four principles – cultural authenticity, mathematical transparency, exploratory construction, and structural decomposition – are offered as a starting theoretical vocabulary for this broader research program, inviting critique, refinement, and empirical testing across diverse settings.

Additionally, there is a pressing need to investigate teacher education. Both pre-service and in-service professional development programs require structured attention to the pedagogical content knowledge and cultural sensitivity needed for effective ethnomathematical instruction (Batiibwe, 2025; Verner et al., 2019). Ultimately, building a cumulative, empirically tested knowledge base through DBR will be essential for scaling up ethnomathematics from locally informed design innovations to widely recognized and systematically implemented educational practices (Anderson & Shattuck, 2012; Design-Based Research Collective, 2003).

REFERENCES

Adam, S. (2004). Ethnomathematical ideas in the curriculum. *Mathematics Education Research Journal, 16*(2), 49–68. DOI: 10.1007/BF03217395

Akdemir, M., & Şengül Akdemir, T. (2024). Kültürel matematiksel bilginin eğitime entegrasyonu: İlköğretim matematik öğretmen adaylarının oluşturdukları etnomatematiksel etkinlikler [Integration of cultural mathematical knowledge into education: Ethnomathematical activities created by middle school mathematics teacher candidates]. *International Journal of New Trends in Arts, Sports & Science Education, 13*(4), 218–226.

Aktuna, H. E. (2013). *Sixth grade students' perceptions of and engagement in ethnomathematical tasks in the area measurement concept* [Unpublished master's thesis]. Middle East Technical University.

Anderson, T., & Shattuck, J. (2012). Design-based research: A decade of progress in education research? *Educational Researcher, 41*(1), 16–25. DOI: 10.3102/0013189X11428813

Arı, A. A., & Demir, B. (2022). Matematiksel modellemenin kültürel perspektifi: Etnomodelleme [Cultural perspective of mathematical modeling: Ethnomodeling]. *Kültür Araştırmaları Dergisi, 12*(12), 253–270. DOI: 10.46250/kulturder.1067167

Asare, B. (2026). Influence of ethnomathematics-based instruction on students' attitudes, participation, and cultural connections in mathematics learning in Ghana. *Cogent Education, 13*(1), 2612395. Advance online publication. DOI: 10.1080/2331186X.2025.2612395

Ascher, M. (1991). *Ethnomathematics: A multicultural view of mathematical ideas*. Routledge.

Bahadır, E. (2021). Göçmen öğrencilerin matematik eğitiminde etnomatematik yaklaşımının kullanılması [Ethnomathematics approach in mathematics education for migrant students]. *Milli Eğitim Dergisi, 50*(1), 577–594. DOI: 10.37669/milliegitim.959829

Bakker, A. (2018). *Design research in education: a practical guide for early career researchers*. Routledge. DOI: 10.4324/9780203701010

Barton, W. D. (1996). *Ethnomathematics: Exploring cultural diversity in mathematics* [Unpublished doctoral dissertation]. University of Auckland.

Batiibwe, M. S. K. (2024). The role of ethnomathematics in mathematics education: A literature review. *Asian Journal for Mathematics Education, 3*(4), 383–405. DOI: 10.1177/27527263241300400

Batiibwe, M. S. K. (2025). Ethnomathematics as a pedagogical tool for mathematics education: Opportunities and challenges. *SN Social Sciences, 5*(221), 1–38. DOI: 10.1007/s43545-025-01260-0

Bishop, A. J. (1988). *Mathematical enculturation: a cultural perspective on mathematics education*. Kluwer Academic Publishers. DOI: 10.1007/978-94-009-2657-8

Cadena, J. R., Machado, J. L., & Muso, M. A. (2024). The formation of critical thinking in mathematics learning through a model of contextualization with the social, political, and historical environment in Ecuadorian education. In C. Nicol, G. Knijnik, A, Peng, M. Cherinda, A. Bose (Eds.), *Ethnomathematics and mathematics education international perspectives in times of local and global change* (pp. 185-212). Springer. DOI: 10.1007/978-3-031-60680-9_9

Çenberci, S., & Horzum, T. (2023). Matematik öğretmeni adaylarının etnomatematik farkındalıklarının incelenmesi [Investigation of ethnomathematics awareness of preservice mathematics teachers]. *Ahmet Keleşoğlu Eğitim Fakültesi Dergisi, 5*(3), 713–732. DOI: 10.38151/akef.2023.80

Cherinda, M. (2024). From ethnomathematical reasoning to mathematical research. In C. Nicol, G. Knijnik, A, Peng, M. Cherinda, A. Bose (Eds.), *Ethnomathematics and mathematics education international perspectives in times of local and global change* (pp. 101-134). Springer. DOI: 10.1007/978-3-031-60680-9_6

Cobb, P., Confrey, J., diSessa, A., Lehrer, R., & Schauble, L. (2003). Design experiments in educational research. *Educational Researcher, 32*(1), 9–13. DOI: 10.3102/0013189X032001009

D'Ambrosio, U. (1985). Ethnomathematics and its place in the history and pedagogy of mathematics. *For the Learning of Mathematics, 5*(1), 44–48.

D'Ambrosio, U. (2006). The program ethnomathematics and the challenges of globalization. *Circumscribere International Journal for the History of Science, 1*(1), 74–82.

Demir, G., & Seçkin Kapucu, M. (2023). Ethnomathematics approaches at middle school textbooks. *Osmangazi Journal of Educational Research, 10*(1), 120–141. DOI: 10.59409/ojer.1269449

Design-Based Research Collective. (2003). Design-Based research: An emerging paradigm for educational inquiry. *Educational Researcher, 32*(1), 5–8. DOI: 10.3102/0013189X032001005

Dumlu, B. Ö., & Alkaş Ulusoy, Ç. (2025). Using an ethnomathematical approach in teaching transformational geometry: the kilim activity. *Çukurova Üniversitesi Eğitim Fakültesi Dergisi, 54*(2), 670-714. DOI: 10.14812/cuefd.1525565

Ergene, Ö., Çaylan Ergene, B., & Yazıcı, E. Z. (2020). Ethnomathematics activities: Reflections from the design and implementation process. *Turkish Journal of Computer and Mathematics Education, 11*(2), 402–437. DOI: 10.16949/turkbilmat.688780

Fouze, A. Q., & Amit, M. (2018). On the importance of an ethnomathematical curriculum in mathematics education. *Eurasia Journal of Mathematics, Science and Technology Education, 14*(2), 561–567. DOI: 10.12973/ejmste/76956

Fouze, A. Q., & Amit, M. (2023). The importance of ethnomathematics education. *Creative Education, 14*(4), 729–740. DOI: 10.4236/ce.2023.144048

Günay, K. (2023). *Etnomatematik uygulamalarının kullanıldığı öğretimin öğrencilerin problem çözme becerileri ve duyuşsal davranışları üzerindeki etkisinin incelenmesi [Examining the effects of instruction using ethnomathematics applications on students' problem-solving skills and affective behaviors]* [Unpublished master thesis]. Sakarya University.

Günay, K., & Takunyacı, M. (2023). Examining the effect of teaching with ethnomathematics on students' problem-solving skills on transformation geometry. *Journal of Interdisciplinary Education: Theory and Practice, 5*(2), 86–106. DOI: 10.47157/jietp.1384160

Hajazizi, F., & Gooya, Z. (2024). Internalized mathematics as intuitive as breathing: an ethnomathematics study of traditional architecture in Iran. In C. Nicol, G. Knijnik, A, Peng, M. Cherinda, A. Bose (Eds.), *Ethnomathematics and mathematics education international perspectives in times of local and global change* (pp. 135-164). Springer. DOI: 10.1007/978-3-031-60680-9_7

Janiola, F. R. (2024). The dynamic components of Janiola's ethnolearning framework. In C. Nicol, G. Knijnik, A, Peng, M. Cherinda, A. Bose (Eds.), *Ethnomathematics and mathematics education international perspectives in times of local and global change* (pp. 235-256). Springer. DOI: 10.1007/978-3-031-60680-9_11

Karademir, A., & Girgin, D. (2024). Eğitimde araştırma ve uygulamanın buluşması: Tasarım tabanlı araştırmaya kuramsal bir bakış [Integration of research and practice in education: A theoretical perspective on design based research]. *Journal of Computer and Education Research, 12*(24), 780–799. DOI: 10.18009/jcer.1522744

Ketenci Tona, F. M., & Akman, B. (2026). Mathematics from a cultural perspective: Systematic review article. *Bayburt Journal of Education Faculty, 21*(49), 424–458. DOI: 10.35675/befdergi.1677972

Kırıkçılar, R. G. (2023). *Matematik merkezli etnosteam ders planlarinin geliştirme ve uygulama süreçlerinin incelenmesi [Examining the development and implementation processes of mathematics-centered ethnosteam lesson plans]* [Unpublished doctoral dissertation]. Marmara University.

Küçük, A. (2014). Ethnomathematics in Anatolia-Turkey: Mathematical thoughts in multiculturalism. *Revista Latinoamericana de Etnomatemática, 7*(1), 171–184.

Kusi, P., & Bonyah, E. (2025). A systematic review and bibliometric report: Ethnomathematics and culturally responsive pedagogy in mathematics education. *European Journal of Education Studies, 12*(9), 113–140. DOI: 10.46827/ejes.v12i9.6187

McKenney, S., & Reeves, T. C. (2012). *Conducting educational design research.* Routledge.

Medina, M. A., Gerofsky, S., & Nicol, C. (2024). Weaving indigenous mathematics: Bringing indigenous ways and stories into conversation with ethnomathematics. In C. Nicol, G. Knijnik, A, Peng, M. Cherinda, A. Bose (Eds.), *Ethnomathematics and mathematics education international perspectives in times of local and global change* (pp. 77-100). Springer.

Meeran, S., & Niranjan, C. (2026). The game of Kgati as an implementation tool in teaching intermediate phase mathematics: A case of two schools in rural Kwazulu-natal, South Africa. *Education Sciences, 16*(2), 294. DOI: 10.3390/educsci16020294

Ministry of National Education [MoNE]. (2024). *Ortaokul matematik dersi öğretim program (5, 6, 7 ve 8. sınıflar) [Middle school mathematics curriculum (grades 5, 6, 7, and 8)].* https://tymm.meb.gov.tr/upload/program/2024programmat5678Onayli.pdf

Ministry of National Education [MoNE]. (2025). *Türkiye yüzyılı maarif modeli öğretim programları ortak metni [Joint text of the Turkish century education model curriculum].* https://tymm.meb.gov.tr/upload/brosur/ortak_metin.pdf

Naresh, N., & Kasmer, L. (2018). Using ethnomathematics perspective to widen the vision of mathematics teacher education curriculum. In Bartell, T. G. (Ed.), *Toward equity and social justice in mathematics education* (pp. 309–328). Springer. DOI: 10.1007/978-3-319-92907-1_18

Nasrum, A., Salido, A., & Chairuddin, C. (2025). Unveiling emerging trends and potential research themes in future ethnomathematics studies: A global bibliometric analysis (from inception to 2024). *International Journal of Learning, Teaching and Educational Research, 24*(2), 206–226. DOI: 10.26803/ijlter.24.2.11

NCTM. (2000). *Principles and standards for school mathematics*. NCTM.

Orey, D. C., & Rosa, M. (2010). Ethnomodeling: A pedagogical action for uncovering ethnomathematical practices. *Journal of Mathematical Modelling and Application, 1*(3), 58–67.

Orey, D. C., & Rosa, M. (2024). Ethnomodelling research: glocalizing educational systems from exclusion to inclusion at local and global levels. In C. Nicol, G. Knijnik, A, Peng, M. Cherinda, A. Bose (Eds.), *Ethnomathematics and mathematics education international perspectives in times of local and global change* (pp. 213-234). Springer. DOI: 10.1007/978-3-031-60680-9_10

Payadnya, I. P. A. A., Wulandari, I. G. A. P. A., Puspadewi, K. R., & Saelee, S. (2024). Student responses to culture-based mathematics learning in the Indonesian and Thailand education curricula. *Jurnal Pendidikan Matematika, 18*(1), 113–128. DOI: 10.22342/jpm.v18i1.pp113-128

Prahmana, R. C. I., & D'Ambrosio, U. (2020). Learning geometry and values from patterns: Ethnomathematics on the batik patterns of Yogyakarta, Indonesia. *Journal on Mathematics Education, 11*(3), 439–456. DOI: 10.22342/jme.11.3.12949.439-456

Rosa, M., D'Ambrosio, U., Orey, D. C., Shirley, L., Alangui, W. V., Palhares, P., & Gavarrete, M. E. (2016). *Current and Future Perspectives of Ethnomathematics as a Program ICME-13 Topical Surveys*. Springer. DOI: 10.1007/978-3-319-30120-4

Rosa, M., & Orey, D. C. (2011). Ethnomathematics: The cultural aspects of mathematics. *Revista Latinoamericana de Etnomatemática, 4*(2), 32–54.

Rosa, M., & Orey, D. C. (2021). Ethnomodelling as the translation of diverse cultural mathematical practices. In Sriraman, B. (Ed.), *Handbook of the mathematics of the arts and sciences* (pp. 2607–2635). Springer. DOI: 10.1007/978-3-319-57072-3_70

Setiaputra, F. I., Subanti, S., Usodo, B., Triyanto, T., Fitriana, L., & Hendriyanto, A. (2025). A systematic review of ethnomathematics research (2019–2023): Cultural integration in mathematics teaching and learning. *Jurnal Pendidikan MIPA*, *26*(1), 195–211. DOI: 10.23960/jpmipa.v26i1.pp195-211

Sharma, T., & Orey, D. C. (2017). Meaningful Mathematics Through the Use of Cultural Artifacts. In Rosa, M., Shirley, L., Gavarrete, M. E., & Alangui, W. V. (Eds.), *Ethnomathematics and its diverse approaches for mathematics education* (pp. 153–182). Springer. DOI: 10.1007/978-3-319-59220-6_7

Štemberger, T., & Cencič, M. (2016). Design based research: The way of developing and ımplementing educational ınnovation. *World Journal on Educational Technology: Current Issues*, *8*(3), 180–189. DOI: 10.18844/wjet.v8i3.621

Supriyadi, E., Turmudi, T., Dahlan, J. A., & Juandi, D. (2024). Development of Sundanese gamelan ethnomathematics e-module for junior high school mathematics learning. *Malaysian Journal of Learning and Instruction*, *21*(2), 139–178. DOI: 10.32890/mjli2024.21.2.6

Verner, I., Massarwe, K., & Bshouty, D. (2019). Development of competencies for teaching geometry through an ethnomathematical approach. *The Journal of Mathematical Behavior*, *56*, 1–14. DOI: 10.1016/j.jmathb.2019.05.002

Yazıcı, E. Z. (2021). *Etnomatematik uygulamalarında kültür ve matematiğin gerçekçi matematik eğitimi çerçevesinde incelenmesi [Examining culture and mathematics within the framework of realistic mathematics education in ethnomathematics applications]* [Unpublished doctoral dissertation]. Marmara University.

ADDITIONAL READING

Nicol, C., Knijnik, G., Peng, A., Cherinda, M., & Bose, A. (Eds.). (2024). *Ethnomathematics and mathematics education: International perspectives in times of local and global change*. Springer. DOI: 10.1007/978-3-031-60680-9

KEY TERMS AND DEFINITIONS

Anatolian Motifs: Traditional decorative symbols and geometric patterns woven into Turkish carpets, kilims, and handicrafts that historically convey societal beliefs, emotions, and expectations while being structurally governed by advanced mathematical rules.

Cultural Artifacts: Objects, tools, textiles, or architectural structures created by members of a specific cultural group that inherently embody the historical, social, and mathematical knowledge of their creators.

Design-Based Research (DBR): An iterative, interventionist research methodology that deliberately blends empirical educational research with the systematic design of learning environments to solve practical classroom problems and advance theoretical understanding.

Emic Approach: A local or insider's perspective in research that focuses on understanding cultural practices and mathematical ideas intuitively from the specific viewpoint of the members within that cultural group.

Ethnomathematics: The study of the diverse mathematical ideas, techniques, and practices that are developed, articulated, and utilized by identifiable cultural groups to understand, explain, and manage their daily realities.

Ethnomodeling: A methodological and pedagogical process that utilizes mathematical modeling to translate and bridge local, culturally embedded mathematical practices with formal academic mathematical systems.

Etic Approach: A global or outsider's perspective in research that utilizes standardized, academic, or universal analytical frameworks to observe, translate, and compare the mathematical practices of different cultures.

Transformation Geometry: A sub-field of geometry emphasized in mathematics education that involves the active manipulation of geometric shapes across a plane, focusing on specific movements such as reflections, translations, rotations.

Chapter 6
Ethnomodeling as a Bridge:
Connecting Local Cultural Knowledge to Academic Geometric Transformations

Seyma Sengil-Akar
http://orcid.org/0000-0002-0032-7439
Kastamonu University, Turkey

Berna Somuncu
http://orcid.org/0009-0006-6408-9085
Hacettepe University, Turkey

Elif Saygi
http://orcid.org/0000-0001-8811-4747
Hacettepe University, Turkey

ABSTRACT

This study explores how 8th-grade students utilize geometric transformations—translation, reflection, and rotation—within cultural contexts, specifically through traditional Turkish rug and ceramic motifs. Grounded in ethnomodeling, the qualitative case study involved nine students and analyzed data from worksheets and recordings using content analysis. The findings reveal that students successfully integrated culturally embedded knowledge into mathematical reasoning. In the "Eşme Rug Problem," they identified pattern repeatability and symmetry to estimate geometric properties. In the "Turkish Tile Plate Problem," students applied rotations and reflections to create original designs using symbolic motifs like crescents and floral patterns. This research underscores ethnomodeling as a pedagogical tool

DOI: 10.4018/979-8-2600-3485-9.ch006

that makes mathematics more relevant and inclusive by connecting local heritage to academic concepts. It demonstrates that cultural context facilitates deeper conceptual understanding of transformational geometry.

INTRODUCTION

Mathematics is a discipline that evolves through cultural practices and is an integral component of humanity's rich heritage (Brandt & Chernoff, 2015; Gerdes, 1998; Putra & Mahmudah, 2021; Rosa & Orey, 2010a, 2016). Diverse cultural groups have developed unique methods of mathematizing reality based on their own value systems (D'Ambrosio, 1990; Orey & Rosa, 2021a; Rosa & Orey, 2006). To identify these culturally embedded mathematical elements, the concept of ethnomathematics was introduced (D'Ambrosio, 1993; D'Ambrosio, 1997; D'Ambrosio, 2001). Ethnomathematics explores how individuals engage with mathematical ideas in their daily lives and represents the intersection of cultural anthropology and mathematics (Alghar & Marhayati, 2023; Bassanezi, 2002; Cortes & Orey, 2020; Rosa, 2000; Rosa & Orey, 2010b; Rosa & Orey, 2013; Rosa & Orey, 2018).

When mathematical modeling is employed within this ethnomathematical framework to solve culturally rooted real-world problems, the approach is conceptualized as "ethnomodeling" (Freire, 1998; Rosa & Orey, 2021a; Rosa & Orey, 2016). Ethnomodeling lies at the convergence of mathematics, cultural anthropology, and mathematical modeling (as shown in Figure 1). This integration allows students to construct mathematical models that are meaningful within the context of their own lived realities and cultural experiences (Alghar & Jamaluddin, 2024; Alghar & Radjak, 2024; Rosa & Orey, 2021b; Rosa & Orey, 2010b; Rosa & Orey, 2019; Umbara et al., 2021).

Figure 1. Ethnomathematics and ethnomodelling as the intersection of three research fields (Rosa and Orey, 2010a)

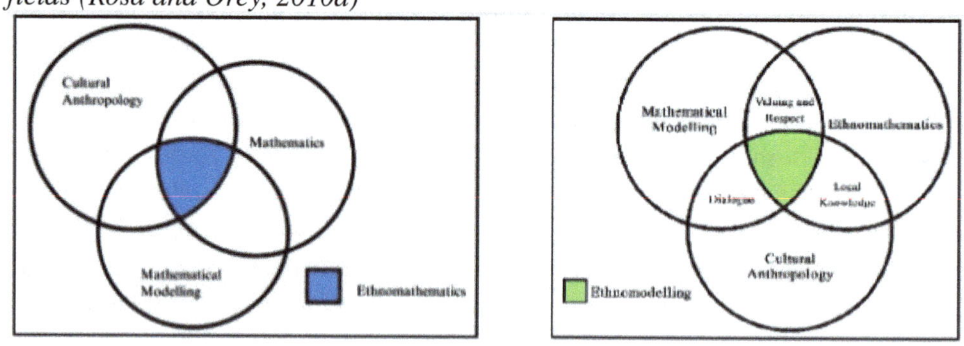

As a pedagogical approach, ethnomodeling opposes the traditional “banking model” of education, emphasizing the co-creation of knowledge rather than mere transfer (Freire, 1970, 1998). It encompasses mathematical practices outside formal school settings, bringing past and present cultural practices into the classroom (Bassanezi, 2002; Rosa & Orey, 2011, 2021a). Ultimately, ethnomodeling fosters connections between these traditional practices and the real-life issues individuals encounter, enhancing the conceptualization of mathematical knowledge through a holistic structure (Cortes & Orey, 2020; Desai et al., 2022).

Significance of the Study

Research has shown that complex mathematical expressions involving geometric principles are embedded in the crafts, architectural designs, cultural activities, and artifacts of many societies (Kalinec-Craig et al, 2019; Rosa & Orey, 2016; Karadag & Karagöz Akar, 2020). In their study, Kalinec-Craig, Prasad, and Luna (2019) argue that Talavera tiles can be used by teachers to make mathematics instruction more culturally meaningful. These tiles serve as concrete representations of mathematical concepts such as symmetry and transformational geometry. Their findings suggest that incorporating Talavera tiles into mathematics lessons can support students in developing a deeper understanding of mathematical ideas while establishing connections to cultural artifacts present in their homes and communities. Another study proposes that the mathematical knowledge employed by lace makers in northeastern Brazil to produce geometric lace patterns can be modeled through ethnomodeling techniques. Karadag and Karagöz Akar (2020) explored the implicit mathematics embedded in traditional crafts, focusing on recurring patterns and examining the concepts of functions and isometries, as well as the relationships between them. Similarly, Topan and Bozkus (2021) found, based on the insights of secondary school mathematics teachers, that five geometric patterns found in Turkish architectural works closely align with the national middle school geometry curriculum. These studies further indicate that students’ awareness of the mathematics embedded in their own cultural heritage enhances their understanding of the nature of mathematics (Rosa & Orey, 2003). While a substantial body of ethnomathematics researchs on mathematics and culture had conducted in Turkey (e.g., Aktuna, 2013; Cenberci& Horzum, 2023; Gürefe, 2025; Sevgi, 2019), studies explicitly addressing the ethnomodeling approach remain limited both nationally (Arı & Demir, 2022; Gürefe, 2025) and internationally (Umbara et al., 2021). Although prior studies have examined cultural artifacts in mathematics education, fewer studies have analyzed, how middle school students use local Turkish cultural knowledge while applying geometric concepts -specifically translation, reflection, and rotation- in ethnomodeling tasks.This study addresses this gap by designing mathematical modeling activities through the integration of cultural contexts. The aim was to enable students to explore mathematical applications related to specific geometry topics and uncover the cultural roots of these applications. Throughout the implementation process, the study sought to create a learning environment where mathematical symbols, concepts, and skills were applied in conjunction with real-life contexts, mediated through ethnomodeling. To realize this goal, two mathematical modeling problems-the Eşme Rug Problem and the Turkish Tile Plates Problem—were implemented. The findings suggest that these tasks provided opportunities for students to connect transformational geometry with culturally familiar designs. The research question guiding the study is as follows: "In ethnomodeling, how do students draw upon local knowledge and practices, using transformations such as translation, reflection, and rotation, to solve geometric problems rooted in traditional Turkish rug and plate designs?"

METHODOLOGY

Research Design

In this study, as argued by Pais and Mesquita (2013), it is assumed that different solution strategies can be better understood when mathematical concepts are taught through problem situations designed within a cultural context. Accordingly, students were encouraged to visualize, elaborate, integrate, and recognize geometric expressions embedded in ethnomodeling problems developed within the framework of Turkish culture. To achieve this aim, a qualitative case study design was adopted. A case study is a distinctive approach used to seek answers to scientific questions by deeply examining a bounded system (Büyüköztürk et al., 2019). Consistent with this methodology, each phase of the study was carefully designed, allowing for an in-depth investigation of how various factors influenced the students' modeling processes within a real-world context (Creswell, 2015; Yin, 2014).

Participants and Setting

In case study research, it is common to work with a group of individuals who interact with each other, share the same environment, and are familiar with one another to examine a phenomenon in depth (Büyüköztürk et al., 2019). The study group consisted of nine 8th-grade students attending a public middle school in Ankara during the second semester of the 2023–2024 academic year. The implementation was structured as an extracurricular, after-school mathematics activity. Group 1 consisted of four students, and Group 2 consisted of five students. Each participant was assigned a code based on their group affiliation, as presented in Table 1.

Table 1. Codes for the participants

Group	Codes of Students in the Group
Group 1	G1S1, G1S2, G1S3, G1S4
Group 2	G2S1, G2S2, G2S3, G2S4, G2S5

A convenience sampling method was utilized; participants were selected on a voluntary basis from the school where the researchers were employed. To foster effective communication and productive collaborative modeling, the students were divided into two groups based on their ability to work harmoniously with one another.

Data Collection Tools and Task Development

The primary data collection method utilized in this study was audio recording of group interactions. Student activity sheets and observational field notes taken by the researcher served as supplementary tools to support and triangulate the data derived from the audio recordings. In parallel with the study conducted by Aydemir and Kubanç (2014), participants were asked to solve mathematical modeling problems while their performances were simultaneously recorded. Two main modeling tasks were used. The first task, the "Eşme Rug Problem," was adapted from Bukova Güzel (2021) and selected for its incorporation of structural elements from Turkish culture. The second task, the "Turkish Tile Plates," was an ethnomodeling problem specifically designed by the researchers to align with the study's geometry objectives, asking students to create original traditional designs. Prior to the main implementation, the researcher-designed task underwent a validation procedure. It was reviewed by mathematics education experts to ensure its alignment with 8th-grade geometry outcomes and cultural relevance. Following expert feedback, a pilot study was conducted with a small group of non-participating students to confirm the clarity and feasibility of the problem instructions.

Implementation and the Role of the Researcher

To ensure methodological transparency, the data collection was carried out systematically across the two modeling activities, with the implementation of each task lasting approximately 70 minutes per group. During the study, one of the authors assumed the role of a teacher-researcher. To prevent any data loss during the audio recording process and to capture the complex group dynamics comprehensively, the teacher-researcher worked with each group separately. The researcher facilitated discussions, ensured task focus, and avoided providing direct mathematical solutions. The audio recordings resulted in a robust dataset of approximately 290 minutes of raw data, which were subsequently transcribed verbatim for analysis.

Data Analysis and Coding Procedure

Content analysis was employed in accordance with the aim of the study to systematically organize and interpret the collected data (Yıldırım & Şimşek, 2021). To address the research questions comprehensively and transparently, a hybrid coding approach—combining both deductive and inductive methods—was utilized to analyze the transcribed audio recordings, student worksheets, and observation notes.

The coding procedure was conducted in multiple iterative stages. Initially, the main thematic categories were predetermined deductively based on the theoretical framework of transformational geometry and ethnomodeling. These predefined main themes included translational movement, reflection movement, rotational movement, and cultural aspects (replacing the initial working term "ethnic situations" to better reflect culturally grounded reasoning). Each theme was constructed as an independent category, with no theme nested within another or treated as a subset.

Following the establishment of these predefined themes, an inductive coding process was applied to generate specific sub-codes directly from the raw data. During this phase, the transcripts and student activity sheets were reviewed multiple times to identify recurring patterns and actions that exemplified how students applied geometric transformations or utilized cultural references. As recommended by Prasad (2008), the coding process began by testing candidate categories and sub-codes on a subset of the data during a pilot phase. Sentences or paragraphs that served as meaningful coding units were marked and systematically assigned to the relevant themes.

Crucially, the coding process was not limited strictly to verbal statements. Students' physical and kinesthetic actions—such as physically rotating the paper to observe changing patterns, using a ruler to measure distances for symmetry, or folding the paper to establish a reflection axis—were also systematically coded as concrete evidence of their geometric reasoning and modeling behaviors. To ensure complete methodological transparency and to illustrate how these inductive sub-codes (including physical actions) map onto the deductive themes, a comprehensive coding framework (Codebook, see App.) detailing the inclusion and exclusion criteria is presented in the Table (see app).

Trustworthiness

In qualitative research, validity and reliability are framed as dependability, credibility, and confirmability (Çelebi, 2022). To enhance credibility, data triangulation was achieved by synthesizing audio transcripts, student activity sheets, and observation notes (Yıldırım & Şimşek, 2021). A clear chain of evidence was established, and direct quotations from participants were frequently incorporated into the findings to reflect authentic student voices. Validity refers to the extent to which a study accurately measures what it intends to measure. Reliability, on the other hand, refers to the consistency and stability of a measurement tool, indicating that it yields similar results under repeated conditions. In qualitative research, these concepts are often framed as dependability and confirmability (Çelebi, 2022). To enhance the validity and reliability of this study, multiple data collection tools were employed. The use of diverse instruments allowed the findings from each data source to corroborate and support one another, creating a richer and more accurate understanding of the research phenomena. This triangulation strategy contributed to a more robust and consistent interpretation of the data. By utilizing multiple data sets, construct validity was also strengthened (Yıldırım & Şimşek, 2021).A clear chain of evidence was established through the integration of various data sources. The data were reviewed repeatedly at regular intervals to ensure consistency. Details regarding how the results were derived are explained in the findings section, and relevant appendices are included to support transparency. In addition, direct quotations from participants, obtained through interviews or observations, were frequently used in the findings section to reflect authentic voices.

To further enhance the trustworthiness of the research, input was sought from external experts and other researchers at various stages of the research process. All procedures were explicitly described, supported by documentation, and presented systematically in a step-by-step manner. Prior to audio recording, informed consent was obtained from all student participants in accordance with ethical research practices.

To address dependability and confirm the robustness of the data analysis, the coding process was not conducted in isolation. Instead of calculating a statistical inter-coder agreement coefficient, the researchers adopted a rigorous consensus-building approach. All researchers jointly reviewed the initial dataset and collaboratively developed the preliminary coding framework. Subsequently, the data were analyzed through iterative group discussions. Rather than coding independently and comparing numerical scores, the research team worked collectively on the dataset, discussing the categorization of specific geometric movements, physical actions, and cultural reasoning until full consensus (100% agreement) was reached for each data segment. This dialogic consensus approach was intentionally chosen over independent statistical calculation because evaluating ethnomodeling tasks—which intertwine abstract geometric transformations with subjective cultural interpretations—requires deep contextual negotiation rather than isolated coding. Finally, all procedures were explicitly documented, and informed consent was obtained from all student participants prior to data collection.

Describing the Findings

The data collected during the study were organized and categorized under specific codes and themes. Sentences or paragraphs that served as the basis for the coding were marked accordingly. The main themes identified in the study include translational movement, reflection movement, rotational movement, and ethnic situations. Each theme is presented with subheadings in the findings section.

The results obtained through content analysis were grouped and interpreted within these thematic categories. Additionally, the characteristics of student group actions were explained, and key cultural elements encountered during the implementation were described. The ethnic references students incorporated into their models were also revealed. In this way, the study demonstrated how local knowledge and cultural practices were integrated into the mathematical modeling process and how this integration contributed to students' mathematical problem-solving. These results provided insight into how students applied their traditional knowledge and cultural experiences in addressing the problems through ethnomodeling.

FINDINGS

This section presents students' approaches to the problems, the strategies they employed, and the skills they demonstrated during the problem-solving process. To ensure clarity and coherence, the findings are organized under separate headings for each problem. The phases that students experienced throughout the implementation were explained individually for each task and supported with excerpts from student dialogue and visuals from their activity sheets.

Findings Related to the Eşme Rug Problem

The first task implemented during the study was the *Eşme Rug Problem.* This problem required students to calculate the price of a rug based on a given rug pattern shown in the example. It was observed that students discussed the problem collaboratively for approximately 30 minutes. They explored different approaches with their peers and collectively sought the most appropriate solution. Although both groups successfully utilized geometric transformations such as translation and reflection, their distinct cultural assumptions led them to construct different mathematical models and, consequently, arrive at different final calculations for the rug's cost. Within this context, the process was examined in detail, and the findings are presented under the relevant subheadings.

Findings on Translational, Reflection, and Rotational Movements in the Eşme Rug Problem

This section presents findings regarding students' understanding and application of translational, reflection, and rotational movements in the context of the Eşme Rug Problem. Students' performances related to geometric transformations were analyzed, and their abilities to understand and apply these movements were observed and evaluated. To support the analysis, sample representations were provided and reinforced with data from student dialogues and activity sheets.

The analysis began with an exploration of how well students recognized and interpreted the patterns featured in the problem. Initially, it was observed that students were able to identify traditional motifs frequently used in the Eşme rugs. For example:

- G1S1 stated, "If we think of these as squares—four squares," and, "The areas of the triangles are equal."
- G1S3 noted, "There are 32 triangles here."
- G1S4 commented, "All of these squares are equal to each other. The triangles here are all equal, too. If we find the area of one square, we can estimate how many there are."
- G2S2 said, "If it's 100 cm^2, then it must be 10×10."
- G2S3 explained, "There are 1, 2, 3, and 5 in the middle, and these parts are equal to those. All of the triangles are the same size."
- G2S5 stated, "All of the triangles are 2.5."

These examples show that students had prior knowledge of the patterns featured in the rug and were able to engage with the design elements analytically. They also considered the repeatability of the patterns, which provided a point of reference for determining the scale and orientation of the patterns. Sample student dialogues related to this issue are presented in **Table 2**. It was determined that the repeatability of the patterns helped students establish a framework for understanding the direction and proportion in which the motifs were applied.

Table 2. Student dialogues demonstrating the use of pattern repetition

Group 1	Group 2
G1S3: There are triangles here—do you think they might be equal? **G1S1:** Actually, since this part repeats after a while, if we just calculate this section and multiply it by two, it'll probably be enough. **G1S2:** Yeah. **G1S3:** Does it divide into two from here? **G1S2:** Exactly. The distances are almost equal.	**G2S3:** This part and that part seem equal. You see those 1, 2, 3, and 5 in the middle—they match with that side, and these match with this side. All the triangles are the same size. **G2S2:** That actually makes a lot of sense. **G2S1:** But we weren't really told that. I mean, that the triangles are equal. **G2S3:** We're going to do it based on the picture. **G2S1:** According to the picture, they do look the same. **G2S2:** But— **G2S1:** But it wasn't stated that the red ones are equal. The ones inside are the same, yes. But do you see the tips of the red ones? There are gaps—so they're not the same. **G2S4:** Look, those same gaps are over here too. These gaps match with those. **G2S2:** So they are the same after all.
Excerpt from Group 1 – Eşme Rug Problem (Solution Phase)	Excerpt from Group 2 – Eşme Rug Problem (Solution Phase)

Note: The dialogues in this table illustrate how students from different groups identified and used the repeatability of geometric patterns as part of their problem-solving strategy.

Findings on Translational Movements

When examining how translational movement was used in the Eşme rug problem, it was observed that students were able to recognize the translation of patterns in a specific direction. Overall, students demonstrated their ability to recognize translational movement by verbally identifying the repeating units and mathematically calculating the pattern's arithmetic continuation (e.g., determining the 4n-2 sequence to solve the problem). It was also noted that, as the process progressed, students were able to apply this type of movement effectively. Several examples of student dialogue supporting these observations are provided below:

G1S1: "Actually, this part repeats after a while. If we just calculate this part and multiply it by two, that will probably be enough."

G1S2: "It looks to me like it continues as a pattern."

The following excerpts are taken from Group 2's discussion on the same pattern-related task.

G2S2: "I found something (referring to the triangles by color)—they increase by four each time. When we divide the yellow ones in half, this one is 2, the next is 6, blue becomes 10, and when blue is 10, red becomes 14. They all increase outward by four. If that one is 10, then white is 14. Look here—when you add these two, the yellow ones total 6, the blue ones 10, the red ones 14, and the white ones 18. Add four more and you get 22."

G2S3: "It totally makes sense to find the pattern—how many repetitions are there?"

G2S1: "The patterns go up by four, one after another."

G2S4: "Then let's connect your 6.25 to the pattern."

In addition to these dialogues, G2S2 and G2S5 identified the general term of the pattern as *4n-2*. G2S5 explained that the given example was part of the rug's overall pattern, stating: "We first found this part (pointing to the sample pattern). The shapes in this section also appear in the rug. That's why this piece must be a multiple of the full pattern."

Findings on Reflection Movements

Regarding reflection skills, students demonstrated their ability to apply symmetry by correctly determining a central axis on the rug and using this visual reflection to halve their calculations, thereby simplifying the problem-solving process. Many students were able to correctly define the symmetry of the patterns and apply reflection transformations accordingly. Visual representations of symmetry applications by the groups are presented in Figure 2. These visuals indicate that students reflected shapes across different axes.

Figure 2. Representations of groups' application of symmetry movements

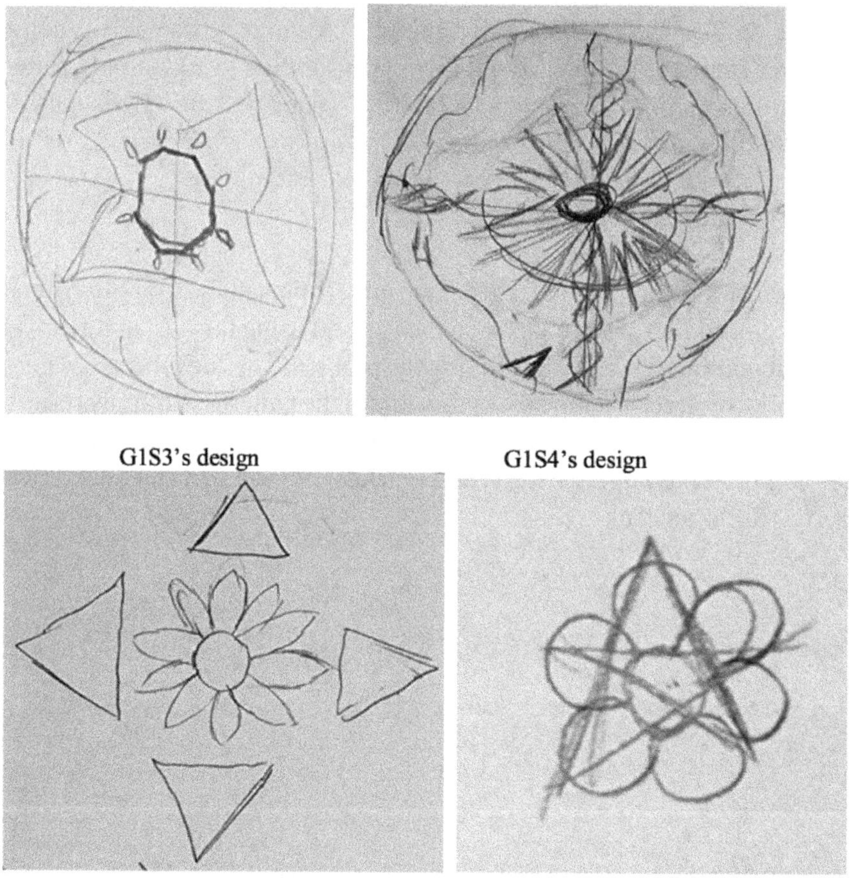

G1S3's design | G1S4's design

G2S2's design | G2S3's design

Evidence confirming students' use of symmetry movements was also found in the audio recordings. The following excerpts from Group 1 demonstrate this:

G1S3: "Does it divide in half from here?"

G1S2: "Exactly—more or less equal distances."

A similar observation is reflected in the following exchange:

G1S2: "The rug isn't evenly aligned from the sides; it's not like a perfect pattern."

G1S1: "I think that's because of the image."

G1S2: "Nothing is definite, but I think we'll go by estimation."

G1S1: "Let's cut it right down the middle... Or think of it this way—there's half of a shape missing here, so this part completes it. It's like the two make one shape."

G1S2: "Then let's go with that. The quickest way is to find one piece. If we isolate this part, we can multiply it by two to get the whole. I think that could work."

G1S1: "Should we split those parts too?"

G1S2: "Yes, if I find this part, I'll multiply it by four to get this section, then by two to get the full pattern. If we find just this part, we'll have the answer."

G1S1: "Okay, let's separate this section."

In *Group 2*, students were also observed applying reflection during the problem-solving process, as illustrated in the dialogue below:

G2S3: "Let's divide the rug in half."

G2S2: "How would that help? What difference does it make?"

G2S4: "If we find half, we can figure out the rest."

G2S2: "How?"

G2S4: "If we draw a line down the center, it becomes perfectly symmetrical."

G2S3: "It's like a reflection."

G2S1: "Exactly—we'll have found half of it."

As seen in the dialogues above, students attempted to simplify the problem by identifying symmetry across an axis and reflecting the patterns accordingly. Applying symmetry helped students recognize structure and order in the patterns. It also allowed them to position the shapes correctly and more easily identify translations within the overall design. Students demonstrated a systematic and thoughtful approach when using both translation and reflection. These movements had a positive effect on their performance. They actively reflected on the patterns within the rug and applied mathematical strategies to analyze them. Furthermore, students considered the relationships among the patterns and their geometric properties while applying these transformations. Another finding was that while some students solved the problem by applying translation and reflection sequentially, others used a more integrated approach, applying both movements simultaneously. By focusing on geometric features such as repeatability, symmetry, and spatial relationships, students were ultimately able to calculate the number of triangles in the rug based on their colors and successfully reach a solution.

Findings on Rotational Movements

Regarding the findings on rotational movements in the Eşme rug problem, it was observed that students attempted to understand rotation by associating it with the patterns in the rug. They followed the visual changes in shapes to grasp how rotation occurred. For example, during the activity, students were frequently seen rotating their papers in order to observe how the patterns changed with movement. It was also noted that some students reflected on how the rotation of the central shape around a fixed point influenced the overall symmetry of the rug.

Findings Related to Cultural Aspects in the Eşme Rug Problem

When considering cultural aspects, it was found that students attempted to describe the shape and design of the rug. During the process, both groups discussed the fact that the given rug shape resembled a trapezoid. This discussion is presented in Table 3.

Table 3. Dialogue between the groups regarding the rug being crooked

Group 1	Group 2
G1S3: I'm not counting the ones that aren't full triangles. **G1S1:** No, you should count the incomplete ones as triangles too. **G1S3:** But we can't really determine their exact value. I think we can include them in the calculation later. **G1S2:** We should count them. **G1S3:** But how can we know if they're actually full triangles or not? **G1S1:** Which ones are you talking about? **G1S3:** Those ones. **G1S1:** Just combine them and count them as a triangle. *(after counting the colored triangles)* **G1S3:** I think there are about 135 yellow ones. **G1S2:** It's not really possible to get an exact number anyway. *(a little later in the discussion)* **G1S1:** By the way, these aren't identical. This part continues over here. **G1S4:** This triangle and that one don't look equal at all. **G1S2:** They are equal, though. (Excerpt from Group 1 – Cultural Reasoning in the Eşme Rug Problem)	**G2S1:** So are we supposed to calculate the cost of the big rug on top, or this one? This part looks like a trapezoid. Are we going to base our calculation on this? It sticks out on the side. **G2S3:** The rugs in our house are all rectangular. **G2S2:** The ones I've seen are rectangular too. **G2S5:** Same here. **G2S4:** Then let's just take it as a plain rectangle. A rectangular rug. (Excerpt from Group 2 – Cultural Interpretation of Rug Shape)

As shown in the dialogue above, students in Group 1 discussed whether partial or incomplete triangles should be included in the total count. While some hesitated due to uncertainty about symmetry and completeness, others argued for including all shapes through approximation or combination. This reflects their engagement with visual estimation and culturally influenced reasoning about the structure of the rug design.

In a related exchange, students in Group 2 questioned the geometry of the rug presented in the problem. Drawing on their personal and cultural experiences, they referenced the typical rectangular shape of rugs in their homes as a basis for interpreting the model. This illustrates how cultural background informed their mathematical reasoning and modeling choices. Furthermore, as evidenced in Group 2's dialogue, the rug—recognized as a traditional element of Turkish culture—was interpreted accordingly by the students. Based on this recognition, they completed the missing section of the rug, demonstrating familiarity with cultural motifs and applying this knowledge to guide their modeling decisions.

Figure 3. Representative example of group 2 completing the rug

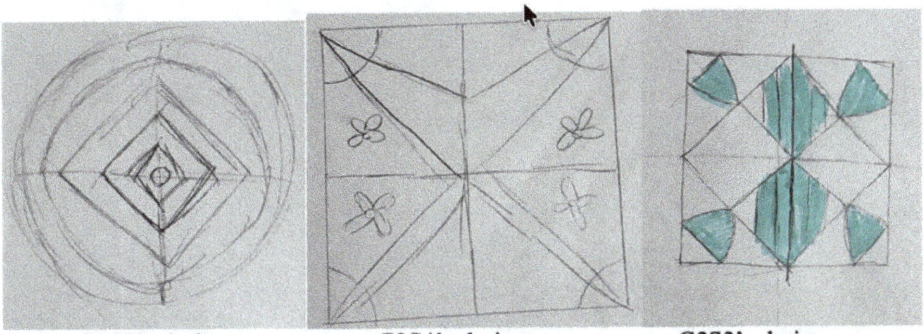

G1S2's design | G2S4's design | G2S3's design

As shown in Figure 3, students completed the missing patterns in the rug design. This demonstrates that they were familiar with rug motifs that are part of Turkish cultural heritage. Group 2 approached the problem by treating the rug as a rectangular shape and added the missing patterns accordingly to reach a solution. Because Group 2 integrated their cultural knowledge into the modeling process—assuming the rug to be a complete rectangle—their final mathematical calculation of the cost differed from that of Group 1. Group 1, relying solely on the visual data provided without applying cultural completion, modeled the rug exactly as pictured. This contrast highlights how culturally grounded reasoning directly influences mathematical modeling decisions and outcomes.

Findings Related to the Turkish Tile Plate Problem

The second problem addressed during the implementation was the Turkish tile plate problem, which involved designing a traditional Turkish tile plate. To support this task, four visual examples of tile plates reflecting elements of Turkish culture were included in the materials.

Initially, students examined the sample plate patterns. They then began working individually on their own designs. After some time, it was observed that students shared their initial ideas with their group members. They discussed their designs collaboratively and began developing mathematical models based on them. During this process, students searched for the most appropriate design that best reflected their ideas. Eventually, each group finalized their work through joint decision-making. As a result, the groups produced models inspired by traditional tile plate patterns. The final designs created by the groups are presented in Figure 4.

Figure 4. Plate designs of the groups

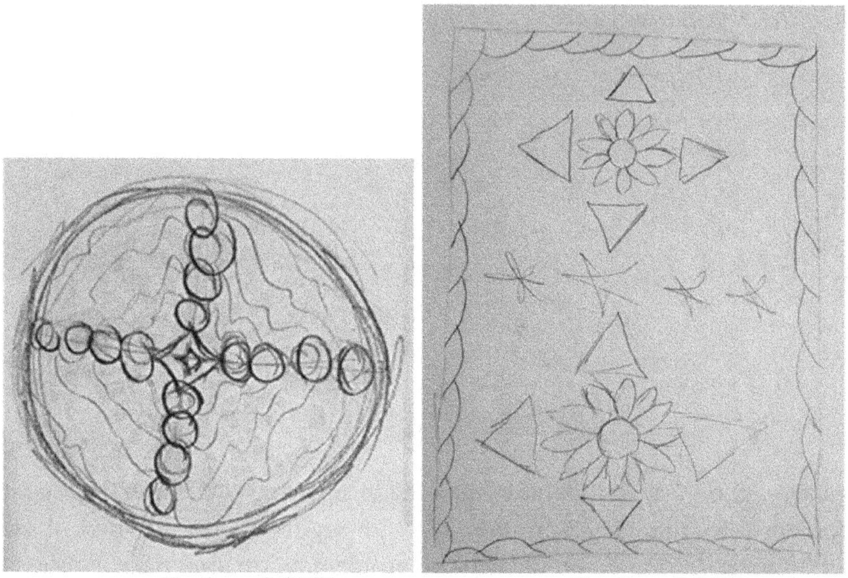

Design of G1S1 Design of G2S3

As seen in Figure 4, although the groups used similar shapes, they created different designs. Group 1 designed their plates on a circular base, while Group 2 created theirs on a rectangular base. In the students' plate designs, a large circle was typically placed at the center, surrounded symmetrically by small geometric shapes or figures. Additionally, more complex patterns—such as polygons and stars—were positioned between these basic elements.

The situations observed during the model design process were examined in detail. The findings obtained from this analysis are presented in the following section under the relevant subheadings.

Findings on Translational, Reflection, and Rotational Movements in the Turkish Tile Plate Problem

When students' performances regarding translation, reflection, and rotation movements in the tile plate design were evaluated, it was observed that their ability to apply these geometric transformations varied across different levels. Students demonstrated an awareness of these transformations during the model design process. In this section, examples related to each transformation are presented, supported by student activity sheets and dialogue excerpts.

Findings on Translational Movements

During the implementation, students' translational skills were analyzed. It was observed that they were able to generate new arrangements by shifting patterns in a specific direction, moving each point of the pattern linearly and at an equal distance. Students demonstrated an understanding of how patterns should be translated and were able to apply this knowledge accurately.

The smooth application of translation allowed for the creation of a continuous visual on the plate surface. By shifting patterns uniformly in a given direction and distance, students produced coherent and repeated motifs. As illustrated in Figure 5, students used translation to construct a motif during the design process. Additionally, by shifting patterns to the left or right, they enabled horizontal movement along the x-axis (see Figure 5).

Figure 5. Sample representations of translational movements in students' plate designs

During the model construction process, as exemplified in the dialogue below, students engaged in discussions regarding the application of translational movement. They considered using translation as part of their design strategy. An example of the dialogues observed in this context is presented in Table 4.

Table 4. Group dialogues on the use of translational movement during the design process

Group 1	Group 2
G1S1: Let's also use translation. **G1S3:** Where exactly would we use translation? **G1S2:** From here to here. *(points to the outer pattern on their model)* **G1S2:** Yeah, but... I'm thinking about how we can apply the math to that... We used pentagons and such. **G1S1:** Maybe we could draw a circle and take the multiplier as "a" from there, then make a proportion. **G1S2:** That's what I was thinking too. **G1S4:** We could use exponential brackets. (Group 1 – Dialogue on Translational Movement)	**G2S5:** Let's draw a star in the middle. **G2S3:** How are we going to draw the star? **G2S4:** In the plates and in mosque patterns I've seen, stars are drawn using triangles. Let's use triangles. **G2S5:** Like a pattern. **G2S3:** Like a floral design. **G2S1:** Then let's draw stars on both sides of the crescents—make it look like a flag. **G2S4:** Should we draw stars from this crescent to that crescent? **G2S2:** That wouldn't create a balanced look… Make it smaller toward the center. *(They begin drawing.)* **G2S5:** It looks like a pattern. **G2S2:** The edges look a bit empty. Maybe we could add some patterns there. **G2S5:** The stars look patterned, but they don't go well with the crescents. We might have to remove the crescents. *(The whole group disagrees—"Let's not remove the crescents.")* **G2S2:** What if we just add some spiral shapes here? **G2S5:** The center feels a bit empty. **G2S4:** All of the plates have some kind of order or pattern. **G2S3:** It doesn't matter if it's empty—as long as there's a pattern. Group 2 – Dialogue on Patterning and Cultural Symbol Use in Tile Design

As shown in Table 4, students did not merely apply translational movement mechanically; rather, they discussed how translation could be used as a design strategy. Their conversations reflect an awareness of both geometric accuracy and aesthetic arrangement. For example, in Group 1, students suggested shifting geometric elements systematically to generate repetition and proportionality. They explored the idea of using translation along with mathematical reasoning such as ratios and variables (e.g., referring to using "a" as a multiplier). This indicates an attempt to connect visual-spatial reasoning with symbolic representations.

Similarly, in Group 2, students discussed the visual balance and placement of symbolic elements like stars and crescents. While they did not explicitly name the transformation as "translation," their discussion of symmetry, repetition, and alignment reflects an intuitive understanding of movement along a horizontal axis. Their decisions about removing or preserving elements, and their effort to maintain a continuous pattern, also suggest engagement with underlying geometric principles. These examples highlight how students blended mathematical transformations with cultural symbolism and visual design considerations to create meaningful models.

Findings on Reflection Movements

Findings regarding students' use of reflection movements are as follows: many students understood the concept of symmetry through reflection along axes and applied it successfully. Representative examples are presented in Figure 6. These visual examples show that students accurately applied symmetry with respect to different axes

Figure 6. Representative examples of reflection in students' tile plate designs

As seen in Figure 6, students applied reflection by physically drawing symmetrical shapes across self-identified x and y axes, ensuring the geometric elements were placed at equal distances from the center. By reflecting shapes across axes, they generated new designs. The objects were repositioned through reflection across the axis of symmetry, resulting in a new appearance of the original shape. Accordingly, students were able to produce new patterns through this process. Throughout the activity, it was observed that reflection was frequently used to analyze, design, and construct symmetrical geometric shapes and patterns. Dialogue excerpts illustrating this process are presented in Table 5.

Table 5. Groups' dialogues on the use of reflection movement

Group 1	Group 2
G1S2: Let's make the base round. Both the base and the shape we draw should be symmetrical, just so we can use the math. **G1S3:** Let's divide it down the middle—make it symmetrical. **G1S2:** It came out a bit small. **G1S3:** Let's make the drawings bigger. *(They begin drawing.)* **G1S4:** We actually made a plate before in design and technology class. **G1S2:** But that one was... different. **G1S4:** Wasn't that a line drawing? **G1S2:** I think what we need to focus on here is that the plate is oval, so everything should be evenly distributed—we need to draw the symmetry carefully. **G1S3:** Let's make the shapes symmetrical across the diagonals. ... **G1S4:** Since we've drawn the plate, let's say: "we'll place it on one side, and the same will appear on the other side using reflection." **G1S2:** Actually, when we divide it into four, everything becomes a reflection of each other. Even your plates are like that—they can be split into four, or two. Either way, all parts are reflections. **G1S1:** Let me see... I'm not sure. Are all of these reflections? **G1S3:** Reflection. **G1S2:** I want to divide the plate into four parts now. **G1S1:** Not yet, I think. Let's decide together first. Group 1 – Dialogue on the Use of Reflection Movements- Turkish Tile Plate Problem	**G2S2:** Let's reflect it, let's reflect it. **G2S5:** We're going to draw a crescent, right? But only part of it should be visible—not the full "C" shape. **G2S2:** We drew an arc here—let's reflect it over the y-axis. Then we can think about what to put inside. Let's use the crescent. **G2S5:** Okay, let's use the crescent. Let's place it on the outside, but not the whole thing—just like this. **G2S2:** But we can't go beyond the border. **G2S5:** We'll erase the excess part. **G2S3:** It'll look like it continues. **G2S5:** We'll draw it the same way on all sides and erase the extra parts. *(draws two crescents symmetrically on the paper)* **G2S2:** Okay, I like the crescent—go ahead, continue. **G2S4:** It has to be symmetrical. Group 2 – Dialogue on the Use of Reflection and Crescent Symbols - Turkish Tile Plate Problem

These dialogues illustrate how students in both groups developed and applied the concept of reflection within a cultural and geometric context. The dialogue from Group 2 demonstrates how students used reflection along the y-axis to incorporate culturally meaningful elements-such as crescents-into their tile designs. Their focus on symmetry, spatial boundaries, and repetition reveals an emerging understanding of geometric transformations within a modeling framework.

The dialogue from Group 1 shows how students engaged with both geometric terminology and visual reasoning to apply reflection in their designs. They frequently discussed dividing the plate into equal parts, recognizing symmetry along multiple axes, and using reflection as a structural tool. Furthermore, audio recordings revealed that students critically evaluated whether a given shape could be divided symmetrically and chose not to use those that could not. This decision-making process highlights their ability to integrate mathematical reasoning with design constraints.

G1S2: Let's draw something here, in every section.

G1S4: Let's draw a pentagon.

(They draw.)

G1S3: How are we going to divide the pentagon symmetrically?

G1S2: Yeah, exactly.

G1S3: I think using a square or circle would make more sense.

G1S1: That would be too classic.

G1S2: Or maybe a star or something else.

G1S3: But you can't divide a star symmetrically either.

G1S1: An octagon?

G1S3: That works.

G1S2: We don't have to stick to those—we could also use a flower.

G1S1: A flower doesn't feel mathematical to me.

G1S2: What do you mean?

G1S4: Let's use the octagon.

G1S2: Whether it's this or that—it doesn't matter. A star-shaped one would be fine.

G1S3: Actually, we could just draw whatever we want in the center first and then divide it into parts.

G1S2: That could work too. It makes sense that it looks the same from all sides, with equal spacing and a balanced design

(Group 1 – Dialogue on Selecting Symmetrical Shapes for Reflection -Turkish Tile Plate Problem)

Group 1 also featured a moment where **G1S1** proposed an idea to achieve symmetry in the plate design by physically folding the paper into quarters and pressing the shape through to trace its reflection. The conversation below captures this suggestion:

G1S1: Wait, I've got a really great idea! What if we cut it and fold it into four equal parts? Then we draw the shape with a pencil and press hard so it goes through to the back. We can trace it from the back.

G1S3: I think that would be too difficult.

G1S1: I think it would actually be easier—it would be perfectly symmetrical.

G1S2: But our pencils don't go through the paper like that. It would be a hassle.
G1S3: We'd have to redraw everything.
G1S2: How will it go through four layers?
G1S1: Then we try with two layers.
G1S2: Two layers makes more sense.
G1S1: So if the two-layer method doesn't work...
G1S2: I don't think it's worth all the trouble.
G1S3: It would be easier to just draw.
G1S4: Let's just keep going like this and then clean it up later.
(Group 1 – Dialogue on Folding and Drawing Symmetrical Reflections- Turkish Tile Plate Problem)

Findings on Rotational Movements

When students' use of rotational movements in their plate designs was evaluated, noteworthy findings were obtained. It was observed that students created new patterns by rotating a design element around a central point on the plate. For instance, in some designs, students started with a flower motif and generated a circular pattern in the center. Examples of such designs are presented in Figure 7. These examples indicate that students understood how to rotate one shape around another and applied this transformation accurately.

Figure 7. Representative examples of rotational movement in students' tile plate designs

Figure 2. Representations of Groups' Application of Symmetry Movements

As seen in Figure 7, students applied rotational movement by sketching identical geometric shapes (such as triangles and petals) at equal intervals around a central focal point, maintaining visual balance. They were able to determine how a shape should be rotated to achieve the desired result and applied this process correctly. Moreover, while implementing this geometric transformation, they also managed to maintain visual balance and compositional harmony. Each step of the rotation contributed to preserving the coherence and aesthetic symmetry of the design. Through these movements, students created tile plate patterns that reflect geometric structure and consistency (see also Figure 3).

This awareness was also evident in their verbal reflections. For example, G1S2 stated, "There is a structure in our design. It includes reflection/symmetry, translation, and rotation." Likewise, G2S2 commented, "We used reflection, made use of patterns, and applied rotation." A dialogue in which this awareness emerged is provided below:

G2S3: Since we're using two crescents, let's make it two.

G2S4: Yes, let's place one here and another one over there.

G2S2: I think just one is enough. On plates, there's usually a large motif in the center.

G2S1: A plate should have order. Nothing should stand out awkwardly.

G2S3: Yes, plates use patterns, reflection, and rotation. Our design should be like that too.

This dialogue shows that the groups not only understood the concept of rotation around a center but also applied it successfully. The process ensured that each part of the design was repositioned in a new location. Repeating the rotation enhanced the symmetry and visual impact of the pattern. Moreover, students appeared to be attentive to the flow and coherence of the design, maintaining unity through rotational symmetry.

Cultural Findings on the Turkish Tile Plate Problem

Significant results were obtained when examining the cultural dimensions of the tile plate problem. It was observed that students had prior knowledge about traditional Turkish tile plates and reflected this understanding in their designs. In particular, students intentionally incorporated traditional motifs, demonstrating familiarity with culturally embedded aesthetic patterns. Examples of observed traditional designs are presented in Table 6.

During the activity, student G2S5 stated: *"In our culture, flower-like shapes are commonly used, especially in the center. There are a lot of floral patterns on plates. We should use them too."* Similarly, G1S2 commented on the importance of symmetry and central design:

"I think we should pay attention to this—the plate is an oval shape, so everything should be distributed evenly. We need to draw the symmetry well."

"Like this, for example. We can use something like this in the very center. In all plates, something is usually placed at the center."

These statements and the designs created by the students show that cultural factors were not only recognized but actively discussed, experimented with, and applied during the modeling process. In short, the findings suggest that students brought prior cultural experiences and knowledge into their mathematical design work.

Table 6. Examples of cultural motifs used by student groups

Motifs	Group 1	Group 2
Border Pattern	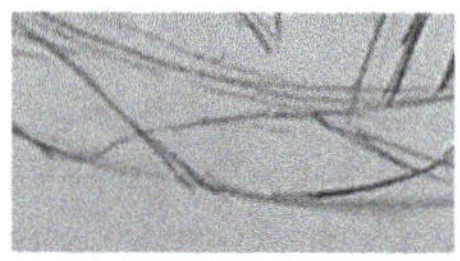	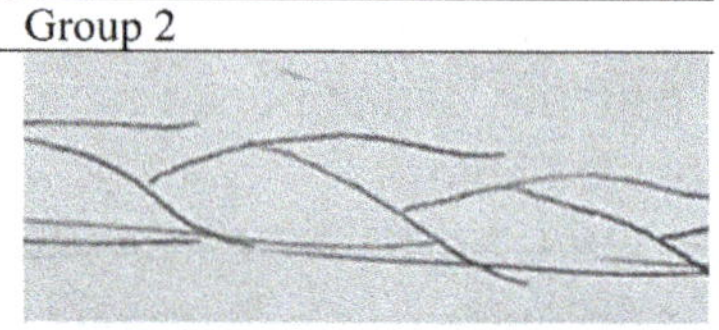 (sarmaşık)
Teardrop	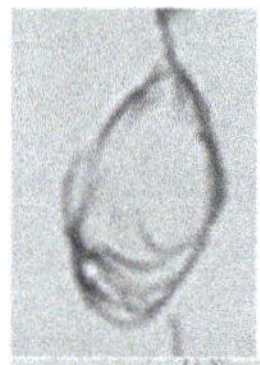	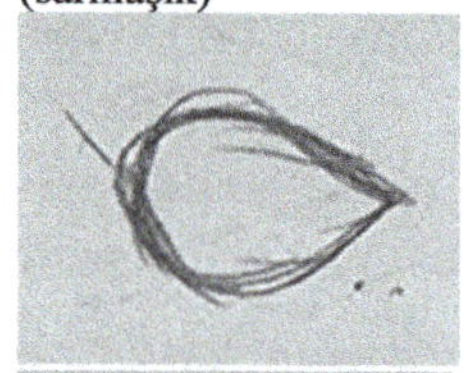
Radial Designs (Rotating around a center/ Star Formation Using Triangles		
Color		
Flovers	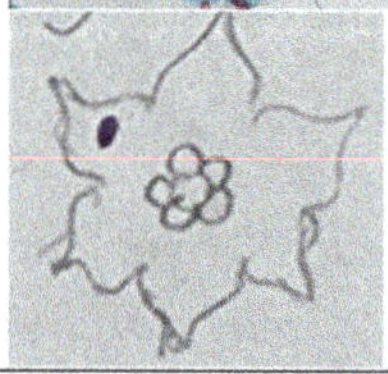	

Table 6 presents shared cultural elements observed across groups. One common motif was the teardrop figure, which was used by both groups. The following dialogue, for instance, illustrates how Group 2 discussed why they chose this motif in their designs:

G2S2: Then let's draw teardrop shapes in the empty spaces.

G2S3: The teardrop is symbolic of Turkish blood.

(They begin drawing the teardrop motifs.)

G2S5: Let's draw a teardrop.

G2S3: Just one.

G2S2: Yes, just one. Otherwise, it'll look too messy.

Group 2 also had a discussion while using triangles and star shapes, during which they explored the use of flowers, symmetry, and centralized patterns in Turkish culture:

G2S5: Let's draw a star in the center.

G2S3: How do we draw a star?

G2S4: In plates and in mosque motifs I've seen, stars are drawn using triangles. Let's use triangles.

G2S5: That would make it patterned.

G2S3: Like a floral pattern.

...

G2S5: Flower-shaped figures are used in our culture—especially in the center. There are many flower designs on plates. Let's use that.

G2S2: Then let's draw a flower in the center. We can make it look like a full moon. Let's draw a full moon and place the flower inside.

G2S3: We can make a star from the flower.

G2S3: Since we're using two crescents, let's keep it to two.

G2S5: Yes, let's place one here and one there.

G2S2: I think only one is enough. On plates, there's usually a single large motif in the center.

(Turkish Tile Plate Problem – Solution – Group 2)

In addition to the examples in Table 6, other cultural features emerged in students' designs. For instance, Group 2 included crescent and moon motifs, which are symbolic in Turkish culture. This was explicitly explained by G2S2, who stated:

> *"We used two crescent designs here. Since they're symbols of the Turkish people, we included both the moon and the star. We were inspired by Ottoman floral motifs."*

In Group 1, the idea of using ebru (marbling) patterns was suggested by G1S2, who said: *"We could also use ebru patterns."* After incorporating floral patterns, which are frequently found in traditional Turkish art, Group 1 also added leaf motifs to complement the floral ones (see Figure 3). It was also observed that students often used circular forms inspired by traditional Turkish dishware. This was evident in the following dialogue from Group 1:

G1S1: Aren't we designing a dinner plate? A regular one?

G1S2: Think of a soup bowl. The sides curve up—like that. Since the inside is oval, the design will appear more clearly.

G1S1: We could even add some splashes of color around the edges—it would look better.

G1S4: That works.

As a result of this discussion, Group 1 decided to design their plates using circular shapes. Later, G1S1 explained:

"We were inspired by plates we've seen before. And since circles are used a lot in tiles, we used a circular pattern."

SUMMARY AND DISCUSSION OF FINDINGS

This study presents an example of how mathematical content can be delivered through the ethnomodelling approach by integrating it with cultural contexts. Ethnomodelling supports students in making meaningful connections between mathematical concepts and their own cultural references (Rosa & Orey, 2021a). This approach aims to enrich school curricula by embedding them within cultural, social, and historical contexts (Cortes & Orey, 2020; Rosa & Orey, 2021b). Accordingly, in this study, mathematical content was implemented by establishing links with students' personal experiences, cultural values, and lived realities. In doing so, students were supported in perceiving mathematics not merely as an abstract discipline but as a comprehensive construct that connects with reality, society, environment, and identity (Bahadır, 2021; Arı & Demir, 2022; Rosa & Orey, 2011; Umbara et al., 2021). In this context, two culturally embedded problems—Eşme rug and Turkish tile plate—were placed at the center of the modelling activities. Through these examples, the study aimed to illustrate how mathematical practices can be contextualized within cultural frameworks (Arı & Demir, 2022). The ethnic motifs, local knowledge, and cultural practices employed by students were examined to understand how they were integrated into the modelling process and how they contributed to mathematical problem-solving. Additionally, students' performances were analyzed with regard to their use of translation, reflection, and rotation within the scope of transformational geometry.

In the first phase of the study, students' performances regarding the Eşme rug problem were examined. Similar to previous studies, students were able to identify geometric transformations (Bassarear, 1995) and explore the structural patterns within the rug design (Britton & Seymour, 1989). Throughout the activity, students demonstrated how frequently used rug motifs could be manipulated through translation, reflection, and rotation. This required them, as Britton and Seymour (1989) emphasized, to carefully analyze all combinations to determine whether the patterns continued consistently. In summary, addressing these three geometric transformations collectively within the context of the Eşme rug problem provided opportunities for students to connect transformational geometry with culturally familiar designs.

Likewise, the evaluation of students' performance on the Turkish tile plate problem revealed that they successfully transformed patterns using translation, reflection, and rotation. These transformations helped students enhance both their understanding and application of geometric concepts, while also strengthening their mathematical modeling competencies. In particular, students utilized reflection to construct symmetrical patterns, in line with the process described by Knuchel (2004). Considering symmetry and geometric relationships during the reflection process (Darvas, 2007) enabled students to develop a more profound conceptual grasp. Throughout the process, students approached the tasks with attention, systematic thinking, and creativity, which allowed them to solve problems effectively and produce original designs. The integrated use of translation, reflection, and rotation supported the flow and visual harmony of their patterns.

During the process of translation, students considered fundamental geometric properties such as repeatability, symmetry, and the relationships between patterns. As in previous studies, geometric transformations were used sequentially throughout the design process (Britton & Seymour, 1989). This allowed students to analyze how patterns change through translation and reflection, explore the relationship between original and reflected designs, and evaluate the effects of such transformations. These cognitive processes significantly contributed to the development of students' mathematical thinking skills (Britton & Seymour, 1989). As the process progressed, students demonstrated improved understanding and application of transformational concepts. This finding is consistent with those of earlier studies (Knuchel, 2004; Conway, Burgiel, & Strauss, 2008). Furthermore, students' efforts to create symmetrical arrangements by focusing on reflection and symmetry supported their awareness of artistic elements—such as balance, proportion, and composition—as noted by Schoevers et al. (2020). This interaction not only addressed aesthetic considerations but also deepened students' comprehension of geometric concepts.

When the problems were addressed within the framework of ethnomodelling, students had the opportunity to associate the mathematical modeling process with cultural, social, and historical contexts. They utilized cultural references as a foundation for building and expressing mathematical understanding (Rosa & Orey, 2021a). Familiarity with the context of a problem increases the likelihood that students will draw on prior experiences during problem solving (Hart, 1993). For example, G1S2 stated, *"Imagine a soup bowl, the sides are curved like that; since the inside is oval, we can get a clearer image here"*, helping Group 1 advance in their plate design. Similarly, G1S1's remark, *"We were inspired by plates we've seen before"*, supported their process. In Group 2, several students (G2S2, G2S3) mentioned that the rugs in their homes were rectangular, and based on this shared recognition, they decided on the rug's shape. These examples suggest that students meaningfully drew on their cultural experiences to support the modeling process (Küçük, 2013; Umar et al., 2019; Alghar & Marhayati, 2023).During the plate design activity, students also discussed culturally embedded patterns such as flowers, stars, and crescents. In Group 2, G2S5 initiated the design by saying, *"In our culture, flower-like shapes are used—especially in the center. There are lots of floral motifs on plates. Let's use one too."* Students employed not only aesthetic elements but also embedded symbolic meanings. Some described star patterns in relation to mosque ornamentation, while others associated the droplet motif with national symbolism. In one exchange, a student mentioned that the droplet "symbolizes Turkish blood." This comment reflects how students linked visual forms to national identity and historical memory, specifically referencing the red color and symbolic imagery of the Turkish flag. Within the context of the study, this interpretation was regarded not as a political or ethnic statement, but as a student-driven expression of cultural heritage. These moments illustrate how students made connections between mathematical modeling and culturally meaningful representations. Ultimately, students found opportunities to link mathematical ideas with cultural and historical narratives, both when analyzing kilim motifs and while designing tile plates. Through expressions such as *"mosque designs"* or *"let's depict it like a full moon"*, students integrated their cultural experiences into the mathematical modeling process in meaningful and reflective ways.

It was observed that symmetrical and translational movements, which are frequently encountered in Turkish motifs and carpet patterns, were easily recognized by the students. The integration of these cultural elements into mathematics education provided a context for students to apply their existing knowledge. In particular, recognizing the symmetry and translation properties embedded in traditional Turkish motifs enabled students to verbalize and visually represent abstract geometric transformations through familiar cultural contexts. Similarly, the application of rotational movements in cultural designs supported students in understanding the mathematical concept of rotation. The activities carried out demonstrated that rotational movement is not merely a theoretical construct but a tangible phenomenon observable in daily life and cultural contexts. Specifically, the Turkish tile plate problem effectively illustrated to students how geometric transformations can be seamlessly integrated with cultural motifs.

Another significant finding of the study was that students enriched their mathematical knowledge through the incorporation of cultural knowledge (D'Ambrosio,

2001). As emphasized in the literature, relating mathematical concepts to cultural and everyday experiences—beyond what is offered in the standard curriculum—enhances both the depth and relevance of learning (Rosa & Orey, 2011; Brandt & Chernoff, 2015). In line with previous research, the present study showed that students formed meaningful connections (Desai et al., 2022), deepened their mathematical understanding (Widada, Herawaty & Lubis, 2018), and developed solid comprehension of mathematical concepts by constructing models (Schoenfeld, 2016). As Rosa and Orey (2021a) explain, students were able to organize and apply prior real-life experiences systematically while solving similar problems. This approach not only promotes a multifaceted perspective in mathematics education but also reveals how mathematical knowledge can be interpreted and adapted across time and place (Rosa & Orey, 2019).

Our research also revealed that examining the geometric properties of Turkish motifs can uncover the hidden dimensions of mathematics. This finding is consistent with the results of Schattschneider's (2004) study and supports the view that ethnomathematical research has the potential to expose the implicit structure of mathematical knowledge (Conway, Burgiel & Strauss, 2008; Rosa & Orey, 2010a; 2011). In this context, the ethnomodeling approach enabled the establishment of connections between local conceptual frameworks and local designs, making it possible to reveal the embedded mathematical structures within these cultural artifacts (Rosa, Orey & Cortes, 2020).

In conclusion, this study demonstrated how elements specific to Turkish culture can be effectively utilized in mathematics education. The findings showed that integrating cultural contexts into mathematics education deepened students' conceptual understanding. Cultural references facilitated students' access to mathematical concepts, while the collaborative nature of the tasks allowed students to exchange ideas and support one another. In this way, students were provided with rich and meaningful learning experiences (Rosa & Orey, 2013). Future research is expected to contribute to the diversification and enrichment of instructional approaches by exploring the relationships between different cultural elements and mathematical concepts from a broader perspective.

RECOMMENDATIONS FOR MATHEMATICS TEACHERS

Based on the findings of this study, the following recommendations are provided for mathematics educators to enhance the integration of cultural elements into their teaching practices:

Incorporate Local Cultural Artifacts: Teachers should utilize culturally familiar objects—such as local rugs, traditional ceramics, or regional architectural motifs—to introduce abstract geometric concepts like translation, reflection, and rotation. Using artifacts that students encounter in their daily lives helps bridge the gap between academic mathematics and lived reality.

Design Open-Ended Modeling Tasks: Instead of using closed-ended geometry exercises, educators should design ethnomodeling tasks that allow students to create their own designs. As seen in the "Turkish Tile Plate Problem," providing students with the freedom to design fosters creativity and deeper conceptual understanding.

Encourage Collaborative Discussion: Implementation should prioritize group work where students can discuss their cultural reasoning and mathematical strategies. These interactions allow students to verbalize their intuitive understanding of symmetry and patterns, transforming the classroom into a space for the co-creation of knowledge.

Validate Intuitive Knowledge: Teachers should recognize and validate students' "local knowledge", such as their understanding of pattern repetition or symbolic meanings, even when it is not expressed in formal mathematical terms. This approach moves away from the "banking model" of education and encourages students to see themselves as mathematical thinkers.

Promote Interdisciplinary Connections: Mathematics lessons can be integrated with visual arts and history by exploring the symbolic and historical narratives behind cultural motifs. For instance, discussing the meaning of floral or celestial symbols in traditional art can make geometric transformations more engaging and meaningful.

Utilize Visual and Physical Modeling: Educators should encourage students to use physical strategies, such as rotating papers or visualizing symmetries through folding, to grasp transformational geometry concepts. These concrete actions support the transition to abstract mathematical modeling

REFERENCES

Aktuna, H. E. (2013). *Sixth grade students' perceptions of and engagement in ethnomathematical tasks in the area measurement concept (Unpublished master's thesis)*. Ortadoğu Teknik Üniversitesi.

Alghar, M. Z., & Jamaluddin. (2024). Ethnomodelling: Fractal geometry on the door ornament of the Sumenep palace using the Lindenmayer system. *Euclid, 11*(1), 1–16. DOI: 10.33603/x5gk7n46

Alghar, M. Z., & Marhayati, M. (2023). Ethnomathematics: Exploration of fractal geometry in gate ornaments of the Sumenep Jamik Mosque using the Lindenmayer system. *Indonesian Journal of Science and Mathematics Education, 6*(3), 311–329. DOI: 10.24042/ijsme.v6i3.18219

Alghar, M. Z., & Radjak, D. S. (2024). Systematic literature review: Implementation of ethnomodelling in mathematics learning. Union. *Jurnal Ilmiah Pendidikan Matematika, 12*(1), 67–81. DOI: 10.30738/union.v12i1.17004

Arı, A. A., & Demir, B. (2022). Matematiksel modellemenin kültürel perspektifi: Etnomodelleme. *Kültür Araştırmaları Dergisi, 12*(12), 253–270. DOI: 10.46250/kulturder.1067167

Aydemir, H., & Kubanç, Y. (2014). Problem çözme sürecinde üst bilişsel davranışların incelenmesi. *Journal of Turkish Studies, 9*(2), 203–219. DOI: 10.7827/Turkish-Studies.6555

Bahadır, E. (2021). Ethnomathematics approach in mathematics education for migrant students. *Milli Eğitim, 50*(Special Issue 1), 577–594. DOI: 10.37669/milliegitim.959829

Bassanezi, R. C. (2002). *Ensino-aprendizagem com modelagem matemática*. Editora Contexto.

Bassarear, T. (1995). *Mathematics for elementary school teachers*. Houghton Mifflin Company.

Brandt, A. L., & Chernoff, E. J. (2015). The importance of ethnomathematics in the math class. *Ohio Journal of School Mathematics, 71*, 31–36.

Britton, J., & Seymour, D. (1989). *Introduction to tessellations*. Dale Seymour Publications.

Bukova Güzel, E. (2021). *Matematik eğitiminde matematiksel modelleme*. Pegem Akademi.

Büyüköztürk, Ş., Kılıç Çakmak, E., Akgün, Ö. E., Karadeniz, Ş., & Demirel, F. (2019). *Bilimsel araştırma yöntemleri* (26. bs.). Pegem Akademi.

Çelebi, M. (2022). *Nitel araştırma yöntemleri*. Pegem Akademi.

Çenberci, S., & Horzum, T. (2023). Matematik öğretmeni adaylarının etnomatematik farkındalıklarının incelenmesi. *Ahmet Keleşoğlu Eğitim Fakültesi Dergisi*, *5*(3), 713–732. DOI: 10.38151/akef.2023.80

Conway, J. H., Burgiel, H., & Strauss, C. (2008). *The symmetries of things*. A K Peters.

Cortés, D. P. O., & Orey, D. C. (2020). Connecting ethnomathematics and modelling: A mixed methods study to understand the dialogic approach of ethnomodelling. *REVEMOP*, *2*, 1–25. DOI: 10.33532/revemop.e202011

Creswell, J. W. (2015). *Educational research: Planning, conducting, and evaluating quantitative and qualitative research* (5th ed.). Pearson.

Darvas G. (2007). *Symmetry: Cultural-historical and ontological aspects of science-arts relations; the natural and man-made world in an interdisciplinary approach*. Basel.

D'Ambrosio, U. (1990). *Etnomatemática*. Editora Ática.

D'Ambrosio, U. (1993). Etnomatemática: Um programa. *A Educação Matemática em Revista*, *1*(1), 5–11.

D'Ambrosio, U. (1997). Ethnomathematics and its place in the history and pedagogy of mathematics. In Powell, A., & Frankenstein, M. (Eds.), *Ethnomathematics: Challenging eurocentrism in mathematics education* (pp. 13–24). SUNY Press.

D'Ambrosio, U. (2001). *Etnomatemática: Elo entre as tradições e a modernidade*. Autêntica.

Desai, S., Safi, F., Bush, S. B., Wilkerson, T., Andreasen, J., & Orey, D. C. (2022). Ethnomodeling: Extending mathematical modeling research in teacher education. *Investigations in Mathematics Learning*, *14*(4), 305–319. DOI: 10.1080/19477503.2022.2139092

Freire, P. (1970). *Pedagogia do oprimido*. Paz e Terra.

Freire, P. (1998). *Pedagogia do oprimido* (25th ed.). Paz e Terra.

Gerdes, P. (1998). On culture and mathematics teacher education. *Journal of Mathematics Teacher Education*, *1*(1), 33–53. DOI: 10.1023/A:1009955031429

Gürefe, N. (2025). Investigation of the effect of geometry teaching with ethnomodeling on students' academic achievement in area measurement. *Ordu Üniversitesi Sosyal Bilimler Enstitüsü Sosyal Bilimler Araştırmaları Dergisi, 15*, 508-535.

Hart, L. C. (1993). Some factors that impede or enhance performance in mathematical problem solving. *Journal for Research in Mathematics Education, 24*(2), 167–171.

Kalinec-Craig, C., Prasad, P. V., & Luna, C. (2019). Geometric transformations and Talavera tiles: A culturally responsive approach to teacher professional development and mathematics teaching. *Journal of Mathematics and the Arts, 13*(1–2), 72–90. DOI: 10.1080/17513472.2018.1504491

Karadag, Z., & Karagöz Akar, G. (2020). El sanatları ve matematik: Cebir dünyasına bakış. *Boğaziçi University Journal of Education, 37*(2), 123–146. https://izlik.org/JA92TY58KB

Knuchel, C. (2004). Teaching symmetry in the elementary curriculum. *The Montana Math Enthusiast, 1*(1), 3–8. DOI: 10.54870/1551-3440.1001

Küçük, A. (2013). Ethnomathematics in Anatolia (in Turkey): Mathematical thoughts in multiculturalism. *Revista Latinoamericana de Etnomatemática, 7*(1), 171–184.

Pais, A., & Mesquita, M. (2013). Ethnomathematics in non-formal educational settings: The urban boundaries project. *Revista Latinoamericana de Etnomatemática, 6*(3), 134–144.

Prasad, D. B. (2008). Content analysis: A method of social science research. In Das, D. K. L., & Bhaskaran, V. (Eds.), *Research methods for social work* (pp. 174–193). Rawat.

Putra, E. C. S., & Mahmudah, F. N. (2021). The implementation of ethnomathematics based-learning for students. *SJME (Supremum Journal of Mathematics Education, 5*(2), 162–169.

Rosa, M. (2000). From reality to mathematical modelling: A proposal for using ethnomathematical knowledge (Unpublished master's thesis). California State University, Sacramento.

Rosa, M., & Orey, D. C. (2006). Abordagens atuais do programa etnomatemática: Delineando-se um caminho para a ação pedagógica. *Bolema, 19*(26), 19–48.

Rosa, M., & Orey, D. C. (2010a). Ethnomodeling: A pedagogical action for uncovering ethnomathematical practices. *Journal of Mathematical Modelling and Application, 1*(3), 58–67.

Rosa, M., & Orey, D. C. (2010b). Ethnomodeling as a pedagogical tool for the ethnomathematics program. *Revista Latinoamericana de Etnomatemática, 3*(2), 14–23.

Rosa, M., & Orey, D. C. (2011). Ethnomathematics: The cultural aspects of mathematics. *Revista Latinoamericana de Etnomatemática, 4*(2), 32–54.

Rosa, M., & Orey, D. C. (2013). Ethnomodeling as a research theoretical framework on ethnomathematics and mathematical modeling. *Journal of Urban Mathematics Education, 6*(2), 62–80. DOI: 10.21423/jume-v6i2a195

Rosa, M., & Orey, D. C. (2016). Humanizing mathematics through ethnomodelling. *Journal of Humanistic Mathematics, 6*(2), 3–22. DOI: 10.5642/jhummath.201602.03

Rosa, M., & Orey, D. C. (2018). The anthropological dimension on ethnomodelling research based on ethnomathematics and modelling. *Open Access Journal of Archaeology & Anthropology, 1*(1), 1–8. DOI: 10.33552/OAJAA.2018.01.000503

Rosa, M., & Orey, D. C. (2019). Ethnomodelling as the translation of diverse cultural mathematical practices. In *Handbook of the Mathematics of the Arts and Sciences*. Springer., DOI: 10.1007/978-3-319-70658-0_70-1

Rosa, M., & Orey, D. C. (2021a). Applying ethnomodelling to explore glocal mathematical knowledge systems. *Acta Scientiae, 23*(1), 199–232. DOI: 10.17648/acta.scientiae.6241

Rosa, M., & Orey, D. C. (2021b). Ethnomodelling as a globalization process of mathematical practices through cultural dynamism. *The Montana Math Enthusiast, 18*(3), 439–468. DOI: 10.54870/1551-3440.1533

Rosa, M., Orey, D. C., & Cortes, D. P. D. O. (2020). Resignifying function concept: A mixed methods study to understand the contributions of the dialogic approach of ethnomodelling. In *Mathematical modelling education and sense-making* (pp. 107–116). Springer., DOI: 10.1007/978-3-030-37673-4_10

Schattschneider, D. (2004). *Escher: Visions of symmetry* (2nd ed.). Harry N. Abrams.

Schoenfeld, A. H. (2016). Learning to think mathematically: Problem solving, metacognition, and sense making in mathematics. *Journal of Education, 196*(2), 1–38. DOI: 10.1177/002205741619600202

Schoevers, E. M., Leseman, P. P. M., & Kroesbergen, E. H. (2020). Enriching mathematics education with visual arts: Effects on elementary school students' ability in geometry and visual arts. *International Journal of Science and Mathematics Education, 18*(8), 1613–1634. DOI: 10.1007/s10763-019-10018-z

Sevgi, S. (2019). Dönüşüm geometrisi etkinliklerine kültürel bağlamın dahil edilmesi sonucu oluşan öğrenci yaklaşımları (Master's thesis, Dokuz Eylul Universitesi (Turkey)).

Topan, B., & Bozkus, F. (2021). Türk mimari eserlerindeki geometrik desenlerin geometri öğretiminde kullanılabilirliğine ilişkin öğretmen görüşleri. *Mehmet Akif Ersoy Üniversitesi Eğitim Fakültesi Dergisi, 58*, 362–389. https://izlik.org/JA38WP94TX

Umar, I. O., Tudunkaya, M. S., & Muawiya, H. U. (2019). Effect of ethnomathematics teaching approach on performance and retention in trigonometry among secondary school students in Zaria Local Government Area Kaduna State, Nigeria. *Abacus (Mathematics Education Series), 44*(1), 104–112.

Umbara, U., Wahyudin, W., & Prabawanto, S. (2021). Exploring ethnomathematics with ethnomodelling methodological approach: How does Cigugur indigenous people use calculations to determine good days to build houses? *Eurasia Journal of Mathematics, Science and Technology Education, 17*(2), 1–19. DOI: 10.29333/ejmste/9673

Widada, W., Herawaty, D., & Lubis, A. N. M. T. (2018). Realistic mathematics learning based on ethnomathematics in Bengkulu to improve students' cognitive level. *Journal of Physics: Conference Series, 1088*(1), 012028. DOI: 10.1088/1742-6596/1088/1/012028

Yıldırım, A., & Şimşek, H. (2021). *Sosyal bilimlerde nitel araştırma yöntemleri* (12. bs.). Seçkin Yayıncılık.

Yin, R. K. (2014). *Case study research: Design and methods* (5th ed.). Sage Publications.

APPENDIX

Eşme Rug Problem

Figure 8.

In the Eşme rug shown in the picture, the colors of the yarns used from the innermost motif to the outer parts and the price per meter of each are given below:

Black yarn: 10 cent, Yellow yarn: 20 cent, Green yarn: 30 cent, Red yarn: 20 cent, White yarn: 40 cent, Dark blue yarn: 50 cent, Light blue yarn: 50 cent, and Cream-colored yarn: 40 cent.

In the 100 cm^2 section shown on the right, the total length of yarn used is 5 meters.

Accordingly, how much would it cost to weave this rug?

Turkish Tile Plates

Turkish tile art is a form of art with a history spanning thousands of years and represents an important part of Anatolia's rich cultural heritage. Turkish tile patterns stand out with their aesthetic richness and symbolic depth. The patterns are often crafted in vibrant colors such as blue, turquoise, green, and red, and may include geometric shapes, plant motifs, animal figures, and abstract compositions. These designs represent the beauty of nature, fertility, and the cycle of life, while also conveying cultural, historical, mystical, and symbolic meanings.

In addition to being an aesthetic form of expression, the patterns used in Turkish tile art reflect the historical, religious, and symbolic legacy embedded in Turkish culture. Each design not only reflects the artist's emotions, thoughts, and beliefs but also conveys a rich story to the viewer. These patterns emphasize the unique and timeless beauty of Turkish tile art and maintain their significance as an essential part of Turkish art and culture. Analyzing the designs on Turkish tile plates offers an opportunity to uncover different dimensions of Turkish culture.

In a tile workshop, the traditional production process of Turkish tile plates is being examined. For this purpose, four images of traditional tile plates are being analyzed. (The craftsman plans to recreate the plates by referring to these historical designs.) Information about the images of the plates in the craftsman's possession is as follows:

Figure 9. A tile believed to have been made using the slip painting technique, continuing the Seljuk tradition in İznik during the early 14th and 15th centuries.

Figure 10. A red clay tile referred to as "Milet ware" from the late 14th century, painted in shades of blue, turquoise, purple, black, and green under a transparent glaze and covered with a whitish cream-colored slip.

Figure 11. A tile created by Azim Tile in approximately 1940.

Figure 12. A modern design found online that is based on traditional Turkish tile culture.

Tile artisans pay particular attention to the diameters and patterns of tile plates during their production. For instance, it has been observed that there is a specific relationship between the diameter of a tile plate (in cm) and the complexity of its design. This Turkish tile plate producer aims to optimize the cost and aesthetic value of the geometric arrangements used on the tile plates. To achieve this, they wish to develop a mathematical model that determines the number and arrangement of basic geometric shapes (such as circles, squares, triangles, etc.) used in these arrangements.

Can you assist the craftsman in designing these plates?

Which geometric arrangements have been used?

Can you model the rule behind the geometric patterns on the plates?

To simplify the model, can you develop sketches by drawing representations of the basic shapes and their positions in the geometric arrangements of the tile plates?

Can you apply your models according to different plate sizes?

Chapter 7
Ethnomathematical Approach to Out-of-School Mathematics Education:
Museums and Heritage Sites

Feyza Aliustaoğlu
http://orcid.org/0000-0001-9262-5216
Kastamonu University, Turkey

ABSTRACT

This study aims to introduce the theoretical frameworks of ethnomathematics and out-of-school learning and to present examples of mathematical activities that can be applied in museums and cultural heritage sites related to middle school mathematics, grounded in these frameworks. First, the theoretical foundations of ethnomathematics and out-of-school learning approaches are discussed in detail. Then, implicit mathematical elements embedded in architectural decorations, handicrafts, and cultural motifs found in museums and heritage sites in Kastamonu are identified. These elements are linked to middle school mathematics learning areas, numbers and operations, algebra, geometry and measurement, data processing, and probability, incorporating them into activity designs. The study provides a theoretical framework for integrating ethnomathematics into out-of-school learning and offers teachers examples of applications related to cultural heritage. It is believed that the proposed activities will contribute to making the cultural and out-of-school aspects of mathematics more visible.

DOI: 10.4018/979-8-2600-3485-9.ch007

INTRODUCTION

In recent years, the field of mathematics education has shown that treating mathematics solely as a discipline learned in the classroom through abstract symbols and operations is insufficient. In this context, the concept of out-of-school learning has emerged. Out-of-school learning is an approach that emphasizes that learning is not limited to formal educational institutions; rather, individuals acquire knowledge through experience in diverse contexts such as museums, heritage sites, science centers, and everyday life (Eshach, 2007). According to Anderson et al. (2002), these settings offer students relevant learning contexts where they can interact with a variety of objects and artifacts that they can analyze and discuss in light of their personal experiences. Out-of-school learning environments, especially museums and cultural heritage sites, offer rich contexts where students can experience mathematical concepts through concrete objects, structures, and cultural artifacts while also supporting the transmission of cultural heritage and values (Cooper, 2011; Kelton, 2021).

Studies in the literature examining the contributions of out-of-school learning environments to mathematics education show that these environments increase students' mathematical achievement (Abdioğlu et al., 2020), reduce math anxiety (Çağlar et al., 2018), and positively influence their attitudes towards mathematics (Duatepe-Paksu et al., 2022; Kelton, 2015). However, it is also stated that teachers do not use these environments sufficiently in mathematics teaching and have difficulty establishing mathematical connections within them (Johnson & Chandler, 2009; Nicol, 2002). When the literature is examined, it is emphasized that out-of-school learning environments are studied intensively, especially in the field of science education (Dierking et al., 2003; Kisiel, 2013), but there are a limited number of studies in the context of mathematics education (Kuş, 2024; Nemirovsky et al., 2017).

On the other hand, the concept of ethnomathematics is gaining importance in mathematics education, particularly in the context of meaningful and authentic learning experiences linked to students' daily lives, cultural experiences, and local contexts. Researchers emphasize that the ethnomathematics approach is an innovative method that can be applied in mathematics teaching (Machaba & Dhlamini, 2021). Ethnomathematics aims to connect school mathematics with these cultural riches by examining the forms of mathematical knowledge, application, and representation that different cultures have developed through historical and social processes (D'Ambrosio, 1985; Rosa & Orey, 2016). This approach requires making students' cultural capital visible and treating mathematics not only as a universal, abstract discipline but also as a cultural human activity (Bishop, 1994; Nasir, 2002).

Studies indicate that ethnomathematical applications, which go beyond traditional classroom practices, increase students' motivation and achievement levels (Barton,

1996; D'Ambrosio, 1985) and contribute to their more meaningful learning of mathematical concepts (Desai et al., 2021; Dumlu & Alkaş Ulusoy, 2025; Umbara et al., 2021). Recent studies increasingly examine ethnomathematics in relation to mathematics learning and cultural contexts (Appelbaum & Stathopoulou, 2023; Khalil, 2023; Mania & Alam, 2021; Pradana et al., 2022). However, ethnomathematics still appears to be used to a limited extent in teaching practices (Bishop, 1994; Rosa & Orey, 2016), and its integration into curriculum and classroom applications remains underdeveloped (Gürefe, 2025; Güreş & Ada, 2025).

Contemporary ethnomathematics research increasingly emphasizes the importance of contextualizing mathematics learning within students' socio-cultural environments in order to support meaningful learning experiences (Hendriyanto et al., 2023). Recent reviews indicate that ethnomathematics research has expanded considerably in relation to cultural artifacts, games, weaving, architecture, measurement practices, and number systems (Batiibwe, 2024; Pradana et al., 2022). Recent studies also emphasize the importance of connecting mathematics learning with cultural contexts, authentic environments, and students' everyday experiences through non-formal learning settings (Williams-Pierce et al., 2024; Falk & Dierking, 2016). Accordingly, museums, heritage sites, and community-based settings are increasingly discussed as valuable spaces for culturally connected mathematics learning.

This study presents ethnomathematical examples illustrating the systematic integration of out-of-school learning environments into mathematics education. The study contributes theoretically, pedagogically, and culturally to the literature. The chapter contributes to the literature in three dimensions:

- Theoretical dimension: The concepts of out-of-school learning and ethnomathematics are discussed within theoretical frameworks, and their interrelationship is explained.
- Application dimension: Teachers are provided with concrete examples of activities to connect middle school mathematics with museums, cultural heritage, and their contexts.
- Student dimension: It is believed that the presented activity examples will contribute to students' ability to relate mathematics to their environment, strengthening their motivation and sense of belonging.

The chapter proposes a conceptual and practice-oriented framework for integrating ethnomathematics into out-of-school mathematics learning environments through context-based and curriculum-aligned activity design.

The aim of this study is to introduce the theoretical frameworks of out-of-school learning and ethnomathematics, to design ethnomathematics-based mathematical tasks that can be carried out in out-of-school learning environments (museums, cultural heritage sites, historical artifacts, etc.), and to reveal and discuss how these tasks can be related to middle school mathematics within the framework of the theory. In line with this general aim, the sub-objectives can be expressed as follows:

- To identify implicit mathematical elements in objects, structures, and applications within museum and heritage site contexts, based on ethnomathematical and out-of-school learning approaches.
- To develop ethnomathematical-based out-of-school learning activity designs that match these elements with mathematics curriculum learning areas (Numbers and Operations, Algebra, Geometry and Measurement, Data Processing, Probability).
- To offer suggestions for teacher training, curriculum development, and in-class/out-of-class applications based on the designed tasks.

Despite the growing body of research on ethnomathematics and out-of-school learning, studies integrating ethnomathematics with museum pedagogy, cultural heritage environments, and curriculum-aligned mathematics task design remain limited. Recent studies also indicate that teachers often lack sufficient pedagogical guidance for designing mathematically rich out-of-school learning activities and for effectively connecting museum environments with mathematics learning outcomes (Casi & Sabena, 2024; Pattison et al., 2017). Although the literature highlights the educational potential of ethnomathematics and out-of-school learning separately, several important gaps remain in mathematics education research. First, there is limited guidance on designing mathematics-specific out-of-school learning activities grounded in ethnomathematical perspectives. Second, the integration of ethnomathematics with museum and cultural heritage pedagogy remains underdeveloped, particularly in curriculum-aligned middle school mathematics contexts. In addition, recent systematic reviews in mathematics education emphasize the need for more studies focusing on the design of effective out-of-school learning environments and mathematics activities rather than only describing existing practices (Eğe et al., 2025). Therefore, the present chapter aims to contribute to the literature by proposing context-based and curriculum-aligned activity examples that integrate ethnomathematics, museums, heritage sites, and middle school mathematics learning areas.

This chapter contributes to the literature by integrating ethnomathematics and out-of-school learning within a common framework for mathematics education and by presenting curriculum-aligned activity examples designed for museums and heritage sites. Unlike many ethnomathematics studies focusing mainly on geometry (Hendriyanto et al., 2023; Kyeremeh et al., 2023), the chapter addresses multiple middle school mathematics learning areas, including numbers and operations, algebra, geometry and measurement, data processing, and probability. Through examples grounded in Kastamonu's cultural heritage, the chapter offers a conceptual and practice-oriented framework for culturally connected mathematics learning in out-of-school settings.

THEORETICAL BACKGROUND

Ethnomathematics

Numerous studies have shown that the development of mathematics is not limited to a single culture but is shaped by contributions from different societies and cannot be considered independently of cultural contexts (Bishop, 1994; Muhtadi et al., 2017). Accordingly, mathematics is viewed not merely as a collection of abstract rules and symbols, but as a human activity that develops in line with social needs, values, and daily life practices (Chaudhary et al., 2023). Examining the historical development of mathematical knowledge reveals that different cultures have actively contributed to this process, and that there is a reciprocal, dynamic, and reinforcing relationship between mathematics and culture (Arı, 2022). This understanding necessitates considering cultural contexts in mathematics education and increasingly highlights the importance of the ethnomathematics approach. Indeed, the increase in studies examining the mathematical accumulations of different cultures in recent years supports this trend (Arı, 2022; Gürefe, 2025; Rosa & Orey, 2016).

Ethnomathematics is defined as a field that examines the relationship between mathematics and culture (D'Ambrosio, 1985; 2001), aiming to understand mathematics through the number systems, measurement methods, spatial arrangements, patterns, and symbolic expressions of different cultures (D'Ambrosio, 1985; D'Ambrosio & D'Ambrosio, 2013). This approach does not limit mathematics to the form taught in schools; it also encompasses mathematical thinking within the practices of artisans, architects, local communities, and everyday life (Rosa & Orey, 2016). Ethnomathematics is also considered a research area that examines how mathematical knowledge is produced, transmitted, and used in daily life (Umbara et al., 2021). It also aims to make mathematics more understandable, accessible, and meaningful by supporting students in establishing meaningful connections between academic mathematics and their daily life experiences (Rosa & Orey, 2016; Umbara et al., 2021).

The ethnomathematical approach offers a broad perspective by examining mathematics within cultural contexts, encompassing mathematical ideas, processes, and practices developed by different societies (Rosa & Orey, 2016, 2020). In this approach, mathematics is viewed as a universal yet non-uniform field of knowledge, manifesting in different forms across cultures. Students begin to perceive mathematics not merely as a theoretical discipline, but also as a cultural product and human activity (Rosa & Orey, 2016, 2020). The inclusion of cultural elements in teaching processes increases student participation, enables a deeper understanding of mathematical concepts, and supports the retention of learning (El Bedewy et al., 2024; Kyeremeh et al., 2023). Ethnomathematics-based applications positively influence students' attitudes towards mathematics, make learning more engaging, and enable them to connect mathematics with daily life (Bishop, 1994; Prahmana & D'Ambrosio, 2020; Rosa & Orey, 2020; Umbara et al., 2021). Teaching without a cultural context, however, can lead to mathematics being perceived as abstract, difficult, and detached from daily life (Dumlu & Alkaş Ulusoy, 2025). Ethnomathematics also fosters cultural awareness, supports critical thinking, problem-solving, and the acquisition of different perspectives (Rosa & Orey, 2016).

Many cultural artifacts encountered in daily life are rooted in mathematical structures. Architectural structures, handicrafts, ornaments, carpet and rug motifs, and embroidery are not merely aesthetically pleasing; they are also structures containing specific information, meaning, and mathematical orders (D'Ambrosio, 1985; 2001; Gerdes, 1994; Kalinec-Craig et al., 2019). Examining these products clearly reveals that mathematics is not merely a theoretical discipline, but rather embodied in cultural and historical contexts. Ethnomathematics also demonstrates that mathematics exists in every culture, regardless of time and place (Arı, 2022; Küçük, 2013). Ethnomathematics studies examine practices from different cultures. Motifs, patterns, architectural structures, and handicrafts belonging to local culture allow students to explore mathematical concepts such as symmetry, patterns, counting, measurement, and spatial relationships (Muhtadi et al., 2017; Rosa & Orey, 2020; Simamora & Saragih, 2019). Prahmana and D'Ambrosio (2020) examined transformation geometry through batik motifs and emphasized that such cultural structures are an effective tool for mathematics teaching. Suryaningsih et al. (2023) analyzed geometric decorations in architectural structures and demonstrated that these structures can serve as teaching materials. Traditional handicrafts such as weaving, basket weaving, pottery, and textiles are among the important areas where mathematical patterns and geometric structures are studied in Europe, Anatolia, Indonesia, and many other cultures (Çenberci & Horzum, 2023). Sevgi (2019) stated that ethnomathematics-based activities enable students to notice mathematics in cultural elements and increase their cultural awareness, while Şahin (2025) stated that students more easily understand geometric concepts in handicraft-based teaching processes.

Ethnomathematics is considered a powerful approach that enriches mathematics teaching with cultural contexts, enabling students to have a more holistic learning experience in both mathematical and cultural terms. An examination of the ethnomathematics literature reveals various thematic areas in which mathematical concepts are addressed across different cultural contexts (Şahin, 2025).

- Traditional Handicrafts and Artistic Productions: Traditional handicrafts such as weaving, embroidery, ceramics, and basket weaving are areas where mathematical structures are intensely observed (Gerdes, 1994). These products incorporate mathematical concepts such as symmetry, pattern, proportion, and geometric transformations. Such cultural products not only serve aesthetic purposes but also contain specific meanings and messages. Therefore, these products are considered a cultural reflection of mathematical thinking (D'Ambrosio, 2001).

- Architectural Structures and Spatial Arrangements: Architectural structures are one of the important cultural areas where mathematical concepts can be concretely observed. Symmetry, proportion, geometry, and spatial arrangements play a significant role in building designs (Bellone et al., 2017). When architectural structures from different cultures are examined, it is seen that mathematical thinking is effectively used in spatial arrangement processes (Chaudhary et al., 2023). This highlights the relationship between mathematics and daily life practices.

- Patterns, Motifs, and Textile Products: Carpet, rug, and fabric patterns are an important theme frequently addressed in ethnomathematics studies. Mathematical concepts such as geometric shapes, symmetry, transformation, and patterns can be clearly observed in these patterns. At the same time, these motifs carry cultural meanings, reflecting individuals' emotions, thoughts, and life experiences (Dumlu & Alkaş Ulusoy, 2025; Küçük, 2013, 2014). Motifs may therefore function as both mathematical and cultural learning resources (Günay & Takunyacı, 2023).

- Games, Stories, and Daily Life Applications: Traditional games, stories, and daily life practices are also considered within the scope of ethnomathematics. These contexts include mathematical processes such as problem-solving, counting, measuring, and strategy development (Nasir, 2002). Situations encountered in daily life, in particular, allow for the natural use of mathematical concepts and help make learning more contextual (Umbara et al., 2021).

- Local Measurement Systems and Traditional Knowledge: Local measurement systems developed by different societies reveal culturally specific aspects of mathematical knowledge. Although these systems differ from standard units of measurement, they are functional and meaningful within their own contexts (Chaudhary et al., 2023). Such practices contribute to students' understanding that mathematics is both universal and culturally diverse (Barton, 1996).

Contemporary ethnomathematics literature increasingly moves beyond defining mathematics as culturally situated knowledge and focuses on how cultural practices can be transformed into teaching and learning resources. Recent systematic reviews indicate that ethnomathematics research commonly examines cultural artifacts, activities, architecture, games, weaving, and local practices as contexts for mathematics learning (Batiibwe, 2024; Hendriyanto et al., 2023; Rosa & Orey, 2020). At the same time, studies focusing on batik motifs, museum ornaments, traditional houses, and architectural structures demonstrate that cultural artifacts may embody mathematical concepts such as symmetry, transformation, congruence, proportional reasoning, and geometric forms (Fatkhurohman et al., 2021; Mariamah et al., 2021; Wahyudi et al., 2021; Sagala & Hasanah, 2023). Similarly, studies on ethnomathematics-based teaching materials emphasize that cultural contexts may support meaningful and contextual mathematics learning experiences (Franita et al., 2024; Wulandari et al., 2024). These studies support the idea that ethnomathematics may help connect school mathematics with students' cultural environments, lived experiences, and meaningful real-life contexts.

In this chapter, activities were considered ethnomathematical not merely because they included cultural objects, but because they aimed to reveal mathematical thinking embedded within cultural practices, artifacts, and local knowledge systems. Therefore, cultural elements were used not only as contextual examples but also as sources of mathematical reasoning, structure, and meaning.

Out-of-School Learning

Learning may occur through formal, informal, and non-formal environments depending on different educational contexts and experiences. Formal learning encompasses structured educational processes within a specific curriculum, while informal learning refers to learning that occurs spontaneously in an individual's daily life (Eshach, 2007). Formal education is structured and usually takes place in schools; informal education is unstructured, and learning occurs spontaneously in places like homes, parks, and gardens within our daily routines. Non-formal education, on the other hand, is a structured form of education that teaches subjects learned in formal education in pre-planned settings (such as museums, zoos, planetariums, or aquariums) (Eshach, 2007). Heimlich (1993) defines non-formal education as planned learning activities carried out in non-school environments. Out-of-school learning therefore emerges as a dynamic learning space located at the intersection of these three types of learning. These environments enable students to access knowledge directly through experience, thereby aligning closely with the constructivist learning approach and supporting the formal learning process (Gerber et al., 2001; Melber & Abraham, 1999). Similarities and differences between formal, informal, and non-formal learning are presented in Table 1.

Table 1. Characteristics of formal, informal, and non-formal learning (Eshach, 2007)

Formal	Non-formal	Informal
Usually at school	At institution out-of-school	Everywhere
May be repressive	Usually supportive	Supportive
Structured	Structured	Unstructured
Usually prearranged	Usually prearranged	Spontaneous
Motivation is typically more extrinsic	Motivation may be extrinsic but it is typically more intrinsic	Motivation is mainly intrinsic

continued on following page

Table 1. Continued

Formal	Non-formal	Informal
Compulsory	Usually voluntary	Voluntary
Teacher-led	May be guide or teacher-led	Usually learner-led
Learning is evaluated	Learning is usually not evaluated	Learning is not evaluated
Sequential	Typically non-sequential	Non-sequential

The National Council of Teachers of Mathematics (NCTM, 2000) emphasizes that every child must develop a strong mathematical foundation, especially at an early age. Therefore, it is important to provide all students with high-quality learning opportunities through formal and informal experiences at an early age (Cooper, 2011). In terms of mathematics education, the field of out-of-school learning appears to have developed later than other educational fields, but it has gained increasing importance in recent years (Kuş, 2024; Nemirovsky et al., 2017). The rise of mathematics museums and interactive exhibitions, in particular, indicates a growing interest in learning mathematics in out-of-school learning environments (Duatepe-Paksu et al., 2022). However, it is noted that teachers lack sufficient experience in these environments (Kayhan Altay & Yetkin Özdemir, 2023; Nemirovsky et al., 2017). In their studies, Nicol (2002) and Johnson and Chandler (2009) found that prospective teachers and teachers experience various difficulties in relating out-of-school learning environments to mathematics. Specifically, Nicol (2002) noted that prospective teachers struggled to connect mathematical content with out-of-school learning, while Johnson and Chandler (2009) emphasized that although prospective teachers positively evaluated their experiences with museums and heritage sites and indicated they would use them in the future, they lacked sufficient skills to transfer these experiences to classroom teaching and faced difficulties in implementation. Similarly, a study by Caniglia (2003) found that teachers attempted to connect mathematics to culture by visiting heritage sites, such as bridges and museums, but this process required systematic guidance. Furthermore, it was stressed that studies on how to design out-of-school learning environments in mathematics education are limited and that more research is needed in this area (Eğe et al., 2025; El Bedewy et al., 2024; Seligmann, 2014). This situation shows that, despite the potential of out-of-school learning environments, there is a significant gap in teacher training (Johnson & Chandler, 2009; Kayhan Altay & Yetkin Özdemir, 2023; Nicol, 2002), and this is an important dimension that needs to be developed.

Previous studies emphasize that the education of prospective teachers can be supported by universities offering courses that help them engage in out-of-school learning (Anderson et al., 2006; Olson et al., 2001). In Turkey, the inclusion of the elective course "Out-of-School Learning Environments in Mathematics Teaching" in the curriculum by the Council of Higher Education (CoHE, 2018) also draws attention to the importance of teacher training. Furthermore, the 2023 Vision Document of the Ministry of National Education (MoNE) emphasizes the importance of integrating out-of-school learning environments into education and training (MoNE, 2018). Out-of-school learning has received increasing attention in recent years. Mathematics education in out-of-school learning environments provides rich contexts for school mathematics, allowing educators to broaden their vision, consider where and how learning can take place outside of school, and engage students in mathematical applications in cultural and realistic contexts (Arcavi, 2002). Anderson et al. (2002) state that these environments allow students to create meaning through objects. Such environments have been shown to positively affect students' mathematical achievement, attitudes, and anxiety levels (Abdioğlu et al., 2020; Çağlar et al., 2018; Duatepe-Paksu et al., 2022; Grothérus & Fägerstam, 2017; Otte et al., 2019). Cooper (2011) states that museum environments offer students the opportunity to explore mathematical concepts such as symmetry, proportion, patterns, and geometric structures in natural contexts. Pumpian et al. (2006) emphasize that mathematics is not only about numbers and operations, but also includes elements such as aesthetic proportions and symmetry as part of mathematical thinking. From this perspective, out-of-school learning environments help students view mathematics from a broader perspective.

Out-of-school learning activities gain a more systematic structure when attention is paid to the content-method-environment relationship (Şen, 2019). The content component answers the question "What?" and expresses the mathematical content to be taught. In the international mathematics education literature, this content is structured around the learning areas of numbers and operations, algebra, geometry and measurement, data processing, and probability (NCTM, 2000). In Turkey, the themes in the middle school mathematics curriculum, updated in 2024, vary from 5th to 8th grade and are addressed as "*numbers and quantities, algebraic thinking with operations, algebraic thinking with operations and changes, algebraic thinking and changes, geometric shapes, geometric quantities, transformation, statistical research process, and from data to probability*" (MoNE, 2024). Furthermore, the curriculum states that out-of-school learning should be included and emphasizes the connection of mathematical topics with daily life and cultural contexts. "*Students may be asked to create original geometric mosaic designs inspired by cultural elements of various cities (examining artifacts in heritage sites and mosaic museums in Turkey)*", and "*They may be made to realize that the patterns of carpets and rugs belonging to our culture in our country vary according to regions. Studies involving the examination of similarities and differences in the reflection transformation in these patterns may be conducted*", are some of the statements included in the curriculum (MoNE, 2024).

The method component answers the question "How?" and encompasses the strategies, methods, techniques, and materials used in out-of-school learning processes (Şen, 2019). In this context, strategies such as experiential learning, problem-based learning, project-based learning, and discovery-based learning stand out (Küçükoğlu & Kaya, 2020). Kelton (2021) states that teachers use both structured and unstructured interaction strategies during mathematics-related museum visits, which enrich students' learning experiences. Furthermore, it is noted that students' curiosity and interactions developed during free exploration have significant effects on learning. Methods and techniques used within the scope of out-of-school learning include "*Field trips, exhibitions, brainstorming, Who am I? What am I?, Guess-observe-explain*" and similar methods and techniques. In addition, the materials used in out-of-school learning are very diverse, including "*visual materials such as worksheets, workbooks, information sheets, maps; audio guides (headphones), mobile learning tools; websites*" (Küçükoğlu & Kaya, 2020).

The environment component answers the question "Where?" and refers to out-of-school learning environments (Şen, 2019). Museums, heritage sites, science centers, zoos and aquariums, botanical gardens, parks, and daily life areas stand out as important out-of-school learning environments (Cooper, 2011; Eshach, 2007). In the next section, museums and heritage sites discussed in this study will be examined in more detail.

Recent studies on non-formal mathematics learning emphasize that mathematical thinking may emerge outside the classroom through interaction, observation, prediction, collaboration, and engagement with meaningful contexts (Williams-Pierce et al., 2024). Learning across settings perspectives also suggest that meaningful mathematical learning may develop through connections established between classroom experiences and non-formal learning environments such as museums and field trips (Kelton, 2021; Falk & Dierking, 2016). In this sense, non-formal learning environments may support students in recognizing mathematics as part of everyday life rather than as a subject limited to formal classroom procedures. However, the literature also emphasizes that these environments require careful pedagogical structuring and teacher guidance in order to transform students' observations and interactions into mathematically productive learning experiences (Casi & Sabena, 2024).

In addition to providing authentic learning environments, effective out-of-school mathematics learning requires carefully designed tasks that meaningfully connect cultural contexts with mathematical concepts and curriculum outcomes (Boaler, 1993; Arcavi, 2002; Nasir et al., 2008). Task design plays a critical role in transforming museum visits and cultural observations into mathematically rich learning experiences. Recent studies on ethnomathematics-based instructional materials also emphasize the importance of systematic design processes, curriculum alignment, and contextual learning in supporting students' engagement and mathematical understanding (Franita et al., 2024; Wulandari et al., 2024). Furthermore, recent reviews indicate that more research is needed on designing effective out-of-school mathematics learning activities rather than merely describing existing practices (Eğe et al., 2025). Context-based and curriculum-aligned task design approaches have gained increasing attention in mathematics education and museum pedagogy research.

Mathematics Education in Museums and Heritage Sites

Pattison et al. (2017) divide out-of-school mathematics learning into two categories: "mathematics in everyday life" and "designed learning environments," placing museums in the latter. Museums and heritage sites hold special significance as out-of-school learning environments. Museums offer rich content that supports mathematical thinking and enable individuals to learn through their own experiences (Cooper, 2011), making them highly interactive out-of-school learning environments (Eshach, 2007). Falk and Dierking (2000) state that museum visitors want to experience a learning-focused entertainment experience. Cooper (2011) found that museums support mathematical learning by allowing students to observe topics such as patterns, symmetry, and ratios in real-life situations. Casi and Sabena (2024) found that mathematics workshops held in museums improve teachers' pedagogical skills and deepen their understanding of mathematics as a cultural phenomenon. This findings suggest that museums have been successful in promoting a holistic, engaging, and context-oriented approach to mathematics education, and that such practices are promising for future educational applications and research. Aydoğdu et al. (2022) demonstrated that the architectural and artistic structures in the Ephesus, Mersin, and Zeugma museums can be related to mathematical concepts. Examples include symmetrical triangular structures in the Laecanius Bassus Fountain in Ephesus, cylindrical columns in the Ephesus Archaeological Site Museum, translational and symmetrical patterns in textile designs exhibited in the Mersin Archaeological Museum, and geometric motifs in the mosaics of the Gaziantep Zeugma Museum. These examples demonstrate how mathematics becomes visible in cultural and historical contexts through art and architecture.

Museums and heritage sites are considered memory spaces carrying both tangible and intangible cultural heritage; they are out-of-school learning environments that make national/urban identity and local culture visible (Falk & Dierking, 2000). Sagala and Hasanah's (2023) ethnomathematical study presents examples of geometric shapes and geometric transformations from the architecture and objects within a selected museum. However, El Bedewy et al. (2024) state that more research is needed across various environments, such as museums, to enrich students' learning opportunities beyond formal learning and enable them to explore innovative mathematical applications. In this respect, the present chapter contributes to mathematics education literature by presenting ethnomathematics-based activity examples designed for museums and heritage sites.

Museum pedagogy literature emphasizes that museums are not only exhibition spaces but also educational environments shaped by interaction, interpretation, inquiry, and visitors' meaning-making processes. Contemporary museum education approaches particularly highlight object-based learning, experiential learning, and inquiry-oriented activities as important pedagogical approaches for supporting deep and meaningful learning experiences (Isa, 2017). In this perspective, authentic cultural objects, architectural structures, and historical artifacts can function as mediating tools that support observation, questioning, interpretation, and discussion. Research also suggests that museum-based mathematics learning may support meaningful connections between classroom mathematics and open-ended exploratory experiences through interaction and contextual engagement (Kelton, 2021; Falk & Dierking, 2016). Furthermore, recent studies demonstrate that art and history museums may support transdisciplinary and context-based mathematics learning by connecting mathematics with architecture, culture, history, and technology in non-formal learning environments (Casi & Sabena, 2024; El Bedewy et al., 2024).

Taken together, ethnomathematics and out-of-school learning provide a complementary framework for designing culturally meaningful mathematics learning experiences. Museums, heritage sites, and local cultural environments can function as authentic learning spaces where mathematical concepts become visible through cultural artifacts, architectural structures, patterns, and daily practices. Accordingly, activity design plays a critical role in connecting curriculum outcomes with culturally situated mathematical experiences. The conceptual framework presented in Table 2 illustrates the interaction between ethnomathematics, cultural heritage environments, out-of-school learning, and curriculum-aligned task design within middle school mathematics education.

Table 2. Conceptual framework integrating ethnomathematics, out-of-school learning, cultural heritage, and curriculum-aligned task design

Ethnomathematics	**Mathematical ideas in culture, daily practices, artifacts, architecture, and traditional designs**
Cultural Heritage and Museums	Museums, heritage sites, handicrafts, historical environments, local culture
Out-of-school Learning	Authentic learning environments, experiential learning, contextual learning
Task Design and Curriculum Alignment	Curriculum-aligned activities, context-based task design, middle school mathematics outcomes

continued on following page

Table 2. Continued

Ethnomathematics	**Mathematical ideas in culture, daily practices, artifacts, architecture, and traditional designs**
Middle School Mathematics Learning Areas	Numbers and operations, algebra, geometry and measurement, data processing, probability
Meaningful and Context-Based Mathematics Learning	Meaningful learning, engagement, mathematical connections, cultural awareness

Building on the theoretical perspectives discussed above, the following section presents the design framework and activity development process for integrating ethnomathematics and out-of-school learning into middle school mathematics education.

DESIGN FRAMEWORK AND ACTIVITY DEVELOPMENT PROCESS

This chapter is conceptual and design-oriented in nature. It does not report empirical classroom implementation, participant-based data collection, or intervention outcomes. Rather, it aims to develop theoretically grounded and curriculum-aligned activity examples within the framework of ethnomathematics and out-of-school learning. The chapter therefore focuses on the pedagogical design of context-based activity examples rather than empirical validation. Therefore, the following sections present proposed activity examples and design principles rather than implemented instructional interventions.

The activity development process generally included the following stages:

- Potential activity contexts for out-of-school learning were first identified. The proposed activity examples were situated within museums and heritage sites located in Kastamonu province, where rich cultural and mathematical contexts are available for ethnomathematics-based out-of-school learning. These museums include the Archaeology Museum, the City History Museum, and the Architect Vedat Tek Culture and Art Center. Heritage sites include the Kasaba Village Mosque, a UNESCO World Heritage Site, historical mansions, handicraft markets, and complexes. The selected museums and heritage sites were chosen purposefully based on their potential to represent rich cultural, architectural, and mathematical elements that could be associated with middle school mathematics learning areas. Particular attention was paid to environments containing patterns, symmetry, geometric structures, measurement opportunities, numerical representations, and culturally significant artifacts.

- Context analysis: During the context analysis process, the researcher examined objects, artifacts, and structures located in the selected museums and heritage sites and identified possible mathematical concepts that could be taught in these environments. Mathematical ideas embedded in cultural artifacts, architectural structures, handicrafts, and decorative patterns were analyzed in relation to curriculum outcomes and mathematical concepts such as symmetry, transformation, ratio, measurement, patterns, and spatial reasoning.

- Matching with learning areas: The identified elements have been linked to the core learning areas of the middle school mathematics curriculum. The NCTM learning areas of numbers and operations, algebra, geometry and measurement, data processing, and probability served as the basis for these connections. The activities were designed in alignment with middle school mathematics curriculum outcomes. They were also informed by theoretical principles of ethnomathematics, contextual learning, and out-of-school learning in order to enhance pedagogical relevance and cultural meaningfulness. Grade-level appropriateness was also considered during the activity design process. Simpler observation, counting, and classification activities were associated with lower middle school grades, whereas algebraic generalization, proportional reasoning, tessellation analysis, and complex transformation tasks were designed for upper middle school levels.

- Task and activity design: The activities were designed by integrating different middle school mathematics learning areas with the theoretical principles of ethnomathematics and out-of-school learning.

EXAMPLES OF MUSEUMS AND HERITAGE SITES LOCATED IN KASTAMONU

In this section, museums, heritage sites, and historical structures included in the proposed activity examples are briefly introduced. Kastamonu was selected because it offers rich examples of cultural heritage, traditional handicrafts, architectural structures, and museums that support the contextualization of mathematical concepts within cultural learning environments. These environments provide opportunities to relate cultural artifacts, ornaments, spatial structures, and traditional products to different middle school mathematics learning areas within the framework of ethnomathematics and out-of-school learning. Information about the venues was compiled and summarized from the official materials provided by the Kastamonu Governorship Provincial Directorate of Culture and Tourism (n.d.).

Kastamonu Archaeology Museum

Kastamonu Archaeology Museum contains stone, metal, terracotta, and glass artifacts from different historical periods. The museum includes artifacts and ornaments that may support mathematical explorations related to geometric forms, measurement, symmetry, and patterns.

Kastamonu City History Museum

Kastamonu City History Museum includes objects and visual materials reflecting the historical, architectural, social, and cultural characteristics of the city. Traditional carpets, handicrafts, and representations exhibited in the museum may be associated with mathematical activities related to patterns, symmetry, representation, and data interpretation.

Architect Vedat Tek Culture and Art Center

The Architect Vedat Tek Culture and Art Center includes sections related to traditional handicrafts, lace, embroidery, and cultural objects reflecting regional culture. In particular, the motifs and hand-crafted products exhibited in the Lace Museum support mathematical investigations related to patterns, symmetry, geometric transformations, and spatial reasoning.

Mahmut Bey Mosque in Kasaba Village

Mahmut Bey Mosque is an important example of Anatolian wooden mosque architecture and contains rich geometric decorations characterized by symmetry, repetition, and spatial organization. These structures provide opportunities for mathematical explorations related to geometric transformations, patterns, and measurement.

Kastamonu Historic Mansions

Kastamonu's historic mansions are important examples of traditional civil architecture. Their spatial organization, proportional structures, and architectural layouts reflect both aesthetic and functional design principles. These structures allow students to explore concepts related to geometry, proportion, and spatial reasoning.

Historical Complexes

Kastamonu contains numerous historical complexes, madrasas, mosques, and tombs from different historical periods. These structures include architectural elements, ornaments, and geometric decorations that reflect various mathematical concepts within ethnomathematical contexts.

Table 3 provides a brief overview of the museums and heritage sites included in the proposed activity examples, together with their cultural/mathematical features and the primary middle school mathematics learning areas associated with the activities.

Table 3. Museums and heritage sites included in the proposed activities

Venue / Heritage Site	Cultural / Mathematical Features	Related Learning Areas in the Proposed Activities
Kastamonu Archaeology Museum	Historical artifacts, geometric forms, ornaments, measurement practices	Numbers and Operations; Geometry and Measurement; Data Processing
Kastamonu City History Museum	Carpets, patterns, traditional objects, representations	Numbers and Operations; Geometry and Measurement; Data Processing;
Architect Vedat Tek Culture and Art Center	Lace, embroidery, motifs, symmetry, transformations	Numbers and Operations; Algebra; Geometry and Measurement; Data Processing; Probability
Mahmut Bey Mosque	Geometric decorations, symmetry, transformations, spatial structures	Geometry and Measurement; Algebra
Kastamonu Historic Mansions	Spatial organization, proportion, architectural structures	Geometry and Measurement; Numbers and Operations
Historical Complexes	Geometric decorations, architectural layouts, ornaments	Geometry and Measurement

EXAMPLES OF ACTIVITIES FOR LEARNING AREAS

This section presents examples of ethnomathematics-based activities for different learning areas in out-of-school learning settings. All photographs presented in this chapter were taken by the author during visits to publicly accessible museums and heritage sites in Kastamonu. The images are used solely for educational and scholarly purposes.

Numbers and Operations

Examples from the Numbers and Operations learning area are presented in Figure 1.

*Figure 1. Images associated with the learning area of numbers and operations**

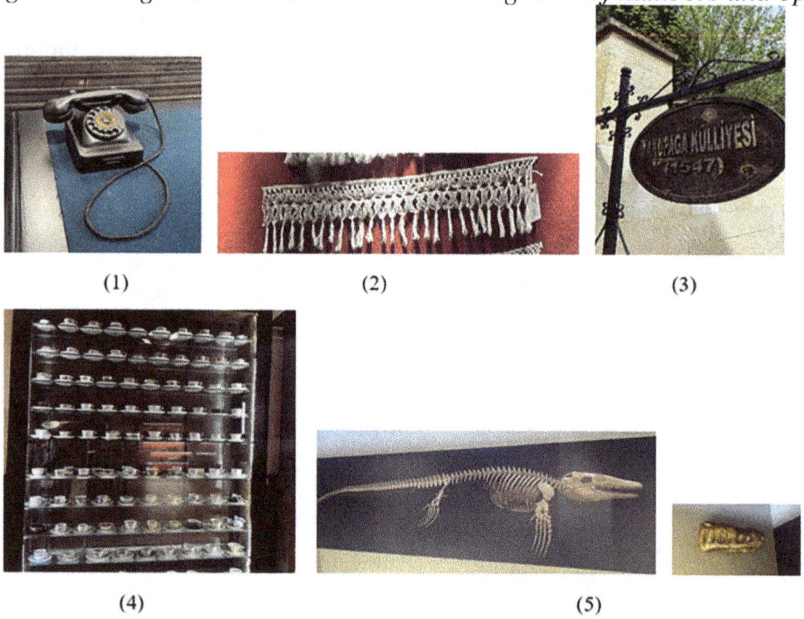

**1: City History Museum - Example of an old-fashioned telephone handset, 2: Architect Vedat Tek Culture and Art Center - Lace Museum - Example of lace, 3: Yakupağa Complex - Plaque showing the year of construction, 4: Architect Vedat Tek Culture and Art Center - Donated Artifacts Section, 5: City History Museum - Example of a fossil*

In Figure 1, using the numbers on the handset telephone shown in image 1, students are asked to find the largest and smallest five-digit numbers that can be formed. The activity helps students understand, compare, and order numbers according to place value. The place value and number comparison activity teaches the concept of numbers through everyday objects, while highlighting the mathematical function of local and historical communication tools for students from an ethnomathematical perspective. In image 2, the number of fringes in the lace is determined, and the number of braids that can be made and the number of remaining fringes are calculated, based on the information that "3 fringes make 1 braid." Through the examination of lace patterns, students may explore division and remainder concepts while analyzing mathematical relationships embedded in traditional handicrafts such as Kastamonu lace motifs. In image 3, students are asked to calculate the structure's age using its construction year. Calculating the age of historical structures may provide opportunities to examine subtraction and time-related calculations through connections with local history and cultural heritage. Problems involving addition, subtraction, and ordering can be created using the number of cups displayed in image number 4. For example, differences can be found by comparing the number of artifacts in a particular section. Questions such as "*How many cups are on each shelf in this section of the museum?*" or "*Find the total number of cups*," can be asked. The arrangement of cultural objects and exhibition layouts may also support contextualized mathematical thinking. Using the Mosasaurs fossil model in image number 5, a ratio can be established between the jaw length and the estimated skull and body length of the Mosasaurs. The actual size can be estimated by scaling. Establishing proportional relationships using fossil models may contribute to the development of ratio, proportion, mathematical modeling, and contextual reasoning skills within historical and natural heritage contexts.

Students may also estimate the total thread length used in a lace sample by examining repeated motif units and applying proportional reasoning. By measuring a single motif and scaling their calculations, students may predict the approximate amount of material required for larger lace designs. Such activities may support estimation, proportional reasoning, and mathematical modeling skills within traditional handicraft contexts. Students may further compare ratios between different body parts and discuss how proportional relationships are used in scientific reconstruction processes. They may also estimate missing dimensions by using scaling strategies and mathematical reasoning. Students may also compare alternative number combinations and justify why certain arrangements produce larger or smaller values. They may further discuss how numbering systems used in historical communication tools reflect mathematical organization in daily life.

These activities aim not only to support counting and operational skills but also to encourage estimation, proportional reasoning, mathematical modeling, and interpretation within cultural and historical contexts.

Geometry and Measurement

Some examples related to the Geometry and Measurement learning area are presented in Figure 2.

*Figure 2. Images related to the geometry and measurement learning area - 1**

(1) (2) (3)

(4) (5) (6) (7)

**1: Architect Vedat Tek Culture and Art Center - Exterior view, 2: Architect Vedat Tek Culture and Art Center - Lace Museum - Lace sample, 3: Kasaba Village Mahmut Bey Mosque - Ceiling section, 4: Kasaba Village Mahmut Bey Mosque - Interior view, 5: City History Museum - Model mansion, 6: Atabey Gazi Mosque - Ceiling decoration, 7: Atabey Gazi Mosque – Door*

The architectural structures and lace pattern given in Figure 2 are examined. Similar and identical shapes are identified. The geometric shapes (triangles, squares, rectangles, circles, etc.) present in the structures are identified. Examining architectural structures and lace motifs helps students recognize geometric shapes, similarity, and symmetry within local cultural products and traditional handicrafts. Some examples related to the Geometry and Measurement learning area are presented in Figure 3. Students may additionally investigate whether decorative floor and ceiling patterns create tessellations and explain which geometric conditions allow tessellation. They may classify motifs according to their symmetry properties and justify their reasoning mathematically.

*Figure 3. Images related to the geometry and measurement learning area - 2**

**1-2: Archaeology Museum - Example of an ancient tombstone, 3: Archaeology Museum - Example of a floor, 4: City History Museum - Example of an antique stove, 5: Kasaba Village Mahmut Bey Mosque - Examples of columns, 6: Kasaba Village Mahmut Bey Mosque - Gate, 7: Architect Vedat Tek Culture and Art Center - Garden - Examples of columns, 8: Archaeology Museum - Garden - Examples of columns, 9: City History Museum - Example of a carpet, 10: Architect Vedat Tek Culture and Art Center - Example of regional embroidery, 11: Nasrullah Kadı Complex*

The translational, reflection, and rotational transformations in the architectural structures shown in Figure 3, as well as in examples such as antique stoves, carpets, and regional embroidery, are analyzed. Alternate interior and exterior angles, as well as corresponding angles, are determined. Parallel structures are examined. Geometric solids are identified. Students are asked to identify the basic elements of a cylinder, such as the base, surface, and axis. Estimated surface area and volume calculations can be performed. It is noted that domes are hemispherical. Investigating carpets, lace, weaving, architectural decorations, and other cultural products may provide opportunities to explore geometric transformations, angles, geometric solids, and spatial relationships within cultural learning environments. Students may also analyze incomplete or damaged motifs and predict missing parts by using symmetry and transformation principles. Such activities may strengthen spatial reasoning, visualization, and mathematical justification skills.

These activities aim to support not only the recognition of geometric forms but also classification, justification, spatial reasoning, transformation analysis, and mathematical explanation within culturally meaningful contexts.

Algebra

Examples related to the algebra learning area are presented in Figure 4.

*Figure 4. Images related to the algebra learning area**

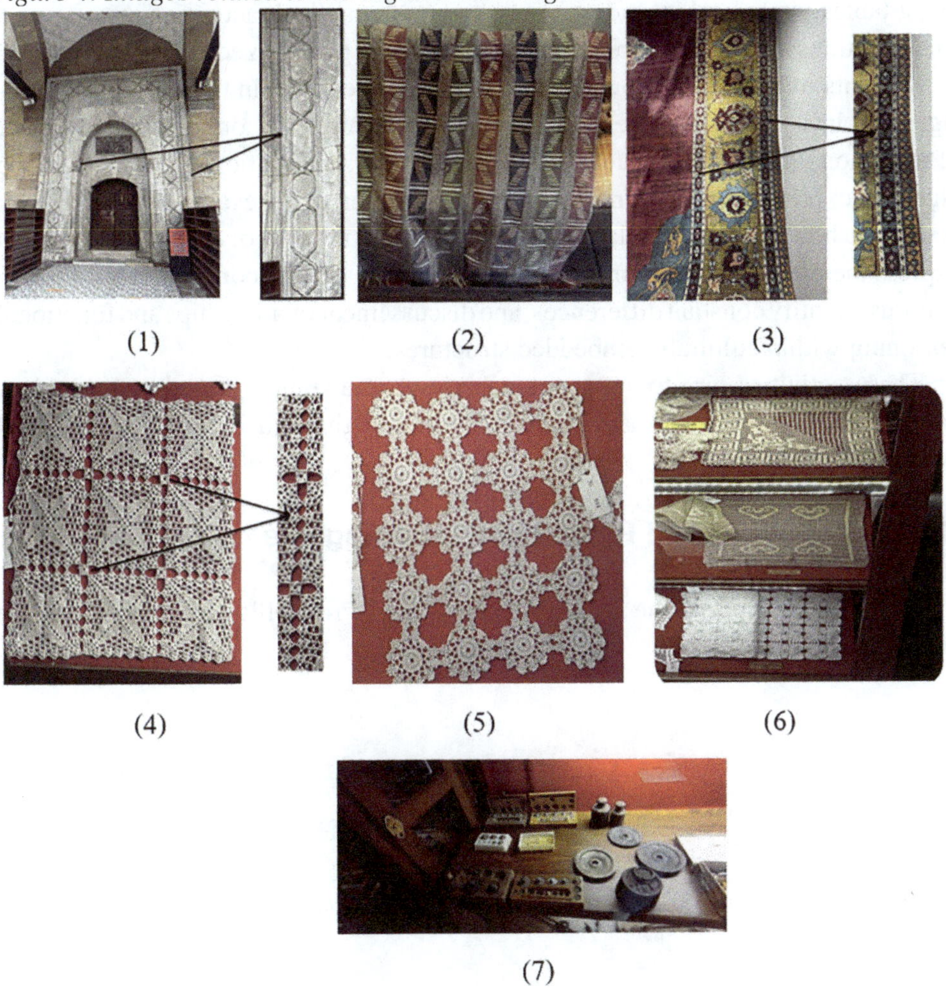

1: İsmail Bey Mosque - Gate, 2: Architect Vedat Tek Culture and Art Center - Handicraft example, 3: City History Museum - Carpet example, 4-5-6: Architect Vedat Tek Culture and Art Center - Lace Museum - Lace examples, 7: Architect Vedat Tek Culture and Art Center - Exhibition Section

The patterns in architectural structures and handicrafts, such as carpets, lace, and weaving, shown in Figure 4, are explored. Exploring repeating motifs and traditional designs supports students' recognition of mathematical patterns and relationships within cultural contexts. Comparisons between different weights may also support discussions related to equality and inequality. The activity may also encourage students to relate algebraic thinking to both historical and contemporary cultural practices.

Students may also be asked to generalize repeating structures and express relationships algebraically. For example, they may investigate how the number of motifs changes according to the number of repetitions and represent these relationships using tables, verbal rules, and algebraic expressions. Such activities may support students' transition from visual pattern recognition to generalized algebraic thinking.

Students may examine how motifs grow and repeat within these traditional designs and investigate how the number of shapes changes according to the number of repetitions. They may create tables, identify recursive relationships, and develop algebraic expressions representing the observed patterns. For example, students may determine how many triangular motifs appear in the nth row of a woven design and express the relationship algebraically. Students may further compare motif growth patterns, identify constant differences, and discuss linear relationships and functional reasoning within culturally embedded structures.

These activities aim to support algebraic generalization, functional thinking, pattern recognition, and mathematical reasoning through culturally embedded designs and structures.

Data Processing and Probability Learning Areas

Examples related to the Data Processing and Probability learning areas are presented in Figure 5.

*Figure 5. Images related to the learning areas of data processing and probability**

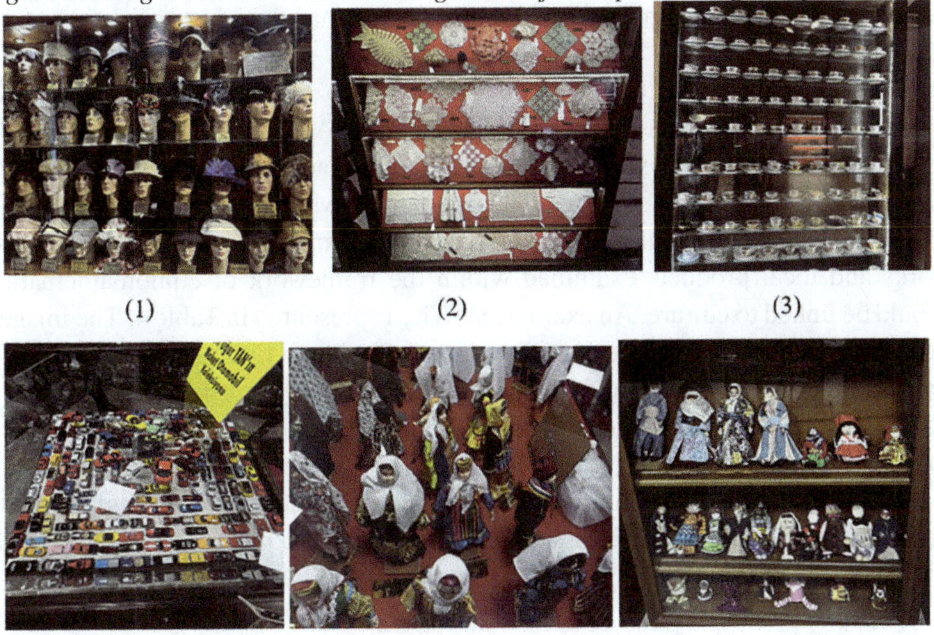

(1) (2) (3)

(4) (5) (6)

1: Architect Vedat Tek Culture and Art Center - Hat Museum - Hat samples, 2: Architect Vedat Tek Culture and Art Center - Lace Museum - Lace samples, 3: Architect Vedat Tek Culture and Art Center - Donated Works Section, 4-5-6: Architect Vedat Tek Culture and Art Center - Doll Museum

Activities such as finding the number of hats, lace, and cups in images 1, 2, and 3 in Figure 5, determining how many hats/lace/cups are in each row, calculating the mean, mode, and median values, and presenting the data with frequency tables and appropriate graphs can be performed. Organizing and interpreting data obtained from cultural objects enables students to examine mathematical relationships within local cultural contexts. The car and regional dolls in images 4, 5, and 6 can be related to probability. Activities such as calculating the probability that the chosen car is red and determining which doll (girl or boy) is more likely to be chosen can be performed. The lace samples and regional dolls encourage connections between mathematics, daily life, and local cultural traditions. In addition, the collections displayed in the museum may provide opportunities to discuss cultural diversity and representation through quantitative data. Beyond simple counting activities, students may classify artifacts according to historical period, geometric structure, or function; compare distributions using percentages, frequency tables, graphs, and statistical measures; interpret trends; and discuss possible cultural and historical explanations for the observed patterns and distributions.

These activities aim to support data interpretation, statistical reasoning, probabilistic thinking, and mathematical justification rather than focusing only on counting and computation.

The examples presented can be expanded upon by considering the NCTM Framework and the topics included in the curricula. Activities should be designed within the scope of out-of-school learning, taking into account the questions "What? (Content), How (Method), Where? (Environment)". Moreover, pay attention to the planning, implementation, and evaluation phases. Furthermore, the architectural works and local products examined within the framework of ethnomathematics should be linked to culture. An example activity is presented in Table 4. The images in Figures 1-5 can be used during this activity. Similar activities may be designed for different learning areas and cultural environments by considering curriculum objectives and local cultural contexts. The following activity example is presented as a proposed design model illustrating how ethnomathematics and out-of-school learning principles may be integrated into mathematics education through context-based and curriculum-aligned task design. The activity is intended as a conceptual and pedagogical example rather than an empirically implemented intervention.

Table 4. Proposed activity design example within the framework of ethnomathematics and out-of-school learning

Activity name	**Exploring Transformations in Cultural Motifs**
Learning area	Geometry and Measurement
Objectives and outcomes	This activity aims to enable students to: - Define basic geometric transformations in the plane (reflection, translation, rotation), - Recognize how these transformations are used in cultural motifs, - Relate mathematical concepts to real-life and cultural contexts, - Recognize the mathematical ways of thinking found in the motifs, patterns, and architectural structures of their culture, and understand that mathematics is a cultural product.
Mathematical reasoning focus	Students are expected to identify, justify, generalize, compare, and explain mathematical relationships embedded in cultural motifs, patterns, and architectural structures.
Out-of-school learning areas	Archaeology Museum, Architect Vedat Tek Culture and Art Center, City History Museum, Kasaba Village Mahmut Bey Mosque (These areas offer unique examples reflecting the aesthetic sensibilities and mathematical knowledge of their respective eras and cultures, thus enabling ethnomathematical inquiry.)
Strategies used	Experiential learning, problem-based learning, project-based learning
Methods and techniques used	Trip, exhibition, brainstorming, what am I, guess-observe and explain, what do we know-what do we want to know-what have we learned

continued on following page

Table 4. Continued

Activity name	Exploring Transformations in Cultural Motifs
Materials used	Worksheets, brochures, flashcards, maps, symmetry mirrors, clear transparencies, greaseproof paper, mobile devices for photography, websites, audio guides, grading rubric.
Before the visit: Teacher's preparation	-The teacher may examine the official websites of the places to be visited to provide preliminary information for the students. -The teacher may collect brochures, digital catalogs, and information cards for museums and heritage sites in advance. -The teacher may prepare worksheets and problem-based scenarios to guide the activity process. -The teacher may also plan transportation, parental permissions, safety precautions, accessibility arrangements, group organization, and museum regulations before the visit. -The teacher may design an analytical rubric for evaluation within the scope of project-based learning. -The teacher may prepare short informational content about the cultural contexts to which the motifs and decorations in the selected places belong.
Before the visit: Student preparation	-Students may examine museum and heritage site websites to gain preliminary information. -Students may conduct brief preliminary research on the question: "Where can geometric transformations be observed in daily life?" - Students may also explore examples related to their local culture and environment.

continued on following page

Table 4. Continued

Activity name	Exploring Transformations in Cultural Motifs
During and after the visit-Implementation	-Students may be divided into groups and may complete the first two parts of a worksheet containing the questions "What do we know? - What do we want to know? - What have we learned?". Their existing knowledge and desire to learn about the activity may be identified. - Before visiting out-of-school learning environments, students are asked to predict what transformations might occur in the places and works of art they will see and to write their predictions in the "Predict-Observe-Explain" worksheet's "Predict" section. - Brainstorming sessions may be held to discuss the following questions: "What mathematical properties make a pattern appear regular?", "Why is symmetry important in daily life?", "How might the motifs used in historical buildings and handicrafts in your environment reflect the aesthetic and cultural understanding of the periods to which they belong?" - Using a map of the locations, students may work in groups under the teacher's guidance in designated areas and may conduct observations. They may observe architectural structures, decorations, and handicrafts (e.g., stonework, wood carving, weaving, and lace) from different periods. They may gain information by reading the information cards in front of the works. They may record their observations on worksheets containing sections on "Transformation Detection, Drawing, and Interpretation," and may take photographs. Where audio guides are available, they may listen to them to learn more. They may also write their observations in the "Observe" section of the "Guess-Observe-Explain" worksheet. - After the field trip and return to the classroom, the teacher may ask students questions related to the visited location within the framework of problem-based learning: "Can the pattern on this floor be continued indefinitely? If so, with which transformation? Continue the pattern on paper," "What transformations are present in the pattern on the antique stove?", "Where is the axis of symmetry of this motif? Can there be more than one axis?" etc. - The "What Am I?" game may be played. Students may be asked to identify which transformation a given statement refers to by providing statements explaining reflection, translation, and rotation transformations. - Students may also be asked to analyze the photographs they took, identify the transformations they noticed, and mathematically explain the properties of these transformations, writing their explanations in the "Explain" section of the "Guess-Observe-Explain" worksheet. They may also be expected to interpret the meanings of the motifs and decorations they examine within the city's cultural context and the purposes for which they were used.

continued on following page

Table 4. Continued

Activity name	Exploring Transformations in Cultural Motifs
Possible misconceptions	Students may confuse reflection and rotation transformations, have difficulty identifying symmetry axes in complex motifs, or focus only on decorative aspects without recognizing the underlying mathematical structures. Therefore, teacher guidance and questioning are important for directing attention toward mathematical reasoning.
Post-visit evaluation	- The teacher may ask students to design a term project as a group. Within this scope, students may create a brochure introducing Kastamonu's cultural heritage by relating motifs from different periods they examined in the places they visited to geometric transformations, or they may create their own original designs inspired by these motifs. The products may be expected to not only show mathematical transformations but also explain elements such as the period to which the motifs belong, their cultural meaning, and their purpose of use. Transparent acetates, symmetry mirrors, and greaseproof paper may be used in the product creation process. The teacher may evaluate the products developed by the students using an analytical rubric prepared beforehand. Evaluation criteria may include students' accuracy in identifying mathematical concepts, quality of mathematical explanations, ability to justify transformations and patterns, creativity in design tasks, participation in collaborative work, and connections established between mathematics and cultural contexts.

CONCLUSION AND IMPLICATIONS

The chapter should be interpreted as a conceptual and practice-oriented contribution rather than an empirical intervention study. The proposed activities are intended to provide examples of how ethnomathematics and out-of-school learning may be integrated into mathematics education through context-based task design. Recent literature also emphasizes that non-formal and culturally contextualized mathematics learning environments may support meaningful mathematical engagement by connecting mathematics with learners' everyday experiences, cultural practices, and authentic environments (Williams-Pierce et al., 2024; Rosa & Orey, 2020). The proposed activities illustrate how ethnomathematics and out-of-school learning may be integrated into middle school mathematics education through curriculum-aligned and context-based task design.

When the theoretical framework and activities presented in this study are considered together, ethnomathematics and out-of-school learning emerge as powerful, complementary approaches to mathematics education. Ethnomathematics helps make mathematics more accessible by revealing how mathematical knowledge is produced and used within cultural contexts, thereby removing it from the realm of an abstract discipline (Rosa & Orey, 2016; Purniati et al., 2021). Out-of-school learning, on the other hand, takes learning outside the classroom by offering authentic learning environments where these cultural contexts can be experienced, creating a multi-dimensional learning process (Arcavi, 2002; Eshach, 2007). Together, these approaches create a more holistic framework for mathematics teaching both pedagogically and conceptually.

The association of cultural elements such as museum objects, architectural structures, carpet-rug-lace motifs, and everyday objects with mathematical concepts, as presented in the activities, is consistent with the literature. Indeed, many studies have emphasized that mathematical concepts can be concretized through cultural patterns, architectural structures, and artistic products (Sagala & Hasanah, 2023; Şahin, 2025; Suryaningsih et al., 2023). Geometric arrangements in architectural structures offer rich content in teaching concepts such as angles, parallelism, polygons, and spatial relationships (Aydoğdu et al., 2022; Bellone et al., 2017; Chaudhary et al., 2023). Especially in the Anatolian region, carpet and rug motifs stand out as natural representations of geometric shapes, symmetry, and transformation concepts (Küçük, 2014; Sevgi, 2019), and such cultural products reflect mathematical thinking (Küçük, 2013).

While the ethnomathematics approach is frequently discussed in the literature through the lens of geometry learning, it is emphasized that it has a broad scope and cannot be reduced to a single area of mathematics (Rosa & Orey, 2016). Indeed, previous studies have shown that different mathematical concepts such as ratio and proportion, counting, patterns, data representations, and measurement can also be addressed through cultural contexts (Johnson & Chandler, 2009; Kayhan Altay & Yetkin Özdemir, 2023). However, it is noteworthy that a significant portion of the applications focus on geometry due to its visual and concrete nature, while other learning areas receive comparatively less attention (Kyeremeh et al., 2023). By designing activities across all middle school mathematics learning areas—numbers and operations, algebra, geometry and measurement, data processing, and probability—contributes to a more comprehensive approach to ethnomathematics.

When evaluated within the context of out-of-school learning, museums and heritage sites are seen to offer rich and multifaceted environments for mathematical learning (Cooper, 2011; Falk & Dierking, 2000). Such environments allow students to interact directly with objects and explore mathematical concepts in real-life contexts (Anderson et al., 2002). Furthermore, museums are reported to offer interdisciplinary learning opportunities and to have the potential to connect mathematics with art, history, and culture (El Bedewy et al., 2024). The activities designed in this study integrate architecture, art, and cultural elements with mathematics, supporting this interdisciplinary approach. In addition, recent studies increasingly emphasize that culturally enriched and context-based mathematics learning environments may support students' mathematical meaning-making processes and strengthen connections between mathematics and everyday life (Williams-Pierce et al., 2024; El Bedewy et al., 2024; Rosa & Orey, 2020).

However, the literature indicates that teachers and prospective teachers experience various difficulties in relating out-of-school learning environments to mathematics teaching (Johnson & Chandler, 2009; Kayhan Altay & Yetkin Özdemir, 2023; Nicol, 2002). In particular, the need for guidance on structuring these environments pedagogically is noteworthy. The literature also suggests that the educational value of museums and heritage environments depends largely on how mathematical tasks and learning experiences are pedagogically structured by teachers (Casi & Sabena, 2024). The activity designs presented in this chapter provide concrete examples that may help address this pedagogical need. Furthermore, the inclusion of different types of museum objects, architectural structures, and traditional products in this study contributes to increasing diversity. Considering that ethnomathematics functions as a bridge between cultures (Küçük, 2013) and that it is important for students to become familiar with mathematical practices across cultures, such studies are also valuable in the international literature. In conclusion, the theoretical framework and activity examples presented in this study demonstrate the potential of integrating ethnomathematics and out-of-school learning to create meaningful learning experiences in mathematics education.

Although the chapter presents examples grounded in Kastamonu's cultural heritage, the proposed framework should not be interpreted as universally representative of all cultural contexts. Cultural artifacts, local practices, and ethnomathematical structures may vary significantly across regions and communities. Therefore, the proposed activities should be adapted according to local cultural characteristics, curriculum expectations, and learner needs in different educational settings.

FUTURE RESEARCH DIRECTIONS

Considering the theoretical foundations and practical implications of this study, it is evident that integrating ethnomathematics into out-of-school learning environments offers a promising yet developing field for mathematics education. Although the literature provides important insights into both approaches individually, there remains a need for more comprehensive, systematic research examining their combined impact on students' cognitive, affective, and sociocultural development. In particular, exploring how culturally enriched out-of-school environments, such as museums, heritage sites, and community spaces, contribute to students' mathematical meaning-making processes can yield deeper insights into effective teaching practices. Furthermore, as education increasingly emphasizes contextualized, student-centered, and interdisciplinary learning, expanding research in this area can play a critical role in bridging the gap between theory and practice. Culturally enriched out-of-school learning environments may contribute to students' deeper understanding of mathematical concepts, their ability to relate mathematics to daily life, and their motivation to learn. The following recommendations can be made within the scope of this study:

- Future studies should employ experimental and longitudinal research designs to examine the impact of ethnomathematics-based out-of-school learning environments on students' academic achievement, mathematical thinking, and affective outcomes.
- Comparative research is needed to evaluate the effectiveness of different out-of-school settings (e.g., museums, heritage sites, cultural spaces) in supporting various mathematical learning domains.
- Design-based research should be conducted to develop and validate sustainable instructional models that integrate ethnomathematics into out-of-school learning and strengthen connections between formal and informal learning environments.
- The integration of digital technologies (e.g., virtual museums, augmented reality) into culturally enriched mathematics learning environments should be explored to expand accessibility and engagement.
- Teacher education and professional development programs should be restructured to equip teachers with the pedagogical knowledge and practical skills necessary to implement these approaches effectively.
- More cross-cultural and interdisciplinary studies are recommended to investigate how ethnomathematics-based practices influence students' cultural awareness, identity development, and their ability to relate mathematics to real-life contexts.

REFERENCES

Abdioğlu, C., Yılmaz, E., & Çevik, M. (2020). 8. sınıf öğrencilerine yönelik fen-matematik temalı bilim kampının değerlendirilmesi: "Gelin tanış olalım; fen ve matematiği eğlenceli kılalım!" projesi. [Evaluation of science-math themed science camp for 8th grade students: The project "let's meet; let's make science and mathematics fun!]. *OPUS International Journal of Society Researches, 15*(22), 1031–1058. DOI: 10.26466/opus.635705

Anderson, D., Lawson, B., & Mayer-Smith, J. (2006). Investigating the impact of a practicum experience in an aquarium on pre-service teachers. *Teaching Education, 17*(4), 341–353. DOI: 10.1080/10476210601017527

Anderson, D., Piscitelli, B., Weier, K., Everett, M., & Tayler, C. (2002). Children's museum experiences: Identifying powerful mediators of learning. *Curator (New York, N.Y.), 45*(3), 213–231. DOI: 10.1111/j.2151-6952.2002.tb00057.x

Appelbaum, P., & Stathopoulou, C. (2023). History of Ethnomathematics: Recent developments. In S. R. Sanchez, A. S. Bayés, P. Appelbaum, & G. Aldon (Eds.). (2023). *The role of the history of mathematics in the teaching/learning process: A CIEAEM sourcebook* (pp. 189-207). Springer Nature. DOI: 10.1007/978-3-031-29900-1_8

Arcavi, A. (2002). The everyday and the academic in mathematics. *Journal for Research in Mathematics Education Monograph, 11*, 12–29. DOI: 10.2307/749962

Arı, A. A. (2022). Ethnomodeling reflections from Turkish culture. *The Journal of Kesit Academy*, *8*(33), 200–223. DOI: 10.29228/kesit.66661

Aydoğdu, A. S. E., Aydoğdu, M., & Aktaş, V. (2022). Matematik dersinde bir eğitim aracı olarak sanal müze kullanımı. [Virtual museum use as an educational tool in math class]. *International Journal of Social Science Research, 11*(1), 51–70. https://dergipark.org.tr/tr/pub/ijssresearch

Barton, B. (1996). *Ethnomathematics: Exploring cultural diversity in mathematics* [Doctoral dissertation, University of Auckland]. University of Auckland ResearchSpace.

Batiibwe, M. S. K. (2024). The role of ethnomathematics in mathematics education: A literature review. *Asian Journal for Mathematics Education, 3*(4), 383–405. DOI: 10.1177/27527263241300400

Bellone, T., Fiermonte, F., & Mussio, L. (2017). The common evolution of geometry and architecture from a geodetic point of view. *International Archives of the Photogrammetry, Remote Sensing and Spatial Information Sciences - ISPRS Archives, XLII-5(W1)*, 623–630. DOI: 10.5194/isprs-archives-XLII-5-W1-623-2017

Bishop, A. J. (1994). Cultural conflicts in mathematics education: Developing a research agenda. *For the Learning of Mathematics*, *14*(2), 15–18. https://www.jstor.org/stable/40248109

Boaler, J. (1993). The role of contexts in the mathematics classroom: Do they make mathematics more "real"? *For the Learning of Mathematics*, *13*(2), 12–17. https://www.jstor.org/stable/40248079

Çağlar, S., Ünal, Y., Çalışkan, B., Gürel, R., & Durmaz, B. (2018). İnformal öğrenme ortamlarının ortaokul öğrencilerinin matematik tutumuna etkisi. [The effect of the mathematıcal learnıng areas to the attıtudes towards mathematıcs]. *Mehmet Akif Ersoy Üniversitesi Sosyal Bilimler Enstitüsü Dergisi*, *10*(23), 11–26. DOI: 10.20875/makusobed.357694

Caniglia, J. (2003). Math in the city: Experiencing mathematics through visiting black historic sites. *Journal of Experiential Education*, *26*(2), 70–74. DOI: 10.1177/105382590302600204

Casi, R., & Sabena, C. (2024). Mathematics in art and history museums: An informal mathematics education case for teachers' in-service training. *Education Sciences*, *14*(5), 489. DOI: 10.3390/educsci14050489

Çenberci, S., & Horzum, T. (2023). Matematik öğretmeni adaylarının etnomatematik farkındalıklarının incelenmesi. [Investigation of ethnomathematics awareness of preservice mathematics teachers]. *Ahmet Keleşoğlu Eğitim Fakültesi Dergisi, 5*(3), 713-732. https://dergipark.org.tr/tr/download/article-file/3187289

Chaudhary, P. R., Panthi, D., & Bhatta, C. R. (2023). Ethnomathematics: Geometry and architecture of the tharu's traditional houses at Chakhoura museum. *Journal of Mathematical Problems. Equations and Statistics*, *4*(2), 9–13. DOI: 10.22271/math.2023.v4.i2a.96

Cooper, S. (2011). An exploration of the potential for mathematical experiences in informal learning environments. *Visitor Studies*, *14*(1), 48–65. DOI: 10.1080/10645578.2011.557628

Council of Higher Education (CoHE). (2018). *İlköğretim matematik öğretmenliği lisans programı* [Elementary mathematics education undergraduate program] https://www.yok.gov.tr/Documents/Kurumsal/egitim_ogretim_dairesi/Yeni-Ogretmen-Yetistirme-Lisans Programlari/Ilkogretim_Matematik_Lisans_Programi.pdf

D'Ambrosio, U. (1985). Ethnomathematics and its place in the history and pedagogy of mathematics. *For the Learning of Mathematics*, *5*(1), 44–48. https://www.jstor.org/stable/40247876

D'Ambrosio, U. (2001). General remarks on ethnomathematics. *Zentralblatt für Didaktik der Mathematik, ZDM*, *33*(3), 67–69. DOI: 10.1007/BF02655696

D'Ambrosio, U., & D'Ambrosio, B. S. (2013). The role of ethnomathematics in curricular leadership in mathematicseducation. *Journal of Mathematics Education at Teachers College*, *4*(1), 19–25. DOI: 10.7916/jmetc.v4i1.767

Desai, S., Kurtz, B., & Safi, F. (2021). Mathematics heritage project: An exploration empowering students' mathematical identities. *Journal of Humanistic Mathematics*, *11*(2), 106–122. DOI: 10.5642/jhummath.202102.05

Dierking, L. D., Falk, J. H., Rennie, L., Anderson, D., & Ellenbogen, K. (2003). Policy statement of the "informal science education" ad hoc committee. *Journal of Research in Science Teaching*, *40*(2), 108–111. DOI: 10.1002/tea.10066

Duatepe-Paksu, A., Kazak, S., & Çontay, E. G. (2022). Okul dışı ortamlarda gerçekleştirilen matematik etkinliklerinin değerlendirilmesi: "Her Yer Matematik Projesi". [Evaluation of mathematics activities performed in out-of-school environments: "Mathematics Everywhere" project]. *Muğla Sıtkı Koçman Üniversitesi Eğitim Fakültesi Dergisi*, *9*(2), 541–558. DOI: 10.21666/muefd.1094581

Dumlu, B. Ö., & Alkaş Ulusoy, Ç. (2025). Using an ethnomathematical approach in teaching transformational geometry: The Kilim activity. *Çukurova Üniversitesi Eğitim Fakültesi Dergisi, 54*(2), 670-714. DOI: 10.14812/cuefd.1525565

Eğe, E. F., Temel, H., & Üzel, D. (2025). Trends in research on out-of-school learning environments in mathematics education: A systematic review. *Balıkesir Üniversitesi Fen Bilimleri Enstitüsü Dergisi*, *27*(2), 802–823. DOI: 10.25092/baunfbed.1659445

El Bedewy, S., Lavicza, Z., & Lyublinskaya, I. (2024). STEAM practices connecting mathematics, arts, architecture, culture and history in a non-formal learning environment of a museum. *Journal of Mathematics and the Arts*, *18*(1-2), 101–134. DOI: 10.1080/17513472.2024.2321563

Eshach, H. (2007). Bridging in-school and out-of-school learning: Formal, non-formal, and informal education. *Journal of Science Education and Technology, 16*(2), 171–190. DOI: 10.1007/s10956-006-9027-1

Falk, J. H., & Dierking, L. D. (2000). *Learning from museums: Visitor experiences and the making of meaning*. AltaMira Press.

Falk, J. H., & Dierking, L. D. (2016). *The museum experience revisited*. Routledge., DOI: 10.4324/9781315417851

Fatkhurohman, F., Ayuningtyas, A. D., Noto, M. S., & Widodo, S. A. (2021). Etnomathematics: Exploration of Geblek Renteng Batik in transformation geometry. *Numerical: Jurnal Matematika dan Pendidikan Matematika, 5*(2), 79-90. DOI: 10.25217/numerical.v5i2.1506

Franita, Y., Hendrastuti, Z. R., Pamungkas, M. D., Chasanah, A. N., Rahmawati, F., 'Adna, S. F., & Wicaksono, A. B. (2024). Development of mathematics teaching materials for junior high school based on ethnomathematics. *Formatif: Jurnal Ilmiah Pendidikan MIPA, 14*(2), 413–418. DOI: 10.30998/formatif.v14i2.28002

Gerber, B. L., Marek, E. A., & Cavallo, A. M. L. (2001). Development of an informal learning opportunities assay. *International Journal of Science Education, 23*(6), 569–583. DOI: 10.1080/09500690116959

Gerdes, P. (1994). Reflections on ethnomathematics. *For the Learning of Mathematics, 14*(2), 19–22. https://www.jstor.org/stable/40248110

Grothérus, A., & Fägerstam, E. (2017). Impact of long-term regular outdoor learning in mathematics: The case of John. In T. Dooley & G. Gueudet (Eds.), *Proceedings of the tenth congress of the European Society for Research in Mathematics Education (CERME10)* (pp. 1074–1081). Dublin City University and ERME. https://hal.science/hal-01936003

Günay, K., & Takunyacı, M. (2023). Examining the effect of teaching with ethnomathematics on students' problem-solving skills on transformation geometry. *Journal of Interdisciplinary Education: Theory and Practice*, 5(2), 86–106. DOI: 10.47157/jietp.1384160

Gürefe, N. (2025). Investigation of the effect of geometry teaching with ethnomodeling on students' academic achievement in area measurement. *Ordu Üniversitesi Sosyal Bilimler Enstitüsü Sosyal Bilimler Araştırmaları Dergisi, 15*(ISRIS 2025), 508-535. DOI: 10.48146/odusobiad.1728168

Güreş, H., & Ada, T. (2025). Revealing and examining students' cultural mathematical knowledge and thoughts through an ethnomathematics activity. *Journal of Mathematics and the Arts, 19*(3-4), 136–159. DOI: 10.1080/17513472.2025.2576831

Heimlich, J. E. (1993). *Nonformal environmental education: Toward a working definition* (ERIC/CSMEE Informational Bulletin No. 2). ERIC Clearinghouse for Science, Mathematics, and Environmental Education. https://files.eric.ed.gov/fulltext/ED360154.pdf

Hendriyanto, A., Priatna, N., Juandi, D., Dahlan, J. A., Hidayat, R., Sahara, S., & Muhaimin, L. H. (2023). Learning mathematics using an ethnomathematics approach: A systematic literature review. *Journal of Higher Education Theory and Practice, 23*(7), 59–74. DOI: 10.33423/jhetp.v23i7.6012

Isa, B. (2017). *Museum pedagogy and learning experiences: An investigation into museum education from institutional perspectives* [Doctoral dissertation, RMIT University]. RMIT Research Repository.

Johnson, D. T., & Chandler, F. J. (2009). Pre-service teachers' fieldtrip to the battleship: Teaching and learning mathematics through an ınformal learning experience. *Issues in the Undergraduate Mathematics Preparation of School Teachers, 2*, 1-9. https://eric.ed.gov/?id=EJ859285

Kalinec-Craig, C., Prasad, P. V., & Luna, C. (2019). Geometric transformations and Talavera tiles: A culturally responsive approach to teacher professional development and mathematics teaching. *Journal ofMathematics and the Arts, 13*(1-2), 72–90. DOI: 10.1080/17513472.2018.1504491

Kastamonu Governorship Provincial Directorate of Culture and Tourism. (n.d.). *Kastamonu e-broşür* [Kastamonu e-brochure] [Brochure]. https://kastamonu.ktb.gov.tr/Eklenti/112738,kastamonu-e-brosurpdf.pdf?0

Kayhan Altay, M., & Yetkin Özdemir, E. (2023). The use of museum resources in mathematics education: A study with preservice middle-school mathematics teachers. *Journal of Education for Teaching, 49*(4), 616–629. DOI: 10.1080/02607476.2022.2150534

Kelton, M. L. (2015). Math on the move: a video-based study of school field trips to a mathematics exhibition. (Publication No. 3712516) [Doctoral dissertation, San Diego State University]. ProQuest Dissertations & Theses Global.

Kelton, M. L. (2021). Mathematics learning pathways on a school fieldtrip: Interactional practices linking school and museum activity. *Visitor Studies, 24*(2), 220–242. DOI: 10.1080/10645578.2021.1939984

Khalil, I. A. (2023). Mathematics teachers' attitudes towards integrating ethnic mathematics in teaching mathematical topics: A mixed methods study. *International Journal of Instruction, 16*(4), 1061–1080. DOI: 10.29333/iji.2023.16458a

Kisiel, J. (2013). Introducing future teachers to science beyond the classroom. *Journal of Science Teacher Education, 24*(1), 67–91. DOI: 10.1007/s10972-012-9288-x

Küçük, A. (2013). Ethnomathematics in Anatolia-I (in Turkey): Geometry perception in multiculturalism. *Educational Research Review, 8*(16), 1360. DOI: 10.5897/ERR012.231

Küçük, A. (2014). Ethnomathematics in Anatolia-Turkey: Mathematical thoughts in multiculturalism. *Revista Latinoamericana de Etnomatemática Perspectivas Socioculturales de la Educación Matemática, 7*(1), 171-184. https://www.redalyc.org/pdf/2740/274030901008.pdf

Küçükoğlu, A., & Kaya, H. İ. (2020). *Kuramdan uygulamaya okul dışı öğrenme ortamları* [Out-of-school learning environments: From theory to practice]. Pegem Akademi.

Kuş, M. (2024). Conceptualization of out-of-school mathematics education. *Yaşadıkça Eğitim, 38*(1), 245–257. DOI: 10.33308/26674874.2024381644

Kyeremeh, P., Awuah, F. K., & Dorwu, E. (2023). Integration of ethnomathematics in teaching geometry: A systematic review and bibliometric report. *Journal of Urban Mathematics Education, 16*(2), 68–89. https://journals.tdl.org/jume/index.php/JUME. DOI: 10.21423/jume-v16i2a519

Machaba, F., & Dhlamini, J. (2021). Ethnomathematics as a fundamental teaching approach. In Luneta, K. (Ed.), *Mathematics teaching and professional learning in sub-Sahara Africa* (pp. 59–76). Springer International Publishing., https://link.springer.com/book/10.1007/978-3-030-82723-6 DOI: 10.1007/978-3-030-82723-6_5

Mania, S., & Alam, S. (2021). Teachers' perception toward the use of ethnomathematics approach in teaching math. [IJEMST]. *International Journal of Education in Mathematics, Science, and Technology, 9*(2), 282–298. DOI: 10.46328/ijemst.1551

Mariamah, M., Muslim, M., Amrullah, A., Prayitno, S., & Kusuma, A. B. (2021). Geometry from Lambitu tribe ethnomathematics. *Journal of Physics: Conference Series, 1778*(1), 012015. DOI: 10.1088/1742-6596/1778/1/012015

Melber, L. H., & Abraham, L. M. (1999). Beyond the classroom: Linking with informal education. *Science Activities, 36*(1), 3–4. DOI: 10.1080/00368129909601027

Ministry of National Education (MoNE). (2018). 2023 vizyon belgesi. [2023 vision document]. https://baklan.meb.gov.tr/www/meb-2023-vizyon-belgesi/icerik/330

Ministry of National Education (MoNE). (2024). *Ortaokul matematik dersi öğretim programı (5, 6, 7 ve 8. sınıflar)* [Middle school mathematics curriculum (Grades 5, 6, 7, and 8)] https://tymm.meb.gov.tr/upload/program/2024programmat5678Onayli.pdf

Muhtadi, D., Sukirwan, S., Warsito, W., & Prahmana, R. C. I. (2017). Sundanese ethnomathematics: Mathematical activities in estimating, measuring, and making patterns. *Journal on Mathematics Education*, *8*(2), 185–198. DOI: 10.22342/jme.8.2.4055.185-198

Nasir, N. I. S. (2002). Identity, goals, and learning: Mathematics in cultural practice. *Mathematical Thinking and Learning*, *4*(2-3), 213–247. DOI: 10.1207/S15327833MTL04023_6

Nasir, N. I. S., Hand, V., & Taylor, E. V. (2008). Culture and mathematics in school: Boundaries between "cultural" and "domain" knowledge in the mathematics classroom and beyond. *Review of Research in Education*, *32*(1), 187–240. DOI: 10.3102/0091732X07308962

National Council of Teachers of Mathematics (NCTM). (2000). *Principles and standards for school mathematics*. Author.

Nemirovsky, R., Kelton, M. L., & Civil, M. (2017). Toward a vibrant and socially significant informal mathematics education. In Cai, J. (Ed.), *Compendium for research in mathematics education* (pp. 968–979). National Council of Teachers of Mathematics.

Nicol, C. (2002). Where's the math? Prospective teachers visit the workplace. *Educational Studies in Mathematics*, *50*(3), 289–309. DOI: 10.1023/A:1021211207232

Olson, J. K., Cox-Petersen, A. M., & McComas, W. F. (2001). The inclusion of informal environments in science teacher preparation. *Journal of Science Teacher Education*, *12*(3), 155–173. DOI: 10.1023/A:1016715127697

Otte, C. R., Bølling, M., Elsborg, P., Nielsen, G., & Bentsen, P. (2019). Teaching maths outside the classroom: Does it make a difference? *Educational Research*, *61*(1), 38–52. DOI: 10.1080/00131881.2019.1567270

Pattison, S., Rubin, A., & Wright, T. (2017). *Mathematics in informal learning environments: A summary of the literature* [Literature review report]. Center for Advancement of Informal Science Education (CAISE). https://www.informalscience.org/mathematics-informal-learning-environments-summary-literature

Pradana, K. C., Putra, R. A., & Rahmawati, Y. (2022). Ethnomathematics on traditional culture: A bibliometric mapping analysis and systematic review on database scopus. *International Journal Corner of Educational Research, 1*(1), 1–8. DOI: 10.54012/ijcer.v1i1.61

Prahmana, R. C. I., & D'Ambrosio, U. (2020). Learning geometry and values from patterns: Ethnomathematics on the batik patterns of yogyakarta, indonesia. *Journal on Mathematics Education, 11*(3), 439–456. DOI: 10.22342/jme.11.3.12949.439-456

Pumpian, I., Fisher, D., & Wachowiak, S. (Eds.). (2006). *Challenging the classroom standard through museum-based education: School in the park.* Lawrence Erlbaum Associates. DOI: 10.4324/9781410617279

Purniati, T., Turmudi, T., Juandi, D., & Suhaedi, D. (2021). Ethnomathematics exploration of the masjid raya bandung ornaments in transformation geometry materials. *Journal of Medives: Journal of Mathematics Education IKIP Veteran Semarang, 5*(2), 235–243. DOI: 10.31331/medivesveteran.v5i2.1639

Rosa, M., & Orey, D. C. (2016). State of the Art in Ethnomathematics. In Rosa, M., D'Ambrosio, U., Orey, D. C., Shirley, L., Alangui, W. V., Palhares, P., & Gavarrete, M. E. (Eds.), *Current and Future Perspectives of Ethnomathematics as a Program* (pp. 11–37). Springer International Publishing., DOI: 10.1007/978-3-319-30120-4_3

Rosa, M., & Orey, D. C. (2020). Principles of culturally relevant education in an ethnomathematical perspective. *Revista de Educação Matemática*, (17), 1–24. DOI: 10.37001/remat25269062v17id306

Sagala, S. A., & Hasanah, R. U. (2023). Ethnomathematics exploration at the state museum of North Sumatra. *Mathline: Jurnal Matematika dan Pendidikan Matematika, 8*(1), 123-136. DOI: 10.31943/mathline.v8i1.364

Şahin, H. B. (2025). Türk-İslam kültür ve mimarisi bağlamında geometri öğretimi: Mimar Sinan'ın eserlerindeki geometrik desenler [Teaching geometry within the context of Turkish-Islamic culture and architecture: Geometric patterns in the works of Architect Sinan]. (Publication No. 933490). [Doctoral dissertation, Bursa Uludağ University]. YÖK Tez Merkezi.

Seligmann, T. (2014). Learning museum: A meeting place for pre-service teachers and museums. *Journal of Museum Education, 39*(1), 42–53. DOI: 10.1080/10598650.2014.11510794

Şen, A. İ. (2019). *Okul dışı öğrenme ortamları* [Out-of-school learning environments]. Pegem Akademi. DOI: 10.14527/9786052419700

Sevgi, S. (2019). Dönüşüm geometrisi etkinliklerine kültürel bağlamın dahil edilmesi sonucu oluşan öğrenci yaklaşımları [Student approaches resulting from the inclusion of cultural context in transformation geometry activities]. (Publication No. 593584) [Master's thesis, Dokuz Eylül University]. YÖK Tez Merkezi.

Simamora, R. E., Saragih, S., & Hasratuddin, H. (2019). Improving students' mathematical problem solving ability and self-efficacy through guided discovery learning in local culture context. *International Electronic Journal of Mathematics Education, 14*(1), 61–72. DOI: 10.12973/iejme/3966

Suryaningsih, Y., Fajriah, N., & Kamid, K. (2023). Geometry exploration for the development of ethnomathematics worksheet based on the ornament of Jingah River Jami Mosque. *Jurnal Teori dan. Aplikace Matematiky, 7*(2), 298–309. DOI: 10.31764/jtam.v7i2.12366

Umbara, U., Wahyudin, W., & Prabawanto, S. (2021). Exploring ethnomathematics with ethnomodeling methodological approach: How does Cigugur indigenous people use calculations to determine a good day to build houses? *Eurasia Journal of Mathematics, Science and Technology Education, 17*(2), 1–19. DOI: 10.29333/ejmste/9673

Wahyudi, H., Widodo, S. A., Setiana, D. S., & Irfan, M. (2021). Etnomathematics: Batik activities in tancep batik. *Journal of Medives: Journal of Mathematics Education IKIP Veteran Semarang, 5*(2), 305–315. DOI: 10.31331/medivesveteran.v5i2.1699

Williams-Pierce, C., Simpson, A., & Eronen, L. (2024). Editorial: Mathematical thinking, practices, and processes in non-formal learning environments. *Frontiers in Education, 9*, 1498414. DOI: 10.3389/feduc.2024.1498414

Wulandari, D. U., Mariana, N., Wiryanto, W., & Amien, M. S. (2024). Integration of ethnomathematics teaching materials in mathematics learning in elementary school. *International Journal of Recent Educational Research, 5*(1), 1–15. DOI: 10.46245/ijorer.v5i1.542

KEY TERMS AND DEFINITIONS

Ethnomathematics: A field that examines the relationship between mathematics and culture (D'Ambrosio, 1985; 2001), aiming to understand mathematics through the number systems, measurement methods, spatial arrangements, patterns, and symbolic expressions of different cultures (D'Ambrosio, 1985; D'Ambrosio & D'Ambrosio, 2013).

Out-of-school learning: A learning approach that takes place outside the traditional school setting, combining elements of formal, informal, and non-formal education. It involves planned instructional activities carried out in out-of-school environments such as museums, science centers, and heritage sites, where learners gain knowledge through direct, hands-on experience and real-world interaction.

Chapter 8
Designing Collaborative Mathematics Tasks as a Professional Learning Space for Teachers:
A Case Study From the Perspective of Islamic Culture During Ramadhan

Zetra Hainul Putra
http://orcid.org/0000-0002-0860-1076
Universitas Riau, Indonesia

Yesi Martha Afrillia
http://orcid.org/0000-0001-5750-5999
Universitas Riau, Indonesia

Muryanti Muryanti
http://orcid.org/0000-0003-3500-061X
Universitas Riau, Indonesia

Yahya Gunawan
Universitas Riau, Indonesia

Hartono Tjoe
http://orcid.org/0000-0002-2728-8484
The Pennsylvania State University, USA

DOI: 10.4018/979-8-2600-3485-9.ch008

ABSTRACT

This chapter aims to investigate the use of mathematical tasks related to the calculation of zakat, fasting duration, food distribution (iftar), ketupat, bedug, and others to teach the basic concepts of mathematics, such as fractions, percentages, time management, and geometry. This social context-based learning aims not only to enhance students' mathematical competences but also to introduce them to social values such as justice, responsibility, and sharing. This study identifies the main challenges in designing context-based tasks, such as task complexity, student acceptance of the social context, and teaching time constraints. This chapter also provides solutions and recommendations to improve task design, as well as proposing future research directions involving learning technology and long-term evaluation in context-based education. Thus, context-based social mathematics learning is expected to introduce social values that can enrich students' lives in society.

INTRODUCTION

Mathematics is a difficult topic to master since it has little relevance to real life and social environment (Dictado & Dagdag, 2025). According to many students consider mathematics as just abstract numbers and, thus, find it hard to relate mathematical concepts to real-life situations (Junger et al., 2026). This, in turn, leads to low involvement in learning. According to Mendoza and Deri (2025), mathematics is typically viewed as something apart from students' social experiences, leading to interpretations built to be less contextual. This is what has led to contextual learning approaches gaining wider attention. This approach connects real-life situations in understanding mathematical concepts, making the contents studied more relatable to students' lives and avoiding the perception of mathematics as an abstract concept. Steen et al. (2007) even emphasize that the essence of good mathematics is that is relevant to students' lives. Rizki and Priatna (2019) found that contextual learning was effective in increasing students' understanding and involvement, especially when the mathematics tasks were designed based on situations close to students, such as the school environment or culture.

Contextual situations that present opportunities for cultural and social integration can strengthen the significance of mathematics learning, one of which is Ramadan. Ramadan presents social and cultural phenomena with practices that contain mathematical concepts, such as measurement, distribution, proportion, and even visual and spatial representation. Ramadan is known as the "holy month," consisting of the rituals of fasting, *zakat* (obligatory alms giving in islam), and sharing blessings with those in need. Various religious practices during Ramadan provide a meaningful context to apply mathematics in real life for many students (Putri & Faradiba, 2025). Learning, which builds upon such social events, corresponds with the ethnomathematics approach that considers mathematical activity as entrenched in cultural practices, social meanings, and community-based modes of reasoning (D'Ambrosio, 1985). Ramadan-related actions such as calculating *zakat*, determining the length of the fast, distributing *takjil* and recognizing the geometry of *ketupat* (rice cake wrapped in woven palm or coconut leaves) and *bedug* (a large traditional drum used in Indonesian Muslim communities), can be seen as culturally contextual mathematical activity (Utari et al., 2025). This view is further supported by Bishop's (1988) concept of mathematical enculturation, especially by the cultural behaviours of counting, measuring, planning and explaining which are evident in the Ramadan-based tasks set in this chapter. As Pinxten (1994) states, mathematics is a cultural language that shapes and informs our thoughts about the world around us.

The mathematical praxeology framework in the Anthropological Theory of the Didactic (ATD) helps understand how mathematical tasks should be designed, the techniques used by students to solve them, the mathematical technologies to justify the techniques, and theories underlying each technology (Chevallard, 2006). This method emphasizes that the purpose of meaningful mathematics learning is to understand the conceptual behinds procedures or algorithms. It also talks about understanding how mathematics is used in a broader social context (Putra, 2017). Ethnomathematics also reveals that the integration of local cultural elements in learning can increase students' engagement and their overall mathematical understanding (Marbun et al., 2026)

At the same time, didactic praxeology emphasizes the teaching strategies that the teachers use to facilitate this learning (Zhu et al., 2025). The right strategies can create social engagement and collaboration in the classroom, which supports the understanding of mathematics as a social activity rooted in the cultural context of the students. Alkan and Ipek (2023) state in their study that the socio-cultural approach is very important in understanding how students build mathematical understanding through social interaction and language use in the learning context. The main objective of this chapter is to identify and analyze the application of mathematical and didactic praxeology in the development of mathematics learning tasks focused on the context of Ramadan. Accordingly, this chapter is guided by the following research questions:

1. How do Master's students in Elementary Education design Ramadan-based mathematics tasks through collaborative learning?
2. What forms of mathematical and didactic praxeology emerge from the developed Ramadan-based lesson plans?

3. What changes about the types of tasks, methods, reasons, and mathematical ideas in the lesson plans when Master's students work together to revise them?

This part is different from other studies. Previous ethnomathematics studies tended to focus on exploring cultural objects and daily activities related to mathematics (Marbun et al., 2026; Motseki et al., 2026; Putra et al., 2024). This study, however, attempts to explore how practices related to Ramadan can be used as a contextual situation for mathematics problems that can be taught through collaborative teacher learning. Using Ramadan practices as a medium for elementary education students to design, revise, and test math assignments through the ATD framework. Using ATD to look at the connection between the type of task, technique, instructional justification, and the underlying mathematical theory in the teacher's lesson planning is what makes this study unique. Therefore, Ramadan is treated as a socio-cultural and professional learning space where teachers negotiate the mathematical meaning and decisions of each design they teach. This chapter also seeks to clarify how these assignments are designed to improve students' comprehension of mathematical concepts, whilst introducing them to important social concepts in everyday life. Therefore, this chapter is meant to guide the educators to build more contextual, relevant and social mathematics learning.

This chapter is structured in several sections. It begins by discussing teachers' professional learning, the design of group tasks, and the context of Ramadan as a social practice. These topics are then framed through the ATD as the theoretical lens. The methods section explains the research design, the participants, and how the data were analyzed based on procedures and data sources. The results section presents the mathematical and didactic practices identified in the lesson plans. The Discussion section connects the findings in this chapter with existing literature. Recommendations, Limitations, and Conclusions conclude this chapter.

THEORETICAL BACKGROUND

In this section, we will discuss: (1) Teacher Professional Learning (Focusing on Master's in Elementary Education), (2) Collaborative in Teacher Education, (3) Task Design in Mathematics Education, (4) Socio-Cultural Perspective of Ramadan and the Connection to Mathematics, and (5) Theoretical Framework: ATD.

Teacher Professional Learning (Focusing on Master's in Elementary Education)

Professional education of teachers, especially those who are focused on Master's programs in primary education, is of great importance for the development of pedagogical and professional abilities of teachers to tackle the problems of contemporary education. Professional learning of instructors requires comprehension of teaching materials, knowledge of teaching methods, knowledge of how students learn, and knowledge of how to tailor the curriculum to different pupils. Effective teacher education allows the development of pedagogical knowledge, which includes knowledge on the teaching materials and comprehension of the socio-cultural contexts of students, as stated by Veliz et al. (2025).

The Master's program in elementary education, especially in the area of mathematics education, seeks to equip teachers with a more profound comprehension of mathematical concepts and the strategies for teaching these concepts successfully to primary school students. According to Evens et al. (2018), this program improves both the teachers' mathematical knowledge and their pedagogical skills to construct relevant mathematics problems that can help students understand. The master's program offers teachers the opportunity to investigate and comprehend the requirements of individual pupils, and how context-based pedagogical strategies can enhance the quality of instruction.

One important aspect of professional learning is the development of teachers' ability to design and adapt tasks in relevant contexts. In mathematics education, this includes designing tasks that teach mathematical concepts effectively. The effective concepts addressed in this learning also introduce students to the social values embedded in mathematics. Mathematical praxeology becomes one of the important tools in this context, as it helps teachers design tasks that focus on mathematical procedures. Teachers are encouraged to emphasize the social and cultural connections of students in each task given. Recent research indicates that the growth of professional learning for teachers, through Master's degree programs in education, can greatly boost their pedagogical and professional competencies. This enables them to better connect theory and practice in teaching, as well as develop more effective teaching strategies that are relevant to the socio-cultural needs of students. Therefore, teachers trained through this program are able to master the teaching material and adjust their teaching to the social context of the students, which is the focus of context-based mathematics learning, as demonstrated in this study by considering Ramadan as a learning context.

The Master's program in elementary education with a focus on mathematics education also provides teachers with the opportunity to deepen their theoretical and practical understanding of mathematical theories that can be applied in task design. For example, this program can help teachers understand how theories of proportion, fractions, and time calculation can be taught in a way that is relevant to students' social lives, such as calculating *zakat* or dividing fasting time in the context of Ramadan. Thus, the professional learning provided through this Master's program offers a strong foundation for teachers to connect mathematics with students' social and cultural lives, and to prepare them to face various challenges in context-based mathematics teaching.

Collaborative in Teacher Education

Collaboration in teacher education can enrich the learning process and their professional development. This collaboration can happen between teachers, students and education practitioners. In mathematics education context, collaboration may give opportunities for teachers to share experiences, knowledge, and skills in designing relevant tasks for their classrooms. This is supported Mishra's (2023) finding which suggests that social interaction can be one of the situations that creates learning opportunities. This exchange may expand their understanding of teaching strategies and their student cultural backgrounds. Prasad (2017) also confirms this in mathematics education context, where he found that collaboration among teachers in mathematics education allows them to exchange ideas, reflect on teaching practices, and provide constructive feedback to each other. This allows teachers to improve their teaching quality because the activity can help them better understand how to connect mathematical theory with practice in the field.

In the context of mathematics education, collaboration among teachers can be a very effective means of introducing context-based approaches, as demonstrated in this chapter with the context of Ramadan. Teachers can work with their colleagues in designing culturally relevant mathematics tasks, such as calculating *zakat* or distributing food during Ramadan, which may benefit students in understanding mathematical concepts. For teachers, they may also integrate social values into mathematics learning using this strategy to build a shared understanding of students' social and cultural needs.

Ebby et al. (2026) also add that teachers who work collaboratively can allow them to identify strengths and limitations in their teaching so that they can improve their teaching practices. This collaboration, which often involves group discussions, allows teachers to develop their teaching skills through hands-on experience and the sharing of successful teaching strategies. In this case, collaboration is beneficial for teachers' sustainable professional development.

Recent research also shows that professional collaboration among teachers plays an important role in creating a more inclusive and effective learning environment. According to Handelzalts (2019), collaboration allows teachers to work together in designing and adapting the curriculum, as well as creating relevant and contextual learning for students. This collaboration helps teachers address the challenges faced in teaching more abstract mathematical concepts more easily and in a manner that aligns with the socio-cultural needs of students. Therefore, this can strengthen mathematical teaching competence, especially when teaching mathematics in contexts rich in social values which involve learning mathematical concepts while introducing values of social justice, sharing, and social responsibility, such as those found in Ramadan practices.

Task Design in Mathematics Education

In mathematics learning, students might face difficulties in understanding mathematical concepts. Teachers can also struggle in explaining them. This is where task design is necessary to assist teachers in shaping students' understanding of mathematical concepts. Effective task design can also benefit classroom in relating the concepts with the real-world applications. To gain these benefits, teachers should examine how these tasks can engage students in learning mathematical concepts while connecting them to real-world applications. As Adleff et al. (2023) explain, the design of mathematical tasks is important because these tasks can influence how students experience their learning. This means that the mathematical tasks the teacher designs can affect their students' understanding and competence of mathematical concepts they learn.

Mathematical tasks should be relevant to students' life to encourage their engagement and understanding (Eka et al., 2025). The tasks should allow students to link abstract mathematical concepts with their experiences. To do so, teachers should consider some aspects when designing the tasks. Firstly, they should align with the learning objectives and be designed to focus on main mathematical concepts such as fractions. The learning tasks should not be designed without referring to the learning objectives that has been set. Besides, the tasks should reflect the real life context, such as calculating *zakat*, determining the duration of fasting, or distributing *takjil* (food) during Ramadan which are relevant to students' cultural practices as muslims. The content and context should also be supported by the quality of the tasks. It should not be too easy or too difficult. Instead, it should be challenging yet still achievable to encourage student engagement.

In this study, Ramadan provides a context for the application of mathematical concepts, where tasks related to *zakat*, food distribution, and the duration of fasting represent the application in the real-life situations. Boaler (1993) defines mathematics as a cultural language that influences how people understand the world around them, which means that the real-world can give meaningful information for learning mathematics. However, designing effective tasks might be challenging, especially in creating meaningful task that can meet high mathematical standards. For example, the problem complexity should move from easy to more complex when designing tasks involving fractions for the distribution of *zakat*. This scaffolding may help teacher develop task that builds on the previously learned knowledge (Putra et al., 2026).

In addition, mathematics assignments in Ramadan context can promote mathematical understanding that can also promote justice and responsibility values. For example, calculating *zakat* can ignite discussions about social justice while they are learning about fractions. Utari et al. (2025) state that effective learning may occur when students get involved in experiential learning. *Zakat* calculation can help create such experiences as it can link mathematical concepts with real-world issues of giving back to community. Such tasks can be used as a tool to help students make connection between the mathematical concepts that they learn with the social values that lies in their culture.

Socio-Cultural Perspective of Ramadan and the Connection to Mathematics

Ramadan is a holy month in muslim culture, where Muslims observe fasting, participate in social activities, give *zakat*, and distribute food to those in need. In ethnomathematical sphere, the month can be referred to as a socio-cultural setting which applies mathematical ideas in its daily practices. This study is based on D'Ambrosio (1985). The development of mathematics stems from the cultural and social activities of society. This can include counting, measuring, distributing, comparing, and explaining. The practices can offer valuable opportunities for teachers to use this context to connect mathematics learning with the students' culture so that the mathematics taught becomes more relevant and meaningful.

Mathematics in a social context can provide students with a more personal experience. Ramadan can be one example of this context where students are invited to use mathematical concepts practically. They can explore social practices that are familiar to them, such as calculating *zakat*, distributing *takjil*, managing fasting times, and shapes like *bedug* which may help them understand how to apply the mathematical concepts. Which is in line with Bishop (1994) Mathematics has evolved from cultural practices such as counting, measuring, locating, and designing

Mathematics tasks related to Ramadan is more than just testing students' mathematical skills of dividing a percentage of total income to ensure fair distribution of *zakat*. When students practice proportionality in calculating the *zakat* distributed to various groups in need, they may also learn to internalize social values like justice and sharing when learning the concept of fractions.

The division of time for fasting also offers a context for applying time concepts in mathematics. In Ramadan, there are two importan times called *imsak* (the predawn cutoff time for eating before the fast begins) and *iftar* (the meal eaten to break the fast at sunset). Students can be invited to calculate the duration of fasting each day by considering the differences in *imsak* and *iftar* times that change daily by converting time from hours to minutes, distributing time throughout the month of Ramadan, and even comparing fasting durations between countries or regions with different time zones. All of this can provide a more applicable concept of time than merely calculating time in theory. Khasanah et al. (2025) also highlight this in the experiential learning theory by saying that effective learning can happen when students are able to relate the material being taught to their direct experiences. When related it to context-based mathematics learning during Ramadan, it can be seen that the mathematical concepts can be a useful tool for solving social problems in their lives. In this case, they can use mathematics for organizing daily life, such as the fair distribution of *zakat* and scheduling fasting times. Apart from the social activities, social objects in Ramadan like *ketupat* and *bedug* can also provide a hands-on experience in learning geometric shapes.

The socio-cultural context of Ramadan can teach students about some social values, such as social justice, sharing, and responsibility toward others. When they calculate *zakat,* they can learn about sharing and responsibility. They can also learn about social justice when deciding who can receive the *zakat* and how much. This can help student internalize their responsibilities as members of society while learning mathematics, as expressed by Padilla et al., (2025). In the socio-cultural perspective, social and cultural contexts is considered influential on how individuals use mathematics in their lives. Ramadan in this perspective can be used as a context for teaching mathematics, where teachers can use mathematical tasks such as *zakat* and food distribution to foster students' understanding of mathematical concepts and introducing them to social values.

In classroom learning, mathematics can become a tool for sharing and supporting social welfare. For example, when students calculate *zakat,* they may apply mathematics in social context, in which they can learn about justice and responsibility. This activity may help them see mathematics as a tool for supporting social welfare. Therefore, Ramadan can provide opportunity for students to experience how mathematics can be used for social purposes.

Theoretical Framework: Anthropological Theory of Didactics (ATD)

The Anthropological Theory of the Didactic (ATD) offers a theoretical framework for analyzing the production, organization, teaching, and transformation of mathematical knowledge in educational settings. ATD, as developed by Chevallard (2006), sees mathematics not only as a set of abstract concepts, but as a type of human activity that is influenced by some aspects, suc as institutional, cultural, and didactic conditions. This is pertinent to the present chapter because Ramadan-based mathematics tasks are treated as designed mathematical and didactic practices through which teachers connect mathematical content, classroom techniques, and socio-cultural meanings (Putra et al., 2025).

A central concept in ATD is praxeology, which describes human activity through the relationship between practice and theory. In mathematics education, praxeology is usually composed of four parts: type of task, technique, technology, and theory (Bosch & Gascón, 2014). The type of task refers to what learners are expected to do, for example, calculating *zakat*, comparing fasting periods, or distributing food for *iftar* fairly. The technique is the method or approach used to solve the problem, such as using fraction operations, percentage calculations, ratio tables, diagrams, or manipulatives. The technology is the explanation or justification that makes the technique meaningful and valid. The theory is the broader mathematical structure that supports the task and its techniques, for example, proportional reasoning, fraction theory, decimal notation, geometry, or measurement. Bosch & Gascón, (2014) point out that ATD allows researchers to analyse mathematical and didactic activities through praxeologies, including their parts, development, and institutional conditions.

In this chapter, ATD is used to analyze two related forms of praxeology: mathematical praxeology and didactic praxeology (Cirade & Crumière, 2019). Mathematical praxeology focuses on the mathematical organization of the tasks designed. For example, a *zakat* task may students to calculate 2.5% of income, determine the direct proportion between income and *zakat* obligation, and report their answer in decimals, fractions, or tables. Meanwhile, a task on fasting duration may involve subtraction of time, converting hours into minutes, and comparing the length of daily fasts. A food-distribution task may involve partitioning objects into equal shares, comparison of fractions, and discussion of proportional fairness. These examples show how Ramadan-related practices can be translated into mathematical type of task supported by specific techniques and theories.

Didactic praxeology, by contrast, focuses on how teachers organize students' access to those mathematical tasks (Pansell, 2023). It includes the instructional techniques, classroom interactions, representations, scaffolding, and justifications used to help students engage with the mathematical content. In the present study, didactic praxeology appears in the participants' use of group discussion, problem-based learning, visual representations, concrete manipulatives, guided questions, and reflection. These didactic techniques are not neutral additions to the task; they shape how students interpret the mathematical meaning of *zakat*, fasting time, food sharing, *ketupat*, *bedug*, and other Ramadan-related contexts. Therefore, ATD allows this chapter to examine not only what mathematical ideas are embedded in

Ramadan practices, but also how teachers transform those ideas into teachable and learnable classroom tasks.

The use of ATD is particularly important because contextual mathematics learning can remain superficial when cultural contexts are used only as decorative settings for routine calculations. Through ATD, the analysis can distinguish between tasks that merely mention Ramadan and tasks that build a coherent mathematical and didactic organization. For instance, a task asking students to "calculate *zakat*" becomes analytically meaningful when the study identifies the type of task, the technique used to solve it, the justification for the technique, and the mathematical theory that underpins it. Similarly, a task about distributing *iftar* food becomes stronger when the teacher clarifies whether students are expected to use equal partitioning, fraction comparison, proportional reasoning, or visual modelling. This distinction helps clarify the contribution of the chapter to research on ethnomathematics, contextual learning, and teacher professional learning.

ATD also provides a useful framework for interpreting the collaborative design process among the participants (Zahedi et al., 2017). In this study, the participants discussed, revised, and justified their task designs, which represent changes in praxeology. A task that was originally based on abstract calculation could be changed by the inclusion of diagrams, manipulatives, or staged questions. A task that was initially too complex for elementary students could be transformed into smaller subtasks with clearer techniques and more accessible representations. From an ATD perspective, such revisions show how teachers' professional knowledge develops through the reorganization of type of task, techniques, technologies, and theories within a collaborative learning space.

Therefore, ATD is used in this chapter as an analytical lens to examine how Ramadan-based mathematics tasks are organized and justified. This framework helps identify two related forms of praxeology. The first is mathematical praxeology, which concerns the relationship between mathematical content, types of tasks, techniques, and the underlying theories. The second is didactic praxeology, which involves how teachers organize instructional tasks, learning techniques, classroom representations, and didactic justifications to support students' mathematical understanding. Through this framework, Ramadan-based practices are treated as structured mathematical and didactic activities.

METHOD

Research Design

A qualitative case study was the approach chosen for this study. This study focuses on the participants' process of designing, discussing, revising, and then implementing mathematics tasks within the context of Ramadan. This was done to examine how the context of Ramadan was integrated with the mathematics tasks, which were analyzed using the ATD framework.

Participants and Context

This study involved seven master's students in the Elementary Education program who were taking the Realistic Mathematics Education (RME) course. Three participants were teachers, and four were recent graduates currently pursuing their master's degrees. Participants were coded as P1 through P7 as pseudonyms. Each participant developed a lesson plan based on the context of Ramadan. The implementation took place in seven elementary schools involving four grade levels: 2nd, 3rd, 4th, and 5th grades. The classes involved followed the teaching context of each respective participant.

Task Design Procedure

The procedure of the task design is based on seven lesson plans, math assignments designed by participants, collaborative recordings, discussion transcripts, classroom implementation results, participant reports, and post-implementation reflections collected via Google Forms. The recordings were used to examine how participants explained, defended, and revised each assignment design decision. Google Forms were used to capture participants' post-implementation experiences regarding assignment execution and student responses,

Data Analysis

The ATD framework was used as the analytical framework in this study. The praxeological aspects namely, type of task, technique, technology, and theory were used to code each lesson plan. The type of task was derived from Ramadan practices such as calculating zakat, distributing *takjil,* or comparing the duration of fasting. The techniques in this study refer to methods of problem-solving, such as fractions, percentages, ratios, or time conversions. Technology refers to the basis for why those techniques are used. Theory refers to the underlying mathematical concepts of the task.

The participants' research data were compared to examine changes in the tasks before and after they were revised. This allowed us to examine how the participants simplified the tasks, clarified the techniques and technologies, and connected mathematical concepts to the context of Ramadan.

Trustworthiness

To ensure the credibility of the analysis, the primary coding process was conducted by a single researcher, and the results were then reviewed through peer checking and discussions among the researchers. This was done to assess consistency and re-examine the evidence of task changes through the lens of praxeology.

Ethical Considerations

This study complies with the ethical guidelines of the University of Riau. Formal Ethics approval was waived because the study was conducted as a course based educational design activity, did not involve experimental treatment, and did not collect sensitive or identifiable personal data. Nevertheless, the research team applied school permission, teacher permission, parental consent, student assent, anonymization and secure data management procedures before classroom implementation

RESULTS

The results are organized according to the ATD framework. Table 1 presents the mathematical praxeology identified from the participants' lesson plans by mapping the relationship among mathematical material, type of task, technique, and underlying theory. Table 2 presents the didactic praxeology by showing how participants planned instructional tasks, selected teaching techniques, and justified the use of Ramadan-related contexts to support students' learning.

Mathematical Praxeology from the Developed Lesson Plan

The application of mathematical theory in task design that results in problem-solving, along with the methods and theories employed to solve those problems, is referred to as mathematical praxeology. Calculating *zakat*, distributing food, and figuring out fasting times are just a few of the mathematical exercises that are intended to help students relate mathematical ideas to their everyday circumstances during Ramadan. The praxeological mathematics used in this study is shown in table 1. Table 1 presents the mathematical praxeology identified from seven participant-designed lesson plans. The code column refers to participant pseudonyms. The table should be read as a mapping of the relationship between mathematical material, type of task, technique, and underlying mathematical theory.

Table 1. Mathematical praxeological

Code	Material	Task	Technique	Theory
P1	Fraction and decimal calculations, geometry.	Calculations, data representation, and pattern analysis.	Visual models, use of manipulatives.	Concepts of multiplication and division of fractions, comparison, and conversion of fractions to decimals. Concept of proportion.
P2	Equivalent fractions using the context of Ramadan.	Division and comparison of fractions.	Drawing visual models, group discussions, and concrete models.	Division of equivalent fractions and comparison of fractions. Converting images into mathematical symbols.
P3	Large whole numbers in decimal notation.	Addition, subtraction, multiplication, and division of large whole numbers with three digits or more.	Writing numbers in decimal notation and the use of diagrams or place-value tables.	Decimal notation, place value, and the concept of large whole numbers.
P4	Simple fractions (1/2 and 1/4) in the context of sharing *iftar* snacks.	Dividing an object into two equal parts and dividing images of *Iftar* snacks as visual fractions and numerical fraction concepts.	The use of visual media such as pictures of food (cakes, fruits, and dates) and paper folding activities with concrete activities.	Simple fractions, dividing objects into equal parts, and the relationship between numbers in fractions. The concept of proportion and percentage.
P5	Simple fractions in the context of sharing food during Ramadan.	Addition and subtraction of fractions, creating story, problems involving fractions.	Visualization of traditional Ramadan foods such as dates, *kolak* (a traditional Indonesian sweet dessert made with coconut milk, palm sugar, and bananas or tubers), fruit salad.	Simple fractions, dividing objects into equal parts, and fraction operations (addition and subtraction).

continued on following page

Table 1. Continued

Code	Material	Task	Technique	Theory
P6	Simple fractions in the context of sharing snacks or food, addition and subtraction of fractions.	Solving fraction word problems, dividing objects like cakes or dates.	Concrete modelling, such as using folded paper and illustrations of food (snacks).	Simple fractions, division into equal parts, and addition and subtraction of fractions.
P7	Fractional calculations, ratios, and scales in the context of *zakat* fitrah (an obligatory charity given by Muslims at the end of Ramadan) and inheritance. Dividing *zakat* fitrah, inheritance distribution based on the ratio between boys and girls.	Fraction calculations, comparisons, and scale applications. Examples: - Solving *zakat* fitrah problems involving addition and multiplication of fractions. - Inheritance distribution with a 2:1 ratio between boys and girls.	Mathematical modelling using real data (for example, rice prices and inheritance amounts). Other techniques involve visual representation using diagrams or comparison tables.	Fraction concepts (addition, subtraction, multiplication, division) as well as comparison and scale.

Table 1 shows that most participants selected Ramadan-related practices as entry points for teaching fractions, proportions, and distribution. However, the mathematical praxeology varied across participants. Some tasks emphasized numerical procedures, while others used visual models, manipulatives, or real-data modelling to support students' understanding.

1. The Given Mathematics Assignment

The assignments given to students aim to develop their understanding of basic mathematical concepts through real-life applications. These tasks focus on some calculations involving fractions, time, and distributions relevant to the context of Ramadan which can be seen as follows:

- Calculating *zakat* based on monthly income and a specified percentage (for example, 2.5% of income).
- Calculating how many hours of fasting are observed each day, taking into account the times of *imsak* and *iftar*. This task involves calculating the time that changes daily based on geographical location.
- Dividing food into equal portions (for example, dividing dates or *takjil* into portions that correspond to the number of people).

2. Techniques Used

To complete the Ramadan-based math assignment, students use visual and practical techniques that help them connect fractions, proportions, and distributions with Ramadan practices. The lesson begins with an introduction led by the teacher using projected Ramadan images in front of the class. As shown in Figure 1.

Figure 1. Teacher introduction through a projected Ramadan visual context

After the introduction, students work in small groups using colored paper, writing tools, and classroom materials. This activity requires students to translate ideas about equal division, grouping, and categorization into concrete models. Figure 2 shows how the use of physical manipulatives supports students in observing how a whole can be divided into parts and how those parts can represent food portions, categories of zakat, or other Ramadan-related distributions.

Figure 2. Students building a visual model of a Ramadan-based math assignment

At this stage, the students demonstrate how mathematical ideas are transformed into visual forms through shapes, colors, and symbolic grouping. As shown in Figure 3, the students attempt to present mathematical ideas about Ramadan in their work. Thus, the students produced a card about mathematics, which was then discussed, presented, and finally carefully corrected.

Figure 3. Student presentation of a group artefact representing Ramadan-related mathematical ideas

3. Logos: Technology and Theory

This task is based on mathematical theories involving the following concepts:

- Proportions and Fractions: Using fractions to calculate *zakat* and food distribution. Students can learn about proportions by dividing *zakat* between two individuals or calculating parts of a whole (e.g., 1/4 and 1/2).
- Time Calculation: Calculating fasting times based on *imsak* and *iftar* times that change daily. The concepts of time and duration are used to measure how long the fasting period lasts.
- Conversion of Fractions to Decimals: Some tasks involve converting fractions to decimals to simplify the *zakat* calculation or food distribution so that it is easier for students to understand the mathematical operations involved.

Didactic Praxeology from the Developed Lesson Plan

Didactic praxeology focuses on the aspects of teaching and didactic techniques used to support students' understanding of the assigned tasks. Context-based teaching is very important, considering the social and cultural values contained in the context of Ramadan, which adds meaning to the given tasks, as it helps students relate their learning to real-life experiences and enhances their engagement with the material. Table 2 illustrates the didactic praxeology of this research.

Table 2 extends the analysis from mathematical praxeology to didactic praxeology by showing how participants planned to support students' engagement with the tasks. The table reports not only what mathematical content was taught, but also how the teacher intended to organize learning and justify the use of Ramadan-related contexts.

Table 2. Didactic praxeological

Code	Didactic Task	Technique	Justification
P1	Building problem-solving skills and understanding the relationship between fractions and decimals.	Problem-based learning and guided discovery.	Connecting mathematics with real-life contexts to make learning more meaningful and relevant.
P2	Understanding the concept of equivalent fractions and building an understanding of justice and responsibility.	Collaborative discussions, problem-based learning, reflection, and open dialogue.	Connecting mathematics with everyday life and mathematics as a tool to achieve justice (such as in the distribution of *iftar* or donations). The concept of fractions and justice in sharing.
P3	Helping students understand how to read, write, and perform operations on large whole numbers. Using decimal notation and understands mathematical terms such as sum, difference, product, and quotient.	Problem-based learning, group discussions, and joint reflections. Presentation of real contexts, such as the number of *iftar* packages distributed during Ramadan.	Connecting mathematics with everyday life contexts, such as sharing *takjil* or *zakat* activities. Practical functions of large whole numbers and place value notation tables.
P4	Building an understanding of the concept of fractions through the activity of sharing *takjil*.	Problem-based learning (for example, sharing cakes or dates), collaborative activities, and the use of visual and manipulative media (for example, folding paper and pictures of *iftar* snacks).	Linking mathematics with real-life contexts (sharing *iftar* during Ramadan), which will help students see the relevance of mathematics in everyday life.
P5	Understanding the concept of fractions and applying fractions.	Group discussions, problem-based learning, and the use of visual aids (such as pictures of food).	Connecting mathematical concepts with real-life contexts (sharing food during Ramadan).

continued on following page

Table 2. Continued

Code	Didactic Task	Technique	Justification
P6	Understanding the concept of fractions in the social and cultural context of Ramadan, dividing objects into equal parts and understanding the meaning of fractions.	Group discussions, problem-based learning, and joint reflection. Interactive and collaborative learning.	Connecting mathematics with real-life contexts. Understanding the application of fractions in fair sharing. The use of mathematics in social situations and emphasizing the values of justice and compassion in line with the spirit of Ramadan.
P7	Helping students understand the concepts of fractions, ratios, and scales in the context of *zakat* fitrah and inheritance. Calculating a part of the whole. Applying the principle of comparison in inheritance distribution.	Group discussions, problem-based learning, and the use of visual representations to help students understand fraction distribution in the context of *zakat* and inheritance.	Connecting mathematical concepts with the real-life contexts of students, such as *zakat* fitrah and inheritance distribution. Studying the concept of fractions and applying it in everyday situations relevant to social and cultural values (such as sharing and justice).

1. Given Didactic Tasks

The didactic purpose in the design of these tasks is to facilitate students in understanding and applying mathematical concepts practically, which include:

- Understanding Fractions and Proportions relevant to social life by understanding how *zakat* is fairly distributed.
- Understanding time management by calculating the duration of fasting from day to day that allow students to practice practical time calculations.
- Understanding fair sharing by dividing food evenly which can introduce the concepts of fairness and proportion.

2. Didactic Techniques Used

The teacher uses various techniques to support students' understanding of these tasks, such as:

- Problem-Based Learning (PBL): teaching mathematics by presenting real-world problems, such as calculating *zakat* or food distribution. The teacher asked the students to solve the problem using the mathematical concepts that have been taught.

"Task 1 (*zakat*): Each group is given data about the number of family members and the price of rice. So, the students have to calculate the total of the zakat that have to be paid by that family."

- Group Discussion: The teacher encouraged the students to work in groups to discuss problem-solving strategies, share their understanding, and learn from each other.

"If some *zakat* is paid by rice and the other part is paid by money, how to calculate it?"

- Direct Instruction and Demonstration: The teacher provides clear and direct instructions on how to complete the task, and demonstrates how to use visual aids and manipulatives.

3. Justification and Reasoning in Task Selection

This task is designed to connect mathematics with students' daily lives. Ramadan is chosen because it is considered as the culture that is familiar to the students who are majority of Muslims. The math assignments are aimed to teach mathematical concepts that contain social and cultural values related to sharing, fairness, and helping others, in line with the concept of Ramadan. The didactic justification for this task selection is to make mathematics learning more meaningful for students by using contexts they are familiar with.

Mathematical Praxeology in Collaborative Discussion

The collaborative discussion between teachers and students focuses on ways to simplify and clarify the application of mathematical theory in those tasks. Some of the topics discussed include difficulty in calculating fractions and using visualization. Some students seemed to have difficulty in dividing *zakat* or calculating fasting durations. Therefore, the discussion was carried out on how to connect the fraction concept with the context of Ramadan in a way that is easier for students to understand. Besides, they also felt challenged in using visualization so that the teacher facilitated discussion on using pie charts or graphs to illustrate the distribution of *zakat* or food portions, which makes it easier for them to understand fair distribution.

1. Tasks Discussed

The discussion is related to various tasks involving the calculation of *zakat*, food distribution, and fasting duration. Students and teachers discussed how to simplify the calculation problems so that they can better align with the students' understanding.

2. Techniques Proposed in the Discussion

The proposed techniques include:

- Application of diagrams and graphs by using pie charts or tables to clarify the proportions of *zakat* or time distribution.

- Application of practical methods by using concrete props, such as pictures of food or money, to help students understand distribution and calculations more easily.

3. The Logos Block: Technology and Theory Emerging in the Discussion
Some concepts that emerged in the discussion are the following:

- Ratios and Proportions: This introduces students to concepts of division and comparison that are relevant in everyday life, such as how to calculate portions of food or money based on ratios and proportions to fairly distribute *zakat* and food.
- Time Analysis: A fasting timer involves understanding the concept of varying durations, which provides students the opportunity to apply time theory in real-world contexts.

Comparison of Praxeology Before and After Revision

Based on the group conversation, the pedagogy may shift throughout task creation and during the participants' reflections. There, participants judged the suitability of the developed exercises to the elementary school students' skills, the clarity of the approaches utilized, and most significantly, whether the Ramadan environment could improve their mathematical knowledge. These tasks were modified from the original procedural tasks. Then, the tasks that just required carrying out arithmetic operations without explaining the underlying concepts were redesigned to be more conceptual and didactically robust. Table 3 shows the changes in the task design before and after revision.

Table 3. Comparing of praxeology before and after revision

Focus Revision	Before Revision	Problem Found	After Revision	Praxeology In ATD
Task complexity	The tasks are designed to require students to perform only a few arithmetic operations. For example, adding up the number of dates, dividing the total, and determining a specific portion of those dates. Example task: A mosque received 25 boxes of dates in the first week and 15 boxes of dates in the second week. How many boxes of dates did the mosque receive in total?	There is too much homework for elementary school students. Students have difficulty understanding the solutions because the problems haven't been broken down into simpler parts. The Ramadan context serves only as a story backdrop and does not help build an understanding of fractions.	The problem is broken down into several simple steps. Students are asked to identify the total number, then determine the part in question, and explain the result of the division as a fraction. Correction: A father bought a box of dates containing 30 dates. He gave one-third of the dates to his grandparents to break their fast. How many dates did he give? Explain how you determined one-third of 30 dates.	The technique used involves dividing 30 into three equal parts. The educational rationale is the relationship between fractions and the practice of sharing in the context of Ramadan.
Problem-solving techniques	Students are asked to complete the assignment involving abstract calculations in their workbooks. As a result, the use of images and concrete objects remains limited.	Participants found that some students had difficulty understanding fractions without a concrete visual representation	The teacher added pictures of dates and a mosque related to the context of Ramadan, as well as folding paper to illustrate fractions. Students were asked to identify the given parts and explain the reasoning behind them	In ATD, this change reinforces the connection between the techniques, technology, and theory of that fraction.
Didactic justification	The initial assignment focuses solely on numerical answers. The reason why Ramadan is used has not yet been explained in their activities.	The social context has not yet been effectively linked to mathematics. For example, the concept of fair sharing is used merely as a story backdrop but has not yet been integrated into mathematical reasoning.	In this section, students are asked to explain why the division must take place and why the shares must be equal, and how a particular division can be considered fair	This is because the students were able to explain the mathematical reasoning behind the division

The results show that the application of mathematical praxeology and didactics in designing Ramadan context-based mathematics tasks can be an effective framework for making mathematics tasks that can enhance students' understanding of abstract mathematical concepts. Based on the results, it can be seen that students can connect mathematical theory with their socio-cultural background because they have experienced the Ramadan practices, such as *zakat*, sharing food, and fasting. However, during the implementation, students found some difficulties in calculating fraction and using the visualization, which they reported in the discussion. The issues were then addressed by the teacher so that the task can give more applicable understanding.

DISCUSSION

This research shows that the use of mathematical praxeology and didactics in the design of Ramadan context-based tasks enriches students' mathematical understanding through direct connections with the social and cultural contexts they experience daily. In this case the socio-cultural perspective plays an important role in providing a deeper meaning to mathematics learning, as the social and cultural contexts relevant to students' lives help them to see the mathematics as a tool that is highly relevant to their daily activities. This finding concurs with research in ethnomathematics that considers mathematics as a practice that arises from surrounding sociocultural activities (Rosa et al., 2016). However, this study goes beyond ethnomathematics by showing how mathematical elements emerge in the context of Ramadan through the ATD framework, rather than just identifying mathematics within culture.

The findings of this study are explained in terms of the four aspects of ATD praxeology. Task types emerged when students calculated zakat, compared fasting durations, distributed takjil, and presented fractions in the context of Ramadan. Techniques were evident through the use of fractions, percentages, and images. The technology aspect emerges when participants explain why a particular technique is correct for example, that the division must be fair or that the representations of the parts must be equal and students understand the relationship between those parts. The theory aspect forms the foundation for tasks involving the concepts of fractions, proportions, and the numerical operations used. Thus, Ramadan becomes meaningful when teachers are able to establish connections among the aspects of task, technique, technology, and theory.

Viewed through ATD, it shows that Ramadan-based task design not only produces a relevant learning context but also shapes the transformation of praxeology. In the early stages, some tasks are still centered on procedural completion, such as calculating fractions, percentages, or time durations. After collaborative discussion, the tasks move towards a more complete praxeology as the completion techniques are supported by visual representations, concrete manipulatives, and social-didactic justifications. Through, the change in assignment design shows the development from a mere mathematical task to a didactic praxeology that considers learning conditions, social meaning, and student accessibility.

Mathematics as a Tool for Solving Social Problems

In the context of Ramadan, the tasks designed, such as the calculation of *zakat* and the distribution of *takjil*, connect mathematical concepts with social values prevalent in society, such as social justice, shared responsibility, and sharing. In a socio-cultural perspective, mathematics learning based on social and cultural contexts can provide deeper meaning for students, as they can see how mathematics is used in real social practices. The results are consistent with research on contextual mathematics learning, which states that mathematical ideas are more meaningful when linked to events relevant to students' daily lives (Putra et al., 2025). This is also consistent with studies on ethnomathematics that emphasize that mathematical activities are rooted in cultural practices, social values, and community-based ways of thinking (D'Ambrosio, 1985). In this way, Ramadan-based projects offer a thematic framework for mathematics learning, and also provide a culturally meaningful context for students to understand how mathematical principles function in social practices.

If viewed through the ATD framework, *zakat* is positioned as a task of type, because it is a type of mathematical task that requires students to determine which part of a whole. The solution techniques can include percentage calculations, fractions, proportions, and structured quantity distribution. Meanwhile, the didactic justification lies in the use of correct mathematical procedures, understanding of justice, social distribution, and their responsibility in allocating resources to those in need. For example, in the calculation of *zakat*, students are not only asked to calculate numbers, percentages, or fractions but are also invited to understand the value of social justice contained in *zakat* itself. Like other studies, this study also confirms that *zakat* can be viewed as a structured mathematical task which involves proportion, percentage, and fair distribution (Marini et al., 2025). The distribution of *takjil* during Ramadan, for example, may teach students that mathematics can be used to determine amounts fairly which contain social justice values. This also show that mathematics can be used to support social justice. Consequently, mathematical concepts can also serve as a way to understand ethical and humanitarian social practices. Overall, the assigned task demonstrates how mathematics can be a tool used to solve social problems, where students learn about the social values embedded in their daily practices, such as *zakat* and sharing food.

Contextual Learning and Social Relevance

Context-based learning in this study reinforces its relevance in mathematics education, as Ramadan is positioned as a condition that shapes how mathematical knowledge is selected, presented, and justified in the classroom, and is no longer just a cultural theme. From a socio-cultural perspective, learning becomes meaningful when mathematical ideas are connected to everyday practices they are familiar with. In this case, Ramadan provides a social environment that tends to be familiar, where students can examine fractions, proportions, then time distribution and allocation while also reflecting on the values of sharing and social responsibility.

Through the ATD framework, the context of Ramadan is interpreted as a condition that influences the formation of mathematical praxeology and didactic praxeology. For example, the type of task that occurs because the teacher designs assignments that ask students to calculate zakat, divide *iftar*, and calculate fasting hours. The aspect of technique occurs when students use calculations of fractions, percentages, ratios, and time tables. In the aspect of technology, techniques must be explained based on procedural accuracy, then understanding social-didactic reasons such as fairness in distribution, measurability in division, and the relevance of mathematics in everyday life. In the aspect of theory, the tasks are based on the concepts of fractions, proportions, number operations, time measurement, and simple mathematical models.

The use of the Ramadan context provides students with the opportunity to understand mathematics in social life. For example, when students calculate *zakat*, they apply percentages so that the calculations relate to the principle of fair distribution. In the activity of distributing *takjil*, in addition to learning fractions, they are also given the opportunity to understand fractions as a tool to ensure equitable distribution. Thus, the concept of mathematics becomes more meaningful through the connection between mathematics and social values in the practice of Ramadan.

Social relevance in learning can be seen through the way math tasks are connected to students' daily experiences. In the activity of calculating fasting duration, for example, it helps students understand the concept of time, its differences, and unit conversions through real-life situations. In the activity of dividing *iftar*, it helps students understand fractions, proportions, and equal parts through the experience of Ramadan practice. Thus, the context of Ramadan increases student engagement because they are given the opportunity to strengthen their mathematical understanding through their social experiences.

Because of this study shows that social relevance in mathematics learning is not sufficiently built through the use of familiar contexts; relevance needs to be supported by task design with a clear praxeological structure. Ramadan-based tasks become meaningful when the relationship between task, technique, technology, and theory can be clearly identified. Through that relationship, mathematics learning helps students understand the concepts of fractions, percentages, and time, and also helps them see how mathematics can be found in social practices that contain aspects of justice and shared responsibility.

Collaboration in Social Discussion and Learning

An important aspect of this study lies in the use of group discussions and collaborative learning as a space to connect mathematical theory with the social context of Ramadan. In a sociocultural perspective, mathematics learning does not take place as an individual activity separate from the student's social experience, but is formed through interaction, negotiation of meaning, and the use of language in the learning community. Through group discussions, participants can test each other's ideas, compare completion strategies, and assess whether the designed mathematics task assignment is in accordance with the students' abilities, learning goals, and social values that they want to build. This finding is consistent with studies on teacher collaboration which show that collaborative work enables teachers to exchange pedagogical ideas, examine classroom artifacts, and reflect critically on the suitability of instructional decisions for students' learning needs (Pischetola et al., 2023)

The group discussions of this study are a didactic milieu in ATD, a didactic space for testing the feasibility of the task, the techniques for completing it, the representations and the didactic justification of the learning before the task is applied in the classroom. Through this process, participants not only discuss whether a problem can be solved mathematically, but also examine whether the techniques used are clear enough, whether visual representations help students' understanding, and whether the context of Ramadan actually reinforces the targeted mathematical meaning. Thus, collaboration is not just an activity of sharing opinions, but a reflective mechanism to improve the quality of mathematical and didactic praxeology.

Discussions about the distribution of zakat, takjil distribution, and the distribution of iftar food raise mathematical and social problems, such as proportion, distribution justice, social responsibility, and sharing values. In this process, participants learn to see mathematics not only as a calculation procedure, but as a tool to explain and solve social problems rationally. For example, when participants discuss the fair distribution of food, they must consider the link among the number of recipients, the size of the portions, and the principle of equality. This process reinforces the understanding that the concepts of fractions, ratios, and proportions have a real social function in everyday life.

Discussion-based learning also helps participants reinterpret the context of Ramadan pedagogically. The values of sharing, caring, and fairness are not placed as moral additions that stand outside of mathematics, but as part of a didactic justification that explains why the task is relevant to the student. In this way, collaborative discussions enrich participants' understanding of the relationship between mathematical concepts, cultural contexts, and learning objectives. As a result, the tasks developed became more directed, more inclusive, and more conceptually robust because they had gone through a joint testing process in a reflective discussion room.

Revision of Tasks Based on Social Feedback

Revising tasks based on student feedback also demonstrates how a socio-cultural approach can be applied in mathematics learning. The task, which was initially too complex, especially for young students, was revised to be more understandable by breaking the problem down into simpler steps. In the perspective of ATD, this revision provides an overview of the praxiological shift from mathematically demanding tasks to structured tasks that are more accessible to elementary school students, supported by clear techniques, visual representations, and didactic justification. This shows that depends not only on teaching techniques but also on the social and cultural needs of the students.

This revision aims to make the mathematical tasks easier to understand and while ensuring that the social values contained in the task are still achieved. For example, although the *zakat* calculation task is simplified, the value of social justice in the distribution is still maintained, ensuring that students understand why *zakat* is important and how to distribute it fairly. Thus, the revision process leads to technical improvements in mathematics teaching that enhances the social dimension of the learning experience.

SOLUTION AND RECOMMENDATIONS

Based on the research results related to the application of mathematical praxeology and didactics in the design of Ramadan context-based tasks, there are several solutions and recommendations that can help address the challenges faced in mathematics learning based on social and cultural contexts. Some issues that have been previously discussed, such as task complexity, student acceptance of social context, and teaching time constraints, require careful adjustments in task design to ensure that students understand mathematical concepts relate those concepts to relevant social values in their lives.

Simplification of Mathematical Task Complexity

One of the main solutions to address the complexity of overly difficult tasks is to simplify the tasks without diminishing the essence of the mathematical concepts being taught. For example, in *zakat* calculation, the tasks assigned should be designed in a gradual manner, starting from a small scale and progressing to more complex situations. To do that, teacher can ask students to calculate *zakat* based on smaller amounts, then moving on to larger ones, but still paying attention to the principles of proportion and percentage.

In some cases, students might face difficulty in understanding a mathematical concept because the ideas of fractions or proportion might seem abstract to them. Teacher can use visual aids and physical manipulatives, such as graphs, pie charts, or real objects like money and food to give concrete form of the concept so that the students can visualize it. For example, teacher can use pie charts to illustrate *zakat* distribution so that students can understand fair distribution in a visual format. This might be easier that using text of oral explanation because students can directly see the chart or touch the real money to illustrate the *zakat* distribution. This may reduce task complexity to facilitate student understanding during the learning process.

Improving Students' Acceptance of Social Context

Even though majority of students in Indonesia are muslims, teacher should acknowledge that there may be some students who are not. This might bring some issues in implementing the topic of Ramadan. To address this, one of the main recommendations is to provide alternative contexts that remain mathematically relevant but are more inclusive for all students. For example, if students do not celebrate Ramadan, teacher can use other social contexts involving fair distribution of resources, such as distributing donations during natural disasters or organizing the distribution of social aid. Thus, teachers can still adopt different context-based mathematics learning as long as it can cover similar mathematical concepts.

Additionally, it is important for teachers to provide a more in-depth explanation of the social context used in the learning process. For example, although not all students observe fasting, teachers can explain the importance of sharing and social justice contained in the practice of Ramadan so that students can see the broader social benefits of the mathematics assignment. This shows that mathematics is also about social responsibility that can be applied in everyday life.

Adjustment of Teaching Time and Curriculum

The limitation of teaching time within a dense curriculum poses a significant challenge in the implementation of context-based learning. One solution to address this issue is by implementing a more flexible curriculum adjustment. Context-based learning, such as that applied during Ramadan, needs to be adjusted to the available time to still provide students with the opportunity to understand mathematical concepts while learning social values.

One way to overcome time constraints is by using problem-based learning (PBL), which allows students to connect mathematical theory with real-world problems in a relatively short time. With PBL, teachers can guide students to work in groups to solve Ramadan-based social problems, such as calculating *zakat* or distributing food, while applying group discussion strategies and collaborative problem-solving. This technique allows students to learn actively and optimize the available time without sacrificing the depth of the mathematics being taught.

Teacher Collaboration in Designing Context-Based Mathematics Tasks

Collaboration among teachers is an important strategy to enrich the design of mathematics tasks based on social context. Through collaboration, teachers can share ideas, teaching strategies, and solutions to the challenges faced in teaching mathematics based on social contexts such as Ramadan. Collaboration among teachers can enhance the quality of teaching by introducing various perspectives in the design of tasks and mathematics instruction. This collaboration is also important to ensure that the social values embedded in mathematical tasks are maintained while ensuring that mathematical concepts can still be well understood by students.

In addition, collaborative discussions among teachers can help in adjusting the material to be delivered. Teachers can share experiences on the best ways to connect mathematics with relevant social contexts, as well as how to adapt math tasks to suit the diverse cultural backgrounds of students. Thus, collaboration among teachers plays a crucial role in creating mathematics learning that is more inclusive and directly related to students' lives.

LIMITATIONS

This study has some limitations. First, the context of Ramadan must be carefully understood, as not all students share the same level of familiarity with it. This is especially true in classes where students have diverse experiences and backgrounds. It is important to note that teachers cannot assume that students have the same level of familiarity with these practices. It is important to remember that math assignments in the context of Ramadan should be framed as a sociocultural context for understanding mathematics. They should not be presented as an invitation or an obligation for students to participate in these practices.

Second, teachers need to ensure that the use of religious contexts remains inclusive and non-coercive. Students who do not celebrate Ramadan should be given access to alternative but mathematically equivalent contexts, such as food donation, disaster relief distribution, community sharing, or social assistance programs. This maintains the mathematical purpose of the tasks while respecting students' diverse backgrounds.

Third, the study involved only seven Master's students in Elementary Education within one institutional context. As a result, the findings are not intended for broad statistical generalization. The classroom implementation was also limited in duration, and some data were drawn from participant reflections, which may contain subjective interpretations. More research with a diverse needed to strengthen the validity of the findings.

Fourth, this study involved only seven master's students in the Elementary Education program at a single educational institution. Therefore, it cannot serve as a basis for broader generalizations. Classroom implementation also took place over a limited period, and some of the data may contain subjective interpretations. Further research involving a larger sample size, more specific grade levels, and a longer duration is needed for future studies.

FUTURE RESEARCH DIRECTIONS

As the approach to mathematics based on social context evolves, particularly in the context of Ramadan, there are several trends and research opportunities that can be pursued to further explore the existing potential and challenges. This research provides important insights into how mathematical praxeology and didactics can be applied to make mathematics learning more relevant and meaningful, but there is still much room for further exploration. Based on the results of this research, several directions for future research may include:

Development of Contextual-Based Learning Models

One of the highly relevant research directions is the development of a social-context-based learning model that can be applied more broadly. The Ramadan context-based learning model tested in this study shows that social context can enrich the mathematics learning experience, but the biggest challenge is adapting this model to be used in various cultural and religious contexts. Therefore, further research could focus on developing a more flexible model, which is based on other social and cultural contexts found in various countries. For example, further research could examine how other social contexts, such as New Year's celebrations, donation distribution during natural disasters, or other charitable activities, can be used as contexts for teaching mathematics based on fair distribution, proportionality, and resource calculation. Such research will open opportunities to create more universal learning models that can be applied in various places while still maintaining relevant social values.

Technology in Mathematics Learning Based on Social Context

With the rapid advancement of technology, the use of technology in mathematics learning based on social context has become a very interesting direction for further research. Learning technologies such as mobile applications can be used to make Ramadan-based math tasks more engaging. Further research can look at how the existing technology can be used to enhance student engagement and their understanding of mathematical concepts in a social context. For example, future researchers can investigate whether web-based applications that simulate the calculation of *zakat* during Ramadan can improve students' understanding compared to traditional methods. Besides, it is also possible for future researchers to develop interactive mobile applications that can help students calculate *zakat* or calculate fasting durations. This can help educators to to create more engaging digital educational resources.

Evaluation of the Long-Term Impact of Contextual Social Learning in Mathematics

So far, this research has focused more on the design and implementation of Ramadan-based mathematics tasks. However, the long-term impact of such approach on students' mathematical abilities and understanding of social values still needs further scrutiny. For future researchers, this research can be extended to exploring how social context-based learning can affect students' their understanding of more advanced mathematical concepts, such as algebra, geometry, or statistics.

Longitudinal studies can also be carried out to follow students' development in terms of mathematical abilities and social engagement over several years after participating in context-based learning. Future research may compare between students who learn with the Ramadan-based approach show and students who follow traditional learning in terms of their math skills and social awareness in their real-life contexts. Such topic may offer valuable insights into the long-term benefits of a context-based social approach.

The Impact of Contextualized Mathematics Learning on Educational Disparities

One of the challenges in education is reducing the gap between different socio-economic groups. Mathematics learning based on social context may have a potential to reduce this gap because it can connect learning with students' real lives so that they can feel more connected to the learning material.

To support this benefit, future research can look at the impact of context-based social learning on reducing this educational gap, as to whether students from lower economic backgrounds feel more motivated and find it easier to understand mathematical concepts when the tasks are connected to social contexts they are familiar with, such as social assistance or charitable activities in their community. This research may help answer if a context-based approach can support educational equity for students, regardless of their socio-economic background.

Research on Collaborative Learning Among Teachers

Collaboration among teachers in designing learning based on social context may have a potential in creating more inclusive teaching that is relevant to students' real-life experiences. Further research can explore the collaborative strategies used by teachers in designing mathematics tasks based on social contexts, as well as how teacher collaboration can enhance the quality of mathematics instruction. This research could focus on the best methods to facilitate teacher collaboration in designing a social-context-based curriculum that is more effective and acceptable to various diverse student groups.

CONCLUSION

This chapter discusses the application of mathematical praxeology and didactics with regards to Ramadan as a context for mathematics learning. This research shows that social and cultural contexts can be effective tools for teaching basic mathematical concepts such as percentages, fractions, and time calculations. During the learning, teacher observed better understanding of mathematical concepts along with a comprehension of social values such as justice, social responsibility, and sharing. This might be due to the relevance of the mathematical tasks designed with students' experiences, such as the distribution of *zakat*, the calculation of fasting duration, and the distribution of *takjil*. This may support the argument that mathematics can serve as a tool to solve social problems.

Even though it has many benefits, applying social context-based mathematics learning still has some challenges that need to be addressed. This study found some issues during the implementation process, such as the complexity of tasks related to the calculation of *zakat* and the duration of fasting, students' acceptance of different social contexts, and the limitations of teaching time. Therefore, this study proposes some solution, such as task simplification, provision of alternative social contexts, and adjustment of teaching time. These solutions may address the identified issues so that the tasks can be implemented effectively to provide meaningful learning for students.

Teacher collaboration is also found important in designing context-based tasks that are relevant to social values because it allows teachers to share their experiences with their colleagues and develop a more inclusive curriculum that can be accepted by students, regardless of their cultural and religious backgrounds. Thus, collaboration among teachers can be considered as one of the key factors in creating learning that is more connected to students' real-life experiences.

Context-based mathematics task still has many aspects that are not yet explored. In future research directions, the development of more inclusive social-context-based learning models and the use of technology in education can be potential opportunities to improve the effectiveness of mathematics learning. Longitudinal investigation can also be considered for further research to elucidate the long-term impacts of social-context-based learning and the reduction of educational disparities in mathematics education system.

REFERENCES

Adleff, A.-K., Ross, N., König, J., & Kaiser, G. (2023). Types of mathematical tasks in lower secondary classrooms in Germany: Statistical findings from a latent class analysis based on general mathematical competencies. *Educational Studies in Mathematics, 114*(3), 371–392. DOI: 10.1007/s10649-023-10254-9 PMID: 37273842

Alkan, S. H., & Ipek, D. (2023). Addressing Language Diversity in Early Years Mathematics: Proposed Classroom Practices through a Live Brief Assessment. *Education Sciences, 13*(10), 1025. DOI: 10.3390/educsci13101025

Bishop, A. J. (1994). Cultural Conflicts in Mathematics Education: Developing a Research Agenda. *For the Learning of Mathematics, 14*(2), 15–18. JSTOR. https://www.jstor.org/stable/40248109

Boaler, J. (1993). Encouraging the transfer of 'school' mathematics to the 'real world' through the integration of process and content, context and culture. *Educational Studies in Mathematics, 25*(4), 341–373. DOI: 10.1007/BF01273906

Bosch, M., & Gascón, J. (2014). Introduction to the Anthropological Theory of the Didactic (ATD). In Bikner-Ahsbahs, A., & Prediger, S. (Eds.), *Networking of Theories as a Research Practice in Mathematics Education* (pp. 67–83). Springer International Publishing., DOI: 10.1007/978-3-319-05389-9_5

Chevallard, Y. (2006). Steps towards a new epistemology in mathematics education. In *Proceedings of the IV Congress of the European Society for Research in Mathematics Education, 1, 21–30.*

Cirade, G., & Crumière, A. (2019). The study of teachers' mathematical and didactic praxeologies as a tool for teacher education. In M. Bosch, Y. Chevallard, F. J. García, & J. Monaghan (Eds.), *Working with the Anthropological Theory of the Didactic in Mathematics Education* (1st ed., pp. 159–168). Routledge. DOI: 10.4324/9780429198168-10

D'Ambrosio, U. (1985). Ethnomathematics and Its Place in the History and Pedagogy of Mathematics. *For the Learning of Mathematics, 5*(1), 44–48. JSTOR. https://www.jstor.org/stable/40247876

Dictado, M., & Dagdag, J. D. (2025). Barriers to Achieving Mathematics Proficiency: Synthesis of Research. *Journal of Interdisciplinary Perspectives*. https://api.semanticscholar.org/CorpusID:280731060

Ebby, C. B., Hess, B., Pecora, L., & Valerio, J. (2026). Facilitating collaborative inquiry into practice around artifacts of mathematics teaching. *Journal of Mathematics Teacher Education, 29*(1), 59–82. DOI: 10.1007/s10857-024-09649-z

Eka, W. D., Tola, T. T., & Negasi, R. D. (2025). Effects of students' mathematics learning strategy and perceived task difficulty on their achievement of mathematics. *PLoS One, 20*(11), e0337115. DOI: 10.1371/journal.pone.0337115 PMID: 41270042

Handelzalts, A. (2019). Collaborative Curriculum Development in Teacher Design Teams. In Pieters, J., Voogt, J., & Pareja Roblin, N. (Eds.), *Collaborative Curriculum Design for Sustainable Innovation and Teacher Learning* (pp. 159–173). Springer International Publishing., DOI: 10.1007/978-3-030-20062-6_9

Junger, A. P., Jacomossi, R. R., Oliveira, V. I. D., Araújo, M. M. P., & Galoni, N. C. (2026). *Digital Technologies And The Overcoming Of Difficulties In Mathematical Learning*. DOI: 10.9790/7388-1601026064

Khasanah, B. A., Prahmana, R. C. I., Adiputra, S., & Arnal-Palacián, M. (2025). The beauty of mathematics in Indonesian culture: An impactful and meaningful context in number patterns learning. *JRAMathEdu (Journal of Research and Advances in Mathematics Education)*. DOI: 10.23917/jramathedu.v10i2.10441

Marbun, L. R. P., Putra, Z. H., Noviana, E., Gunawan, Y., Afrillia, Y. M., Anriana, R., & Ahmad, S. (2026). Ethnomathematics exploration of tandok of the Batak Toba tribe as geometry learning resources for elementary school. *Ethnography and Education*, •••, 1–24. DOI: 10.1080/17457823.2026.2641176

Marini, M., Tohir, M., Muhasshanah, M., & Anam, A. C. (2025). Mathematical Calculations in Determining the Amount of Zakat Mal and Zakat Fitrah as an Effort to Empower the Community Economy. In Alfarisi, R., Mason, J., Abdullah, N. B., Margunayasa, I. G., & Chen, H. H. (Eds.), *Proceedings of International Conference on Neuroscience and Learning Technology (ICONSATIN 2025)* (Vol. 980, pp. 355–375). Atlantis Press SARL. DOI: 10.2991/978-2-38476-525-6_37

Mendoza, J. E. D., & Deri, R. A. (2025). Real-World Context in Mathematics Teaching: Teachers' Insights. *International Journal of Innovative Science and Research Technology*. https://api.semanticscholar.org/CorpusID:277686467

Mishra, N. R. (2023). Constructivist Approach to Learning: An Analysis of Pedagogical Models of Social Constructivist Learning Theory. *Journal of Research and Development (Srinagar), 6*(01), 22–29. DOI: 10.3126/jrdn.v6i01.55227

Motseki, P. D., Jojo, Z. M. M., & Gumbo, M. T. (2026). Transforming mathematics teaching through Ndebele art: An ethnomathematics perspective. *Diaspora, Indigenous, and Minority Education*, *20*(1), 1–7. DOI: 10.1080/15595692.2025.2451239

Padilla, M. O., Velasco, N. Y. G., & Ortiz, S. A. (2025). Analysis of pedagogical content knowledge for teaching mathematics in early childhood education. *Journal of Infrastructure Policy and Development*, *9*(1), 10135. DOI: 10.24294/jipd10135

Pansell, A. (2023). Mathematical knowledge for teaching as a didactic praxeology. *Frontiers in Education, Volume 8-2023*. DOI: 10.3389/feduc.2023.1165977

Pischetola, M., Møller, J. K., & Malmborg, L. (2023). Enhancing teacher collaboration in higher education: The potential of activity-oriented design for professional development. *Education and Information Technologies*, *28*(6), 7571–7600. DOI: 10.1007/s10639-022-11490-x PMID: 36532797

Prasad, R. (2017). Trainee Teachers' Feedback and Reflection on Math Lesson Study. *International Journal of Advanced Research*, *5*(12), 237–242. 10.21474/IJAR01/5959

Putra, Z. H. (2017). Pre-service elementary teachers' knowledge of comparing decimals based on the anthropological theory of the didactic. *HAL Open Science*, 01948872.

Putra, Z. H., Ramiati, Zufriady, Hidayat, R., Jismulatif, Hermita, N., & Sulistiyo, U. (2024). Development of computational thinking tasks based on Riau Malay culture: A study of fifth-grade public school students in Pekanbaru, Indonesia. *Education 3-13, 52*(8), 1387–1397. DOI: 10.1080/03004279.2022.2150063

Putra, Z. H., Alim, J. A., Melihayatri, N., Dahnilsyah, , Afrillia, Y. M., & Aljarrah, A. (2026). Project-based learning in teacher education: A reflection on analysing prospective elementary teachers' knowledge of math, didactics, and technology. *Journal on Mathematics Education*, *17*(1), 179–198. DOI: 10.22342/jme.v17i1.pp179-198

Putra, Z. H., Hermita, N., Alim, J. A., Dahnilsyah, D., & Gunawan, Y. (2025). The role of project-based ethno-didactics of mathematics learning toward pre-service elementary teachers' perspectives, mathematical, and didactic knowledge. *Frontiers in Education*, *10*, 1568366. DOI: 10.3389/feduc.2025.1568366

Putri, R. A., & Faradiba, S. S. (2025). *Kajian Literatur Integrasi Aritmetika Sosial dengan Nilai-Nilai Fiqih dalam Zakat pada Pembelajaran Matematika*. Mandalika Mathematics and Educations Journal., DOI: 10.29303/jm.v7i3.9900

Rizki, L. M., & Priatna, N. (2019). Mathematical literacy as the 21st century skill. *Journal of Physics: Conference Series, 1157*, 042088. DOI: 10.1088/1742-6596/1157/4/042088

Rosa, M., D'Ambrosio, U., Orey, D. C., Shirley, L., Alangui, W. V., Palhares, P., & Gavarrete, M. E. (2016). *Current and Future Perspectives of Ethnomathematics as a Program*. Springer International Publishing., DOI: 10.1007/978-3-319-30120-4

Steen, L. A., Turner, R., & Burkhardt, H. (2007). Developing Mathematical Literacy. In Blum, W., Galbraith, P. L., Henn, H.-W., & Niss, M. (Eds.), *Modelling and Applications in Mathematics Education: The 14th ICMI Study* (pp. 285–294). Springer US., DOI: 10.1007/978-0-387-29822-1_30

Utari, R. S., Putri, R. I. I., Zulkardi, , & Hapizah, . (2025). Seeing data differently: Developing a local instructional theory for culturally relevant statistical literacy in Indonesian teacher education. *Journal on Mathematics Education, 16*(3), 915–936. DOI: 10.22342/jme.v16i3.pp915-936

Zahedi, M., Tessier, V., & Hawey, D. (2017). Understanding Collaborative Design Through Activity Theory. *The Design Journal, 20*(sup1), S4611–S4620. DOI: 10.1080/14606925.2017.1352958

Zhu, J., Huang, X., & Trouche, L. (2025). The development of didactic praxeologies in lesson design through social resource interactions: Evidence from two novice mathematics teachers. *Educational Studies in Mathematics, 119*(3), 487–513. DOI: 10.1007/s10649-025-10403-2

ADDITIONAL READING

Bosch, M., & Gascón, J. (2014). Introduction to the Anthropological Theory of the Didactic (ATD). In Bikner-Ahsbahs, A., & Prediger, S. (Eds.), *Networking of Theories as a Research Practice in Mathematics Education* (pp. 67–83). Springer International Publishing., DOI: 10.1007/978-3-319-05389-9_5

Pansell, A. (2023). Mathematical knowledge for teaching as a didactic praxeology. *Frontiers in Education, 8*, 2265977. DOI: 10.3389/feduc.2023.1165977

Zhu, J., Huang, X., & Trouche, L. (2025). The development of didactic praxeologies in lesson design through social resource interactions: Evidence from two novice mathematics teachers. *Educational Studies in Mathematics, 119*(3), 487–513. DOI: 10.1007/s10649-025-10403-2

KEY TERMS AND DEFENITIONS

Anthropological Theory of the Didactic, ATD: A theoretical framework used to analyze how mathematical knowledge is produced, organized, taught, and transformed within educational institutions. In this chapter, ATD is used to examine how *Ramadan*-based practices become teachable mathematical tasks.

Collaborative Teacher Learning: A professional learning process where teachers or prospective teachers design, discuss, revise, and justify instructional tasks together.

Context-Based Mathematics Learning: Mathematics learning that uses real-life situations to support conceptual understanding. In this chapter, *Ramadan* functions as a context for learning fractions, proportions, percentages, time, and geometry.

Didactic Justification: The teacher's reasoning for selecting a task, representation, or teaching strategy. In this chapter, didactic justification links mathematical goals with *Ramadan*-related social values.

Didactic Praxeology: The organization of teaching activity that helps students access mathematical ideas. It includes instructional techniques, classroom representations, scaffolding, teacher questions, group work, and didactic justification.

Ethnomathematics: A perspective that views mathematics as embedded in cultural practices, community activities, social meanings, and everyday ways of reasoning.

Fair Distribution: A mathematical and social idea involving equal sharing, proportional allocation, and justice in resource distribution, especially in tasks related to *zakat* and food sharing.

Mathematical Praxeology: The organization of mathematical activity through four connected elements: task type, technique, technology, and theory. In this study, examples include calculating *zakat*, comparing fasting duration, dividing *takjil*, and representing fractional quantities.

Praxeological Revision: The improvement of a task after discussion or classroom feedback by changing its task type, technique, representation, justification, or theoretical focus.

Ramadan-Based Mathematics Tasks: Mathematical tasks developed from *Ramadan* practices, such as *zakat* calculation, fasting duration, iftar food distribution, ketupat geometry, and bedug representation.

Social Relevance: The degree to which a mathematics task connects with students' lived experiences, social practices, and community values.

Task Type: The form of mathematical work expected from students, such as calculating, comparing, dividing, converting, representing, or explaining quantities in *Ramadan*-related contexts.

Technique: The method used to solve a mathematical task. Examples include fraction operations, percentage calculation, ratio tables, visual diagrams, concrete manipulatives, and time conversion.

Technology: The explanation or justification that supports a technique. In this chapter, technology includes mathematical reasoning and social reasoning, such as why a proportional distribution is fair.

Theory: The broader mathematical structure behind a task, such as fraction theory, proportional reasoning, decimal notation, measurement, geometry, and time calculation.

Zakat-Based Task: A mathematical task that uses *zakat* as a context for learning percentage, proportion, quantity, and social responsibility.

Chapter 9
Solid Geometry in Rumah Lontiok Architecture of Kampar:
An Ethnomathematical Analysis and Pedagogical Framework for Elementary Education

Fuji Lestari
http://orcid.org/0009-0001-2302-8853
Universitas Riau, Indonesia

Mahdum Mahdum
Universitas Riau, Indonesia

Zetra Hainul Putra
http://orcid.org/0000-0002-0860-1076
Universitas Riau, Indonesia

Hermandra Hermandra
http://orcid.org/0000-0002-2002-059X
Universitas Riau, Indonesia

Hartono Tjoe
http://orcid.org/0000-0002-2728-8484
The Pennsylvania State University, USA

DOI: 10.4018/979-8-2600-3485-9.ch009

ABSTRACT

This chapter uses Rumah Lontiok, a traditional elevated timber house of the Kampar Malay community in Riau Province, Indonesia, as the cultural basis for an ethno-mathematical pedagogical framework. The authors conducted content analysis at three heritage sites with laser-assisted field documentation, and then validated the resulting framework through a two-round expert panel of five specialists (overall validity score 3.76 out of 4.00). The study identified five solid geometry domains: the tiang pillar network, the lantai panggung floor platform, the cosmologically organized interior rooms, the curved bumbungan melentik ridge, and the integrated composite solid of the whole building. Domains 1 to 3 align with Kurikulum Merdeka of Phase C core outcomes, whereas domains 4 and 5 serve as enrichment. Van Hiele's (1986) theory provides the structural backbone for the four-stage framework. Bishop's (1988) and Gerdes' (1988, 1994) contributions extend it further. The framework was then translated into six classroom learning activities for grades 5 to 6.

INTRODUCTION

Somewhere in Kampar Regency, Riau Province, a ten-year-old girl stands beneath a *Rumah Lontiok*. Massive timber columns rise on all sides, and they hold a floor platform that she must climb to enter. The rooms inside are divided by cosmological rules that she cannot yet name: *alam berkawan, alam bersamak,* and *alam semalu.* These are zones that her grandparents' generation understood without explanation. Above her head, a ridge beam sweeps upward at both ends in a curve that has always reminded people of the prow of a sailing vessel. She almost certainly does not know that she is standing inside a solid geometry lesson. Every measurable feature of that house corresponds to something in her school mathematics curriculum. Nobody has ever formally made that correspondence available to her teachers.

The research reported in this chapter grew out of that gap, the distance between mathematical knowledge encoded in traditional Kampar architecture and the solid geometry instruction in nearby classrooms. *Rumah Lontiok* is a traditional elevated timber dwelling of the Kampar Malay people. Its *bumbungan melentik,* the dramatically upswept ridgepole, resembles the hull of a *lancang* (a traditional Malay sailing vessel), which is why the structure is also called *Rumah Lancang* or *Rumah Pencalang*. Built on grids of wooden columns and organized according to a sophisticated cosmological philosophy, the house has held national Intangible Cultural Heritage recognition since 2017 (Faisal, 2019; Kementerian Pendidikan dan Kebudayaan, 2017; Munjiatun et al., 2022).

Solid geometry is, by most accounts, among the hardest topics to teach well in the elementary mathematics curriculum. Battista (2007) traced the central difficulty to what he called the 2D-to-3D representational gap. Students are asked to reason about three-dimensional objects from two-dimensional diagrams, and that mismatch exceeds the spatial visualization capacity of many elementary-age learners. Indonesian national assessment data from 2022 show that most elementary students have not reached minimum numeracy benchmarks. Spatial reasoning and solid geometry remain among the weakest subdomains (Pusat Asesmen Pendidikan, 2023), and international meta-analyses point in the same direction. Atit et al. (2022) documented a robust correlation between spatial skills and mathematics performance. Hawes et al. (2022) found that spatial training produces measurable transfer to mathematical outcomes, and Gilligan-Lee et al. (2022) treat spatial thinking as a systematically underdeveloped component of mathematics curricula.

D'Ambrosio (1985) proposed ethnomathematics as a principled response to problems like this one. The idea is to root abstract mathematical content in cultural artifacts that students already know, inhabit, and care about. Research across multiple national contexts has shown that instruction built on ethnomathematics improves engagement and deepens understanding (Nicol et al., 2024; Rosa et al., 2016; Wildfeuer, 2022). Indonesian scholarship in this area has expanded considerably over the past decade. Yet Batiibwe's (2024) review of 61 published articles arrived at an uncomfortable finding: most studies catalog the mathematical features of cultural objects and then stop, without ever producing validated, classroom-ready instructional resources. That shortfall, between documenting what mathematics lives in a cultural artifact and knowing how to teach it, is what this chapter takes on.

The work was organized around three research goals. We first set out to identify and document the solid geometry concepts embedded in *Rumah Lontiok*. That involved laser-assisted field measurement and ethnographic interviews at three heritage sites. A second goal was to validate that content analysis through a structured expert panel covering mathematical, cultural, architectural, and pedagogical expertise. The third goal was to turn the validated analysis into a four-stage teaching framework, drawing on Clements and Battista's (1992) reading of Van Hiele (1986) and developed through six learning activities aligned to the Indonesian curriculum known as *Kurikulum Merdeka* in Phase C outcomes (Grades 5 to 6). The most sophisticated geometry classroom may already exist in timber and curvature. This chapter is about learning to recognize it.

BACKGROUND

Ethnomathematics: Origins, Theoretical Foundations, and the Descriptive-Prescriptive Gap

When D'Ambrosio (1985) published his foundational essay, the central provocation was epistemological. Every human community produces mathematical knowledge through cultural engagement with the world. Western academic traditions are not the only source. Mathematics, on that account, is not context-free. Presenting it as universal has always involved a choice, one that has systematically excluded the mathematical practices of communities whose heritage lies outside the Western academic canon. The consequences for students from those communities are real.

Rosa and Orey (2013) and Rosa et al. (2016) responded to this diagnosis by developing ethnomodelling. Their methodology translates cultural mathematical practices into school-applicable models through documentation, abstraction, and curricular adaptation. What sets ethnomodelling apart from purely descriptive ethnomathematics is its orientation toward use. The question is not only what mathematical knowledge exists in a cultural artifact, but how to make that knowledge accessible to students in formal educational settings. Wildfeuer (2022) reached a related conclusion: when teachers incorporate ethnomathematics and make culturally responsive pedagogical choices, students experience their own community's knowledge as substantive mathematics rather than as a passive inheritance, which supports both their cultural and mathematical identity.

In Indonesia, ethnomathematics scholarship has expanded considerably over the past decade. Studies have examined mathematical structures in *batik*, *gamelan*, traditional agriculture, and regional architecture (Hendriyanto et al., 2023; Karina et al., 2022; Sunzuma & Umbara, 2025). Work focused on Malay traditional architecture across Riau has appeared (Cahyani et al., 2025), and some researchers have begun weaving digital technologies into ethnomathematical learning resources (Rachman et al., 2025). Even so, Batiibwe (2024) found that the literature catalogs mathematical concepts in cultural artifacts without producing validated, classroom-ready frameworks. The authors characterize this as the descriptive-prescriptive gap. Closing it, with the rigor and curricular specificity the field seldom reaches, is the aim of this work.

Ethnomathematics in a Wider Theoretical Perspective

Ethnomathematics is not a single-framework enterprise. The theoretical landscape is richer and more contested than introductory accounts suggest, and this chapter draws selectively on four distinct strands as complementary analytical resources.

First, Bishop (1988, 1990) offered a structural taxonomy of six universal mathematical activities found across all human cultures, though enacted differently in each: counting, locating, measuring, designing, playing, and explaining. At least four of these operate at once in *Rumah Lontiok*. The proportional relationships among *tiang* (pillar), *lantai panggung* (stage floor), and roof structure are governed by measuring. Locating organizes the cosmologically motivated tripartite spatial arrangement of *Alam Nan Tigo*. The structural and aesthetic requirements across all five geometric domains documented in this chapter are coordinated through designing. Counting operates through the regular four-by-six grid of the pillar array. Bishop (1988) argued that mathematical knowledge consolidates most durably when multiple activities operate in concert, a condition vernacular architecture routinely satisfies and classroom mathematics rarely does.

Gerdes (1988) gave ethnomathematics the concept of frozen mathematics, that is, knowledge embedded in cultural artifacts through generations of accumulated practice, no longer consciously articulated by those who carry it. His method of defrosting such knowledge, which involves recovering geometric and computational structures through deliberate analytical scrutiny, is essentially what the field documentation phase of this study did. The *bumbungan melentik* is a precise illustration. Kampar Malay master carpenters produced a curved ridgepole through a series of graduated wooden segments, each positioned slightly higher than the last. Through empirical practice, they achieved what analytical scrutiny now recognizes as a frustum-series approximation of a continuous curve, without ever naming it as such. The mathematics was frozen into the procedure, and the present analysis thaws it.

Knijnik (2012), whose work has been extended in recent international scholarship (Nicol et al., 2024), brought the political stakes of ethnomathematical instruction into focus. Incorporating cultural mathematical knowledge into formal schooling is never a neutral act. Either the knowledge is treated as genuinely mathematical, as a source of insight in its own right, or it gets reduced to illustrative material propping up an unchanged Western curriculum. This chapter takes the first position. We treat the architectural knowledge of Kampar Malay builders as a substantive mathematical achievement. That choice shapes every instructional design decision. The activities are framed so students recognize the sophistication of traditional practice, rather than learning standard geometry with cultural scenery around it.

Rosa and Orey (2013) and Rosa et al. (2016) described bidirectionality as a core principle. Cultural practices shed light on mathematical models, and those models in turn help explain the cultural practices. The *bumbungan melentik* illustrates this clearly. The frustum-series analysis is not simply a mathematical translation of an architectural form. It also explains why the traditional construction technique works: the graduated segments produce a visually continuous curve through a discrete-step procedure that carpenters without formal calculus training could actually execute.

This two-directional engagement distinguishes substantive ethnomathematical work from the superficial use of cultural examples to dress up a pre-existing curriculum.

Each strand has a distinct role in what follows. Bishop supplies the analytical inventory. Gerdes legitimates the recovery procedure. Knijnik shapes the rhetorical orientation of the activities. Rosa and Orey underwrite the two-directional translation. Working from all four together is what produces the methodological robustness Batiibwe (2024) identifies as the field's most persistent unmet need.

Solid Geometry in Elementary Education: Cognitive Demands and Instructional Challenges

When asking most primary school teachers which mathematics topic is hardest to teach well, solid geometry comes up regularly. Battista (2007) traced the root difficulty to what he termed the 2D-to-3D representational gap: the fundamental mismatch between the two-dimensional medium of most geometry instruction (textbook diagrams, worksheets, blackboard drawings) and the three-dimensional nature of what students are actually supposed to learn. A child looking at a printed diagram of a cube must reconstruct a three-dimensional object from a flat image, and that reconstruction depends on spatial visualization skills many elementary students have not yet consolidated (Pittalis & Christou, 2010; Sinclair & Bruce, 2015).

Pittalis et al. (2010) studied Grades 5 to 9 students in Cyprus and identified six dimensions of 3D geometric thinking: constructing and recognizing nets, representing three-dimensional objects, structuring cube arrays, identifying shape properties, computing volume and surface area, and comparing shape properties across figures. Pittalis and Christou (2010) grouped these into four broader reasoning types: representation, spatial structuring, property conceptualization, and measurement. These dimensions correspond closely to the five geometric domains identified in *Rumah Lontiok*, which helps explain why the architectural domains work as Phase C instructional targets rather than simply as cultural illustration. The same research showed that students develop unevenly across dimensions: a child who handles volume calculation reliably may still struggle with net construction, and each dimension needs to be addressed on its own.

A few specific errors appear with such regularity in the literature that they can almost be anticipated in advance. Students undercount the edges of rectangular prisms, typically reporting eight rather than twelve, because they miss the edges connecting the top and bottom faces. Net construction tasks defeat many students, since they cannot mentally unfold and refold solids. When computing composite solid volumes, students commonly double-count overlapping regions. Across all these tasks runs the same procedural-conceptual disconnect: correct computation of $V = l \times w \times h$, but no ability to say what the resulting number means spatially (Battista, 2007; Clements & Battista, 1992).

Several pedagogical requirements follow from this body of research. Students benefit from genuine three-dimensional physical experience alongside or before symbolic reasoning, not from diagram-based instruction alone (Hawes et al., 2022; Sinclair & Bruce, 2015). Calculations should also be embedded in situations with real meaning, so that numerical results carry spatial reference (Clements & Battista, 1992). Geometry instruction works best when it connects to spaces students know and inhabit (Bishop, 1988). *Rumah Lontiok* satisfies these conditions, which is why it functions as a genuine pedagogical resource rather than as a motivating anecdote.

Van Hiele's Theory of Geometric Thinking

Van Hiele's (1986) theory remains the most widely cited developmental account of how geometric understanding grows. It provides the structural backbone for the four-stage framework presented in this chapter. The theory holds that geometric understanding develops through qualitatively distinct levels, and that progression depends on instruction quality, not age. Students who receive well-sequenced, level-appropriate teaching move forward; those who receive instruction pitched above their current level stall at rote memorization (Crowley, 1987; Mainali, 2021).

Van Hiele's original Dutch formulation numbered the levels 0 through 4, with Level 0 as visualization. American mathematics education, including Crowley (1987) and Mason (2002), renumbered these as Levels 1 to 5. Clements and Battista (1992) added a pre-recognition stage below Level 0. This chapter uses Van Hiele's original 0 to 4 numbering throughout, consistent with Mainali (2021) and Sinclair and Bruce (2015). Pre-recognition is acknowledged but not engaged here, since Phase C students have typically moved past it.

At Level 0, students recognize shapes by overall visual appearance through gestalt recognition, without property analysis. Shown a rectangular prism, a Level 0 student will say it looks like a box, not that it has six rectangular faces or twelve edges. Students at Level 1 identify shapes through their properties: faces are counted, edges labeled, and vertices distinguished. Level 2 introduces relational reasoning; students at this level understand that properties imply other properties and that geometric figures relate hierarchically to one another. Levels 3 and 4 involve formal proof and axiomatic rigor, both beyond the elementary range. The framework presented here targets Levels 0 through 2.

One practical consequence has been documented repeatedly: most elementary students arrive at solid geometry instruction at Level 0, yet most textbook instruction is pitched at Level 1 or above. The result is rote performance masquerading as understanding (Clements & Battista, 1992; Mainali, 2021). Every stage of this framework begins where students actually are.

The Indonesian Educational Context

Indonesian primary schooling (*Sekolah Dasar*, SD) covers Grades 1 through 6 for children typically aged 7 to 12. The 2022 *Kurikulum Merdeka* reform, introduced under Permendikbudristek No. 56/M/2022, reorganized the primary curriculum into three developmental phases: Phase A (Grades 1 to 2), Phase B (Grades 3 to 4), and Phase C (Grades 5 to 6). Moving from grade-by-grade specification to phase-based outcomes was intended to give teachers more room to pace instruction and to acknowledge the genuine developmental variability present in any cohort (Kemendikbudristek, 2022). Mathematics in *Kurikulum Merdeka* covers five content strands: numbers, algebra, measurement, geometry, and data analysis.

Phase C is the level this framework targets. Its solid geometry outcomes cover systematic work on cubes, rectangular prisms, and composite solids; volume and surface area calculation; spatial visualization through plan views; and application to real-world contexts. These outcomes correspond directly to four of the five geometric domains identified in *Rumah Lontiok*. The triangular prism and frustum content from Domains 4 and 5 is positioned as enrichment, consistent with those topics formally belonging to Phase D. Students at Phase C are also at the developmental stage where Van Hiele Levels 0 to 2 can be productively engaged through instruction. *Kurikulum Merdeka's* Phase C also foregrounds contextual, project-based, and culturally grounded learning, which creates the curricular space this kind of integrated unit requires.

Running alongside the subject-area outcomes are the six dimensions of the *Pancasila* Student Profile (*Profil Pelajar Pancasila*). The framework directly engages three of them: *Berkebhinekaan Global* (cultural diversity and heritage appreciation), *Bernalar Kritis* (critical reasoning), and *Kreatif* (creativity). These dimensions are not tacked on as afterthoughts; they emerge naturally from the learning activities, particularly Stage 4.

Rumah Lontiok: Architecture, Cosmology, and Cultural Heritage

Among the traditional dwelling types of the Malay world, *Rumah Lontiok* is notable for one structural feature that has no precise counterpart in standard solid geometry taxonomy: the *bumbungan melentik* (see Figure 1). Faisal (2019), who produced the most thorough architectural documentation of the tradition, describes how the ridge achieves its dramatic upswept curvature. The form is not achieved by bending continuous timber, which available wood species and traditional tools would not support, but through a graduated series of wooden segments, each positioned slightly higher than the one before it. Centuries of empirical refinement produced a form that is architecturally singular and pedagogically rich. Cahyani et al. (2025) recently situated *Rumah Lontiok* within the broader landscape of Malay heritage structures across Riau, identifying shared structural principles and features unique to the Kampar tradition.

Figure 1. Rumah Lontiok at the primary research site in Kuok Village, Kampar Regency, displaying the characteristic bumbungan melentik, the elevated lantai panggung, and the tiang pillar array

Faisal (2019) also documents the structural logic of the elevated floor system. Post-and-beam construction with carefully proportioned member dimensions achieves load-bearing capacity without engineered calculation in the modern sense. Al Husaini et al. (2023) confirmed that the original construction logic remains legible at the *Kendil Kemilau Emas* site despite decades of weathering. The timber pillars (*tiang* in Kampar Malay) appear in multiple cross-sectional profiles across documented examples of the tradition. Square (*segi empat*), hexagonal, heptagonal, octagonal, and nonagonal forms are all recorded, each carrying specific symbolic meaning in Kampar tradition (Faisal, 2019; Karina et al., 2022). Mathematical structure in *Rumah Lontiok* is a culturally articulated achievement.

Karina et al. (2022), through ethnographic fieldwork at the touristic village of Pulau Belimbing, documented the cosmological organization of the interior into three named zones: *alam berkawan* (the communal front space where village members gathered), *alam bersamak* (the middle family room), and *alam semalu* (the private kitchen). In the broader Kampar Malay *adat* tradition, these zones are realized architecturally as the *rumah induk* (main house body), the *anjungan* (an elevated zone within the *induk* footprint), and the *dapur* (a rear extension). The cosmological tripartition motivates the decomposition of the building's interior into a composite solid with culturally differentiated components.

Rumah Lontiok has held national Intangible Cultural Heritage recognition since 2017 (Kementerian Pendidikan dan Kebudayaan, 2017). Modernization, urbanization, and the gradual loss of master carpenters with traditional joinery knowledge all pose documented preservation threats. Using this building as a mathematics education resource is simultaneously ethnomathematical research and an act of cultural documentation and advocacy, consistent with the dual orientation D'Ambrosio (1985) and Knijnik (2012) share toward mathematical justice and cultural survival.

RESEARCH DESIGN AND METHODOLOGY

The authors combined qualitative ethnomathematical content analysis (Krippendorff, 2018; Rosa & Orey, 2013) with the validation phase of educational design research (Plomp & Nieveen, 2013). Neither approach was sufficient by itself. Ethnomathematical content analysis provides procedures for recovering mathematical knowledge embedded in cultural artifacts (Gerdes, 1988). Design research validation supplies the quality standards needed to ensure that the resulting instructional resource is accurate, culturally credible, and usable in actual classrooms. Without the second, the analysis produces an interesting description that teachers cannot act on. Without the first, validation produces a polished product disconnected from cultural reality.

The work proceeded through three sequential phases. Field documentation across three *Rumah Lontiok* sites produced the empirical foundation, including laser-verified architectural measurements and nine ethnographic interviews. Content analysis and curricular alignment then turned those data into a structured map connecting each identified architectural-geometric domain to specific Phase C competency statements. The third phase was expert validation: two rounds evaluated the content map and pedagogical framework against eight criteria through a five-member panel.

Rumah Lontiok in Kuok Village, Kampar Regency, served as the primary site. It was selected because all five geometric domains are present in canonical form, it is publicly accessible for verification, and it holds formal regional heritage recognition that guarantees structural integrity. Pulau Belimbing and Koto Masjid villages served as secondary sites, providing cross-validation of dimensional measurements and confirming that the dimensional patterns at *Rumah Lontiok* represent the tradition rather than a single idiosyncratic building.

Field Documentation Procedure

The authors conducted three site visits using a Leica DISTO X4 laser distance measurer (manufacturer accuracy ±1.5 mm at 10 m) and a Bosch GCL 2-50 CG cross-line laser level (±0.3 mm/m at 10 m). The choice of laser instrumentation over tape measures was deliberate: tape-measure documentation in vernacular architecture studies introduces imprecision that, for ethnomathematical applications, undermines the credibility of every calculation built on those measurements (cf. Al Husaini et al., 2023; Cahyani et al., 2025).

All linear dimensions were measured in triplicate across three visits on separate days, and the mean of the three readings is the reported value. Where readings varied by more than 1% from their mean, a fourth reading was taken and the outlier excluded. *Tiang* cross-sections were measured at base, mid-height, and near-top to confirm constant cross-section along each pillar's full length, confirming that each qualifies as a rectangular prism rather than a frustum.

Nine semi-structured ethnographic interviews were conducted. At each of the three sites, the authors spoke with two *adat* custodians (*penghulu adat*) and one master carpenter (*tukang tuo*). The questions covered five areas: the cultural and cosmological significance of the *Alam Nan Tigo* spatial organization; the symbolic meaning of different pillar cross-sectional profiles; the traditional construction technique of the *bumbungan melentik*; the proportional system governing the building's dimensional relationships; and current preservation conditions. Interviews were conducted in Bahasa Indonesia with elaboration in Bahasa Melayu Kampar where participants preferred it, audio-recorded with informed consent, and fully transcribed.

Measurements and ethnographic data were triangulated iteratively. When the *tukang tuo's* account of *bumbungan melentik* construction implied a frustum series, the physical measurements were re-examined to confirm or disconfirm that geometric structure. When measurements revealed the regular four-by-six pillar grid, the interview record was consulted for its cultural rationale. Neither data source was treated as primary; each constrained the interpretation of the other.

Geometric Content Analysis and Curricular Alignment

Coding proceeded deductively first, then inductively. The deductive phase searched the field data for instances of standard Phase C solid geometry concepts: rectangular prisms, cubes, triangular prisms, and composite solids. The inductive phase then searched for geometric structures not covered by Phase C content, principally the frustum-series of the *bumbungan melentik*. Running both phases in sequence preserved the dual requirement of curricular connection and architectural fidelity.

Two researchers coded independently before comparing results. Initial agreement was 92%, with Cohen's $\kappa = 0.86$, representing substantial reliability by established interpretive thresholds (Krippendorff, 2018). Disagreements were resolved by returning to source measurements. The main point of disagreement concerned the *bumbungan melentik*: whether to categorize the curved ridge as a triangular prism with approximation, as a frustum series, or as a hybrid of both. A hybrid characterization was adopted because it best preserves both mathematical accuracy and architectural fidelity.

Each identified domain was then matched to specific competency statements from Permendikbudristek No. 56/M/2022, that is, the concrete outcome statements teachers are accountable for, not curriculum themes. Each domain was also assigned a Van Hiele level conservatively. Placement at a given level required that a typical Phase C student could meaningfully engage there using accessible methods.

Two-Round Expert Validation

Five validators were selected to cover four relevant areas of expertise: mathematics education with elementary geometry specialization; elementary mathematics curriculum with active *Kurikulum Merdeka* involvement; cultural heritage architecture with *Rumah Lontiok* expertise; Kampar Malay cultural studies with published scholarship on Kampar material culture; and mathematics teacher education at the elementary level. A five-member panel suits the requirements of content validity assessment (Aiken, 1985).

Eight criteria structured the assessment: mathematical accuracy of geometric identification; correctness of property identification; alignment with *Kurikulum Merdeka* Phase C; Van Hiele level-appropriateness; pedagogical sequencing; classroom feasibility; cultural accuracy; and translatability to other regional contexts. Validators rated each criterion on a four-point scale (1 = Not Valid, 4 = Highly Valid), with acceptance thresholds established before data collection: 1.00–1.75 Not Valid; 1.76–2.50 Less Valid; 2.51–3.25 Valid; 3.26–4.00 Highly Valid (Aiken, 1985).

After Round 1, validators provided ratings alongside written justifications. All specific corrections were integrated into a revised framework before Round 2.

Round 2 scores constitute the final validity assessment. Table 1 presents the results across both rounds.

Table 1. Expert validation results across two rounds (n = 5 validators; scale 1–4)

Validation Criterion	Round 1 Mean	Round 2 Mean	Interpretation
Mathematical accuracy of geometric identification	3.40	3.80	Highly Valid
Geometric property identification correctness	3.35	3.75	Highly Valid
Curricular alignment with *Kurikulum Merdeka* Phase C	3.45	3.80	Highly Valid
Van Hiele level-appropriateness of framework stages	3.30	3.70	Highly Valid
Pedagogical sequencing and progression	3.40	3.75	Highly Valid
Feasibility of learning activities in Grade 5 classrooms	3.50	3.80	Highly Valid
Cultural accuracy of Kampar Malay representation	3.55	3.85	Highly Valid
Translatability to other regional contexts	3.40	3.70	Highly Valid
Overall Mean	3.42	3.76	Highly Valid

Round 2 produced an overall mean of 3.76, which falls within the upper portion of the Highly Valid range (3.26–4.00). Individual criterion scores ranged from 3.70 to 3.85, and no criterion fell below the Highly Valid threshold. The 0.34-point improvement from Round 1 to Round 2 reflects substantive revision in response to panel feedback, not cosmetic adjustment.

Solid Geometry Domains in Rumah Lontiok Architecture

Five solid geometry domains emerged from the field documentation and content analysis. Each domain links to specific Phase C outcomes and can be engaged at one or more Van Hiele levels. Table 2 summarizes the verified architectural dimensions.

Table 2. Verified architectural dimensions across five geometric domains (Rumah Lontiok, Kuok Village)

Domain	Component	Dimensions	Quantity
1. *Tiang* system	Square-section timber pillars	0.20 × 0.20 × 1.75 m	24 (4×6)
2. *Lantai panggung*	Main floor platform	10.4 × 7.2 × 1.75 m	1
3. Interior zones	*Rumah induk* (Main area)	Net floor area × 3.5 m height	1
	Anjungan	2.4 × 3.6 × 3.0 m	1

continued on following page

Table 2. Continued

Domain	Component	Dimensions	Quantity
	Kitchen	3.6 × 5.8 × 3.4 m	1
4. Roof body (enrichment)	Triangular prism approximation	Span 10.4 m, height 2.1 m, depth 9.8 m	1
***Bumbungan melentik* (enrichment)**	Frustum series (curved ridge)	3 segments × varying dimensions	2 (each end)

Domain 1: The Tiang System

Twenty-four square-section timber pillars stand in a four-by-six grid at *Rumah Lontiok* (see Figure 2). Pillar counts vary considerably across examples of the tradition. Karina et al. (2022) document 6 primary pillars at Pulau Belimbing village, and Faisal (2019) records 38 *tiang* at *Rumah Lontiak Majo*, a different exemplar not included in this study's three-site sample. All figures here pertain specifically to the verified primary site. Each pillar measures 0.20 m × 0.20 m in cross-section and 1.75 m in height, making it a square prism with six faces, twelve edges, and eight vertices. One pillar occupies $V = 0.20 \times 0.20 \times 1.75 = 0.07\ m^3$. All 24 together occupy 1.68 m^3, roughly the usable timber from three to four mature tropical hardwood trees, making concrete to students the forest resources a single house demands.

Figure 2. Adjacent Rumah Lontiok structures showing the tiang construction system with square cross-sections and the characteristic carved wooden window panels

Grid spacing of approximately 2.08 m along the long axis and 2.4 m along the short axis reflects empirical optimization over generations: enough distance to preserve open interior space, with enough density to distribute structural loads reliably (Faisal, 2019; Munjiatun et al., 2022). Bishop's (1988) counting and locating activities operate simultaneously here. Phase C outcomes covered include identification of solid figures by properties, rectangular prism classification, and aggregate volume across multiple identical instances, at Van Hiele Levels 0 and 1.

Domain 2: The Lantai Panggung

At 10.4 m × 7.2 m × 1.75 m, the elevated main floor platform is a single large rectangular prism (see Figure 3). The volume of the space beneath it is $V = 10.4 \times 7.2 \times 1.75 = 131.04\ m^3$. That number comes to life when placed alongside the air volume of a standard Indonesian elementary classroom, approximately $7 \times 8 \times 3 = 168\ m^3$. The under-platform space of one *Rumah Lontiok* is nearly three-quarters of a classroom. Once students see the comparison, they no longer treat $131.04\ m^3$ as an abstract computational result; it becomes something spatially real.

Figure 3. The carved entrance staircase and elevated lantai panggung of Rumah Lontiok, showing the 1.75 m elevation above ground level

Why 1.75 m? The Kampar River floods seasonally, and a raised floor at that height gave meaningful protection. The sheltered ground-level space below was used for livestock, storage, and craft work. The geometry here is practical, not decorative. Students who understand why the elevation is 1.75 m gain something beyond calculation: they encounter the reasoning of people who faced real engineering problems and solved them with geometry. That connection is precisely what *Kurikulum Merdeka's* Phase C outcomes mean when they call for authentic, real-world application.

Domain 3: Interior Spatial Zones (Composite Solids and the Anjungan Subtraction)

Karina et al. (2022) document three zones at the touristic village of Pulau Belimbing: *alam berkawan* (communal front space), *alam bersamak* (family middle room), and *alam semalu* (private kitchen). Architecturally, these correspond to the *rumah induk* (main area, interior height 3.5 m), the *anjungan* (elevated zone within the *induk* footprint, 2.4 × 3.6 m, height 3.0 m), and the kitchen (rear extension, 3.6 × 5.8 × 3.4 m).

The geometry here requires care. The *anjungan* footprint sits inside the main floor area, so a naive sum of the three component volumes would count the *anjungan* column twice. The correct computation starts with the area of the main floor: (10.4 × 7.2) – (2.4 × 3.6) = 74.88 – 8.64 = 66.24 m^2. The volume is therefore 66.24 × 3.5 = 231.84 m^3. Adding the *anjungan* (2.4 × 3.6 × 3.0 = 25.92 m^3) gives 257.76 m^3 for the combined interior space. Adding the kitchen (3.6 × 5.8 × 3.4 = 70.99 m^3) brings the aggregate to 328.75 m^3. The cultural tripartition of *Alam Nan Tigo* is not just a narrative frame; it is the actual reason the decomposition takes the form it does.

Domain 4: The Roof Body (Triangular Prism with Approximation)

Setting the curved ridge aside for a moment, the roof body approximates a triangular prism: span 10.4 m, apex height 2.1 m above the wall plate, depth 9.8 m. Treated as a triangular prism, $V = ½ \times 10.4 \times 2.1 \times 9.8 = 107.02$ m^3. The word approximation carries real weight here, because the actual form departs from a pure triangular prism where the *bumbungan melentik* curves upward at both ends.

The departure is pedagogically useful. Students working with Domain 4 encounter something rarely addressed in standard Phase C textbooks: mathematical models of physical objects are idealizations. Making that explicit by identifying where the model departs from reality, and whether it is adequate for the task at hand, is itself a form of mathematical reasoning (Rosa & Orey, 2013; Rosa et al., 2016). Formal triangular prism instruction belongs to Phase D. What Phase C students gain here is the half-of-rectangular-prism relationship and, more importantly, the metacognitive experience of evaluating a principled approximation.

Domain 5: The Bumbungan Melentik (Frustum Series Approximation)

No feature of *Rumah Lontiok* is architecturally or mathematically more unusual than the *bumbungan melentik*. Traditional Kampar Malay carpenters could not bend continuous timber into a curve. Available wood species and hand tools did not allow it. Their solution was a graduated series of segments, each placed slightly higher than the previous one. The result is a form that reads visually as a continuous upward sweep (Faisal, 2019). At *Rumah Lontiok*, each upswept end consists of three segments of progressively decreasing cross-sectional area, ending in the characteristic upturned tip.

Each segment is geometrically a frustum, that is, a truncated rectangular pyramid with rectangular bases of different sizes. Three sequential frustums together approximate what a contemporary structural engineer would specify with a parametric curve equation. In Gerdes' (1988) terms, this is frozen mathematics in its clearest form: a continuous-curve problem is solved through a discrete-step procedure, refined over generations without recourse to calculus. The frustum analysis exceeds the Phase C scope. Yet students can productively engage with the visual observation that the curved ridge is built from graduated segments, and with the insight that those segments approximate something smooth.

A Four-Stage Van Hiele-Grounded Pedagogical Framework

With five domains documented, the empirical foundation was set. The next task was turning that foundation into structured, sequenced instruction that teachers could actually use. Each of the four stages targets a Van Hiele level and a set of Phase C outcomes, with specific learning activities carrying the work. Stages are not interchangeable. Each one prepares the ground for the next, and Stage 4 pulls together what the first three developed.

Stage 1: Visual Recognition (Van Hiele Level 0)

At Level 0, students recognize geometric forms by overall visual appearance, before they can analyze properties. Stage 1 works at that level. Photographs, scale models, and site visits (where feasible) give students real experience of the five domains before any formal analysis begins. Everyday, descriptive language is not just permitted at this stage; it is the goal. A student who says the roof looks like a smile or that the floor is floating above the ground is doing exactly the right cognitive work for Level 0. Formal vocabulary is held back on purpose.

The sequencing reflects a principle Knijnik (2012) argued for: ethnomathematical instruction should treat cultural knowledge as substantive in its own right, not as packaging for formal content. Starting with the cultural meaning of the building, that is, what it is, who built it, what the rooms mean, and why the ridge curves, before asking what geometric shapes it contains, is not a nicety but a pedagogical commitment. Phase C outcomes at this stage are primarily orientational, covering cultural appreciation, curiosity about geometric structure, and the experiential foundation that later stages will analyze formally. Among the three Pancasila Student Profile dimensions this framework engages, *berkebhinekaan global* is most directly activated here, through sustained encounter with the building's cultural and cosmological significance before any formal analysis begins.

Stage 2: Property Analysis (Van Hiele Level 1)

At Level 1, shapes are no longer recognized by appearance alone; they are analyzed through their properties. Stage 2 is where formal geometric vocabulary enters: face, edge, vertex, base, height, prism. The terms are introduced not in the abstract but attached to *Rumah Lontiok* components that students now know by sight and name. With scale models or carefully prepared 2D representations, students count and label faces, count and label edges, distinguish vertices, measure dimensions, and record their findings on structured worksheets.

Stage 2 is organized around a predict-verify-reflect cycle. Students predict how many faces a *tiang* has, count systematically to verify, and reflect on why the prediction was right or wrong. This structure has a consistent track record for surfacing and engaging misconceptions (Battista, 2007; Pittalis & Christou, 2010; Sinclair & Bruce, 2015). The most common Level 1 error, confusing edges with vertices or undercounting the twelve edges of a rectangular prism, tends to appear here, and catching it early is instructionally valuable. Phase C outcomes include identification and classification of solid figures by properties, with attention to rectangular prisms, cubes, and triangular prisms.

Stage 3: Relational Reasoning (Van Hiele Level 2)

Stage 3 moves into Level 2, which involves reasoning about relationships between properties and between figures. Composite solid analysis enters here. Students decompose complex three-dimensional forms into identifiable components and then aggregate their volumes. Domain 3's interior zone analysis is the natural setting: the *Alam Nan Tigo* tripartition supplies a culturally grounded rationale for the decomposition, because the zones are not arbitrary but spiritually and functionally distinct (see Figure 4).

Figure 4. Interior carved doorway connecting the cosmologically organized zones of Rumah Lontiok, marking the spatial transition between zones of Alam Nan Tigo

Two particularly demanding operations appear at Stage 3. The first is net construction and unfolding: students must mentally and physically transform 3D solids into flat representations and back again. This is one of the most reliable diagnostic tasks for spatial visualization capacity (Pittalis & Christou, 2010; Sinclair & Bruce, 2015). The second is principled approximation, encountered through the triangular prism model of the roof, which introduces the idea that mathematical modeling of real objects involves deliberate idealization. This is a meta-mathematical concept that standard Phase C curricula rarely address explicitly (Rosa & Orey, 2013; Rosa et al., 2016). Phase C outcomes addressed here are composite solid volume and surface area, recognizing decomposability, and real-world application. *Bernalar kritis* (critical reasoning) is most directly exercised at this stage through the demands of relational reasoning, principled approximation, and systematic decomposition planning.

Stage 4: Integrated Application (Levels 1–2 Synthesis)

Stage 4 is a project-based unit. Students design and construct a scaled paper model of *Rumah Lontiok*. They compute dimensions, volumes, and surface areas, present their reasoning to classmates, and document the process in a structured portfolio. The Pancasila Student Profile dimensions addressed here are genuinely embedded in the mathematical work: *berkebhinekaan global* comes through cultural heritage engagement; *bernalar kritis* is exercised through integrative geometric reasoning across the five domains; and creativity emerges from the design-and-build format.

Structuring Stage 4 as a Project-Based Learning unit (*Pembelajaran Berbasis Proyek*, PjBL) is consistent with *Kurikulum Merdeka* guidance. What raises it above a culminating craft project is the simultaneous activation of Bishop's (1988) multiple mathematical activities. Counting happens through pillar enumeration, locating through spatial layout decisions, measuring through dimensional specification, and designing through aesthetic and structural coordination. Bishop argued that mathematical knowledge consolidates most durably when several of these activities operate together; Stage 4 sets up that condition by design.

Six Validated Classroom Learning Activities

Six learning activities carry the four stages into the classroom. Each was designed for Phase C and reviewed through the two-round expert procedure. Table 3 presents the full sequence.

Table 3. Six validated classroom learning activities for Phase C solid geometry instruction

No.	Activity Title	Stage / Level	Duration	Primary Outcomes
1	Shape Detectives at a Rumah Lontiok	Stage 1 / Level 0	2 × 35 min	Holistic recognition; cultural orientation
2	Counting the Pillars: Faces, Edges, Vertices	Stage 2 / Level 1	2 × 35 min	Property analysis; systematic enumeration
3	Measuring the Lantai Panggung: Volume and Meaning	Stage 2 / Level 1	2 × 35 min	Volume calculation; unit conversion
4	Unfolding the Roof: Nets and Approximation (enrichment)	Stage 3 / Level 2	3 × 35 min	Net construction; principled approximation
5	Decomposing the Interior: Alam Nan Tigo as Composite Solid (enrichment)	Stage 3 / Level 2	2 × 35 min	Composite solid decomposition
6	Junior Architect: Designing a Rumah Lontiok	Stage 4 / Levels 1–2	5 × 35 min	Project synthesis; design integration

Two Illustrative Lesson Vignettes

Vignette 1: Activity 1 — Shape Detectives at a Rumah Lontiok (Stage 1, Level 0)

Session 1. A large photograph of *Rumah Lontiok* goes on the projector. The teacher asks two questions only: What do you notice? What does it remind you of? Responses at Level 0 are predictably holistic: it looks like a boat, the roof is smiling, the floor is floating. The teacher listens and records without correction or formal vocabulary. The goal of Session 1 is a mental image, not a property list.

Five large-format photographs are then placed at five numbered stations, one for each geometric domain. Pairs rotate every four minutes, recording what the component looks like in everyday terms, what shapes from earlier lessons it calls to mind, and the Kampar Malay name from the vocabulary card provided. After rotation, pairs regroup into fours to compare observations, and each group identifies one surprising observation to share. Those observations are recorded for Session 2.

Session 2. The class chart from Session 1 reappears, and formal geometric names are introduced beside the Kampar Malay terms. *Tiang* is a rectangular prism. The *anjungan* column is a rectangular prism. The roof body approximates a triangular prism. Each pair receives a small 3D model and is asked to find three things their component shares with their neighbour pair's component. Students then write one sentence, in their own words, explaining why their component carries its geometric name. The session ends by naming what shifted: from seeing a building to noticing its properties.

Vignette 2: Activity 2 — Counting the Pillars: Faces, Edges, Vertices (Stage 2, Level 1)

Session 1. Opening question: How do we know every rectangular prism has the same number of faces, edges, and vertices, without checking every one that has ever existed? Predictions are collected and recorded without comment, including wrong ones. Scale models of *tiang* (5 cm × 5 cm × 43.75 cm at 1:4 scale, or 2 cm × 2 cm × 17.5 cm at 1:10 scale, cardboard or wood) are distributed with investigation record sheets. Pairs count and label each face, edge, and vertex using sticky notes to prevent double-counting. The teacher circulates asking: how do you know you haven't missed a face? After counting, pairs compare strategies. Final class consensus: 6 faces, 12 edges, 8 vertices. Predictions are revisited without blame.

Session 2. Bridge question: if one *tiang* requires 70,000 cm^3 of timber, what does the entire pillar system require? Students compute via repeated addition and via multiplication, then discuss what each method feels like cognitively. A follow-up question asks which properties of a single *tiang* (volume, or the face-edge-vertex profile) change when the number of pillars increases, and which stay the same. That distinction, between properties that scale with quantity and properties that remain constant, is the conceptual seed of Level 2. The session names it explicitly and points forward to Activity 3.

Anticipated Student Reasoning and Common Misconceptions

One practical mark of a well-designed instructional framework is that it tells teachers where students are likely to get stuck before those moments arrive. Drawing on Battista (2007), Clements and Battista (1992), Pittalis and Christou (2010), and Sinclair and Bruce (2015), this section maps the predictable misconceptions domain by domain and sketches productive instructional responses.

Domain 1 (Tiang System): Holistic-Shape and Property-Counting Errors

At Level 0, students often treat all elongated rectangular solids as equivalent boxes, without distinguishing cubes, square-based prisms, and non-square rectangular prisms. A more specific difficulty appears at the Level 0 to Level 1 boundary: students count edges or vertices without reliably distinguishing between them, so edge counts get reported as vertex counts and vice versa. Requiring physical contact with the model before any count begins (touching and labeling each face, edge, and vertex before writing anything down) helps students bridge visual recognition and kinesthetic confirmation. The most predictable counting error is reporting eight edges rather than twelve, because students miss the four edges that connect the top and bottom faces. Working through that error by systematic enumeration is instructionally productive in its own right.

Domain 2 (Lantai Panggung): Formula-Procedure and Unit-Conversion Errors

The procedural-conceptual disconnect that Clements and Battista (1992) identified as the central failure of conventional volume instruction surfaces here reliably. Students compute $V = l \times w \times h$ correctly, but cannot explain what the result means spatially. Unit conversion is a companion problem: students working in centimetres who shift to metres without adjusting produce results off by factors of a million. The *lantai panggung* volume of 131.04 m^3 is pedagogically useful precisely because it is large enough to demand spatial interpretation. When students compare it to their classroom volume, the number becomes experientially real. Asking students to predict the spatial meaning of a calculation before executing it, then identify a comparison object afterward, addresses both errors at once.

Domain 3 (Interior Zones): Double-Counting Error in Composite Solids

The canonical composite solid error appears here: students add all component volumes without accounting for nested or overlapping regions. The *rumah induk–anjungan* configuration is set up to surface this error. Students who sum V(*induk*) + V(*anjungan*) + V(kitchen) without subtracting the *anjungan* footprint from the main floor area will double-count that column. Making the subtraction explicit and asking students to explain in their own words why it is necessary has diagnostic value. A student who performs the subtraction without explaining its rationale is still at Level 1. A student who can explain it is crossing into Level 2.

Domain 4 (Bumbungan Melentik): Over- and Under-Idealization Errors

Two opposing errors appear in this domain. Some students accept the triangular prism approximation uncritically, failing to notice that the ridge curves and that the model therefore misrepresents the actual form at both ends (over-idealizing). Others reject the approximation as simply wrong because the roof is not a triangle, missing the fact that principled mathematical simplification is a legitimate and widely used analytical strategy (under-idealizing). A productive response makes the idealization choice explicit and evaluative: compute the volume under the approximation, identify visually where the model over- and underestimates the real form, and decide whether the approximation is adequate for the purpose at hand. The sequence introduces the concept of principled idealization at a level few Phase C curricula reach.

Domain 5 (Complete Structure): Systematic-Decomposition Error

Students who attempt to compute the total composite solid volume often try to do it in one step, generating errors that are difficult to trace and correct. Others show what might be called formula preference, a strong pull toward a single clean calculation over structured sequential decomposition. The capacity to decompose systematically is diagnostic of the Level 1 to Level 2 transition: students who plan before computing are moving into Level 2; students who cannot are still consolidating Level 1. The framework does not prevent these errors from occurring; its value lies in giving teachers the advance information they need to engage those errors substantively, rather than meeting them unexpectedly in summative assessments.

SOLUTIONS AND RECOMMENDATIONS

Theoretical Contributions to Ethnomathematics Research

Batiibwe (2024) diagnosed the descriptive-prescriptive gap as the field's most persistent problem. This chapter offers a concrete counter-instance: the distance between cultural documentation and validated classroom design can be crossed. The process described here shows how, and the demonstration itself is a contribution. The claim is not about *Rumah Lontiok* specifically, but about what ethnomathematics research can produce when it follows the full methodological sequence from field documentation through expert-validated instructional design.

A second contribution is the theoretical integration. Bishop (1988), Gerdes (1988), Knijnik (2012), Rosa and Orey (2013), and Rosa et al. (2016) have each been drawn on separately in previous work. When the four strands are coordinated explicitly, with each one assigned a specific analytical function, the result is a more rigorous theoretical apparatus than any single framework provides. Hendriyanto et al.'s (2023) systematic review found that Indonesian ethnomathematics studies most commonly engage a single theoretical orientation at a time. The chapter presented here works on four at once.

A third contribution is geographic. The ethnomathematics literature remains disproportionately concentrated in Latin American, African, and Western European contexts (Nicol et al., 2024). Indonesian Malay-architectural scholarship is underrepresented in that literature. The *Rumah Lontiok* case adds substantively to a body of work whose regional scope needs to expand if its claims to universality are to hold.

Practical Recommendations for Classroom Teachers

The full instructional sequence is designed to be usable without extensive preparatory reworking. It includes verified field measurements, six validated activities, and explicit Phase C outcome alignment. The misconception map gives teachers advanced warning of where students are likely to struggle, rather than leaving those difficulties to surface in summative assessment. The Pancasila Student Profile dimensions are genuinely embedded in the mathematical work. The framework demonstrates in practice that cultural responsiveness and rigorous content instruction can coexist, which *Kurikulum Merdeka* calls for but has been difficult to implement.

For Kampar Malay students, the framework communicates something else: their cultural heritage is a source of genuine mathematical achievement. Wildfeuer (2022) argued that culturally responsive pedagogical choices incorporating ethnomathematics support students' cultural and mathematical identity simultaneously. When students see their own community's knowledge treated as substantive mathematics rather than as a decorative backdrop, research consistently links that shift to increased engagement and achievement among students from historically underrepresented backgrounds (Kong et al., 2023; Nicol et al., 2024).

Implementation Considerations and Adaptive Adjustments

Teachers unfamiliar with *Rumah Lontiok* will need professional development, including both content study and, where geography permits, site visits. Material resources such as scale models, photograph sets, and structured worksheets must be available in sufficient quantities for paired work. The activity descriptions specify requirements in enough detail to support pre-unit preparation. Schools outside Riau Province can adapt the cultural framing to locally relevant vernacular architectural traditions, keeping the methodological structure but substituting culturally proximate content.

The translatability criterion scored 3.70 in Round 2, placing it in the Highly Valid range. This suggests real potential for replication beyond Riau. *Rumah Gadang* in West Sumatra, *Mbaru Niang* in East Nusa Tenggara, *Tongkonan* in South Sulawesi, *Joglo* in Java, and *Bale* in Bali each present a distinct geometric vocabulary and cultural context that the same methodological approach can handle. Coordinated replication across Indonesian universities could produce a national network of validated ethnomathematics frameworks, which currently does not exist.

One last practical note concerns assessment. The framework does not prescribe a specific summative instrument. Teachers implementing it should align their assessment tasks with the Van Hiele level targeted at each stage. Level 0 performance is most reliably captured through observation and oral response. Level 1 is best assessed through property identification and enumeration tasks. Level 2 calls for composite solid problems that require written decomposition plans and relational justifications. The most common mismatch in conventional instruction is assessing Level 0 work with Level 1 instruments, which results in underreporting of student understanding and undermines the developmental logic on which the framework is built.

Future Research Directions

The work suggests four directions for further research. First, classroom implementation research is the most urgent. Expert panel validation establishes a necessary condition for deployment, but it does not establish sufficiency. Whether the framework actually improves student learning must be tested in Grade 5 classrooms through quasi-experimental or design-based studies. Outcomes worth measuring include mathematical learning on standardized geometry assessment, cultural identity on validated scales, student engagement through classroom observation, and mathematics self-efficacy through self-report instruments. One particularly important sub-question concerns whether effectiveness differs between Kampar Malay and non-Kampar Indonesian students, which bears on when cultural specificity in instruction functions as support and when it functions as constraint.

Longitudinal research on Van Hiele progression addresses a question this study cannot answer: does the framework produce durable advancement through Levels 0 to 2 over the full Phase C span, or only short-term performance gains? Cross-sectional measurement at the Grade 5 and Grade 6 endpoints, relative to conventionally taught comparison groups, would provide the needed evidence (Mainali, 2021; Pittalis & Christou, 2010).

A third direction is methodological replication across other Indonesian vernacular traditions. Coordinated work targeting *Rumah Gadang, Tongkonan, Mbaru Niang, Joglo,* and *Bale*, each aimed at different mathematical content strands and different *Kurikulum Merdeka* phases, could, if sustained across multiple universities, transform Indonesian ethnomathematics from a collection of individual studies into a coherent and cumulative research program.

The fourth direction is technology-enhanced extensions. Rachman et al. (2025) showed that interactive digital media can be productively integrated with ethnomathematics instruction. Three-dimensional virtual models of *Rumah Lontiok* that students can decompose interactively, augmented reality overlays annotating geometric features onto photographs, and mobile platforms distributing the validated activities regardless of geographic location all address the resource-distribution constraints that limit the paper-based framework's reach, particularly in rural areas of Riau and beyond.

CONCLUSION

A traditional Kampar Malay building became the empirical foundation for an elementary solid geometry curriculum. The path from one to the other was methodologically demanding: it required field documentation at three heritage sites, ethnographic interviews with nine participants, geometric content analysis with established inter-coder reliability, curricular alignment at the level of specific competency statements, and two rounds of expert validation against eight criteria. What emerged was a four-stage pedagogical framework, six classroom-ready learning activities, and an overall validity score of 3.76 out of 4.00.

Set against the descriptive-prescriptive gap, the persistent tendency of ethnomathematics scholarship to document and then stop, this chapter makes a methodological argument through demonstration. The gap can be closed. Closing it requires the full sequence described above, not a subset of it. Each step was necessary. The expert panel confirmed this, and the 0.34-point improvement from Round 1 to Round 2 shows what deliberate revision across two rounds of structured feedback actually produces.

The framework is more than a solid geometry unit for Phase C students in Riau. It demonstrates a replicable methodology. A researcher working with *Rumah Gadang, Tongkonan, Mbaru Niang*, or any other vernacular architectural tradition can follow the same sequence of field documentation, content analysis, curricular alignment, and expert validation to produce their own validated instructional resources. The methodological template carries across contexts. That is the broader contribution.

The ten-year-old girl standing beneath the *bumbungan melentik* is surrounded by geometry she does not yet have names for. The ridge above her is a discrete-step approximation of a continuous curve. The pillars around her form a four-by-six grid of rectangular prisms together amounting to 1.68 m^3 of tropical hardwood. The floor she is about to climb encloses a volume nearly three-quarters the size of her classroom. With this framework, a teacher who has studied *Rumah Lontiok* can show her all of it. In doing so, the teacher communicates something that most mathematics classrooms never manage to say: her community's way of building is mathematics, not a metaphor for it.

REFERENCES

Aiken, L. R. (1985). Three coefficients for analyzing the reliability and validity of ratings. *Educational and Psychological Measurement, 45*(1), 131–142. DOI: 10.1177/0013164485451012

Al Husaini, M. A., Hidayat, W., Mashuri, Djuniati, S., & Nawawi, A. (2023). Study of existing conditions as preservation Rumah Lontiok of Kendil Kemilau Emas Foundation. [MARCOPOLO]. *Indonesian Journal of Interdisciplinary Research in Science and Technology, 1*(8), 719–730. DOI: 10.55927/marcopolo.v1i8.6255

Atit, K., Power, J. R., Pigott, T., Lee, J., Geer, E. A., Uttal, D. H., Ganley, C. M., & Sorby, S. A. (2022). Examining the relations between spatial skills and mathematical performance: A meta-analysis. *Psychonomic Bulletin & Review, 29*(3), 699–720. DOI: 10.3758/s13423-021-02012-w PMID: 34799844

Batiibwe, M. S. K. (2024). The role of ethnomathematics in mathematics education: A literature review. *Asian Journal for Mathematics Education, 3*(4), 383–405. DOI: 10.1177/27527263241300400

Battista, M. T. (2007). The development of geometric and spatial thinking. In Lester, F. K. (Ed.), *Second handbook of research on mathematics teaching and learning* (pp. 843–908). Information Age Publishing.

Bishop, A. J. (1988). *Mathematical enculturation: A cultural perspective on mathematics education.* Kluwer Academic Publishers. DOI: 10.1007/978-94-009-2657-8

Bishop, A. J. (1990). Western mathematics: The secret weapon of cultural imperialism. *Race & Class, 32*(2), 51–65. DOI: 10.1177/030639689003200204

Cahyani, A. L., Putra, Z. H., Witri, G., & Ahmad, S. (2025). Exploring ethnomathematics in a Malay traditional house and architecture on Meranti Island as a geometry learning resource for elementary school. *Journal of Teaching and Learning in Elementary Education, 8*(2), 105–117. DOI: 10.33578/jtlee.08.02.105.117

Clements, D. H., & Battista, M. T. (1992). Geometry and spatial reasoning. In Grouws, D. A. (Ed.), *Handbook of research on mathematics teaching and learning* (pp. 420–464). Macmillan.

Crowley, M. L. (1987). The van Hiele model of the development of geometric thought. In Lindquist, M. M., & Shulte, A. P. (Eds.), *Learning and teaching geometry, K–12* (pp. 1–16). National Council of Teachers of Mathematics.

D'Ambrosio, U. (1985). Ethnomathematics and its place in the history and pedagogy of mathematics. *For the Learning of Mathematics*, *5*(1), 44–48. https://www.jstor.org/stable/40247876

Faisal, G. (2019). Arsitektur Melayu: Identifikasi Rumah Melayu Lontiak Suku Majo Kampar. *Langkau Betang: Jurnal Arsitektur*, *6*(1), 1–12. DOI: 10.26418/lantang.v6i1.31007

Gerdes, P. (1988). On culture, geometrical thinking and mathematics education. *Educational Studies in Mathematics*, *19*(2), 137–162. DOI: 10.1007/BF00751229

Gilligan-Lee, K. A., Hawes, Z. C. K., & Mix, K. S. (2022). Spatial thinking as the missing piece in mathematics curricula. *npj Science of Learning, 7*(1), 10. DOI: 10.1038/s41539-022-00128-9

Hawes, Z. C. K., Gilligan-Lee, K. A., & Mix, K. S. (2022). Effects of spatial training on mathematics performance: A meta-analysis. *Developmental Psychology*, *58*(1), 112–137. DOI: 10.1037/dev0001281 PMID: 35073120

Hendriyanto, A., Priatna, N., Juandi, D., Dahlan, J. A., Hidayat, R., Sahara, S., & Muhaimin, L. H. (2023). Learning mathematics using an ethnomathematics approach: A systematic literature review. *Journal of Higher Education Theory and Practice*, *23*(7), 59–74. DOI: 10.33423/jhetp.v23i7.6012

Karina, A., Faizah, H., Elmustian, & Syafrial. (2022). Filosofi rumah adat Lontiok khas Desa Wisata Pulau Belimbing Kuok Kabupaten Kampar Provinsi Riau. *Jurnal Pendidikan dan Konseling, 4*(6), 10188–10202. DOI: 10.31004/jpdk.v4i6.10007

Kemendikbudristek. (2022). *Permendikbudristek No. 56/M/2022 tentang pedoman penerapan kurikulum dalam rangka pemulihan pembelajaran*. Kementerian Pendidikan, Kebudayaan, Riset, dan Teknologi Republik Indonesia.

Kementerian Pendidikan dan Kebudayaan. (2017). *Penetapan warisan budaya takbenda Indonesia: Rumah Lontiok*. Kementerian Pendidikan dan Kebudayaan Republik Indonesia.

Knijnik, G. (2012). Differentially positioned language games: Ethnomathematics from a philosophical perspective. *Educational Studies in Mathematics*, *80*(1–2), 87–100. DOI: 10.1007/s10649-012-9396-8

Kong, J. E., Arizmendi, G. D., & Doabler, C. T. (2023). Implementing the science of math in a culturally sustainable framework for students with and at risk for math learning disabilities. *Teaching Exceptional Children*, *56*(1), 44–51. DOI: 10.1177/00400599221127385

Krippendorff, K. (2018). *Content analysis: An introduction to its methodology* (4th ed.). SAGE Publications.

Mainali, B. (2021). Representation in teaching and learning mathematics. *International Journal of Education in Mathematics. Science and Technology*, *9*(1), 1–21. DOI: 10.46328/ijemst.1111

Mason, M. (2002). The van Hiele levels of geometric understanding. In Jane, L. (Ed.), *Professional handbook for teachers, geometry: Explorations and applications* (pp. 4–8). McDougal Littell.

Munjiatun, K., O., Guslinda, Zufriady, Santika, Y., & Rahayu, S. K. (2022). Typology of Rumah Lontiok architecture in Kuok Kampar Riau. In *Proceedings of the 5th International Conference on Language, Literature, and Education (ICLLE-5 2022)* (pp. 420–425). Atlantis Press. DOI: 10.2991/978-2-494069-85-5_45

Nicol, C., Knijnik, G., Peng, A., Cherinda, M., & Bose, A. (Eds.). (2024). *Ethnomathematics and mathematics education: International perspectives in times of local and global change*. Springer. DOI: 10.1007/978-3-031-60680-9

Pittalis, M., & Christou, C. (2010). Types of reasoning in 3D geometry thinking and their relation with spatial ability. *Educational Studies in Mathematics*, *75*(2), 191–212. DOI: 10.1007/s10649-010-9251-8

Pittalis, M., Mousoulides, N., & Christou, C. (2010). Students' 3D geometry thinking profiles. In V. Durand-Guerrier, S. Soury-Lavergne, & F. Arzarello (Eds.), *Proceedings of the Sixth Congress of the European Society for Research in Mathematics Education* (pp. 816–825). Institut National de Recherche Pédagogique.

Plomp, T., & Nieveen, N. (2013). *Educational design research: An introduction*. Netherlands Institute for Curriculum Development (SLO).

Pusat Asesmen Pendidikan. (2023). *Laporan hasil Asesmen Nasional 2022: Rapor Pendidikan Indonesia*. Kementerian Pendidikan, Kebudayaan, Riset, dan Teknologi Republik Indonesia.

Rachman, K., Mas'ud, A. I., Vesya, S. R., Arbiyanti, A. P., & Rasyid, I. (2025). Development of digital puzzle based on Indonesian traditional houses as virtual manipulative learning media for plane geometry. *Unnes Journal of Mathematics Education*, *14*(2), 167–178. DOI: 10.15294/ujme.v14i2.29951

Rosa, M., D'Ambrosio, U., Orey, D. C., Shirley, L., Alangui, W. V., Palhares, P., & Gavarrete, M. E. (2016). *Current and future perspectives of ethnomathematics as a program*. Springer., DOI: 10.1007/978-3-319-30120-4

Rosa, M., & Orey, D. C. (2013). Ethnomodelling as a research lens on ethnomathematics and modelling. In G. A. Stillman, G. Kaiser, W. Blum, & J. P. Brown (Eds.), *Teaching mathematical modelling: Connecting to research and practice* (pp. 117–127). Springer. DOI: 10.1007/978-94-007-6540-5_10

Sinclair, N., & Bruce, C. D. (2015). New opportunities in geometry education at the primary school. *ZDM Mathematics Education*, *47*(3), 319–329. DOI: 10.1007/s11858-015-0693-4

Sunzuma, G., & Umbara, U. (2025). Ethnomathematics-based technology in Indonesia: A systematic review. *Asian Journal for Mathematics Education*, *4*(1), 129–153. DOI: 10.1177/27527263241305812

Van Hiele, P. M. (1986). *Structure and insight: A theory of mathematics education.* Academic Press.

Wildfeuer, S. (2022). Ethnomathematics and cultural identity to promote culturally responsive pedagogical choices for teachers in early childhood and elementary education. In Danesi, M. (Ed.), *Handbook of cognitive mathematics* (pp. 1–15). Springer. DOI: 10.1007/978-3-030-44982-7_1-1

ADDITIONAL READING

D'Ambrosio, U. (1985). Ethnomathematics and its place in the history and pedagogy of mathematics. *For the Learning of Mathematics*, *5*(1), 44–48. https://www.jstor.org/stable/40247876

KEY TERMS AND DEFINITIONS

Bumbungan Melentik: The upswept curved ridgepole distinguishing Rumah Lontiok from other traditional Malay house forms, achieved through graduated wooden segments rather than bent continuous timber and analyzed in this chapter as a frustum-series approximation of a continuous curve.

Descriptive-Prescriptive Gap: Term used in this chapter, following Batiibwe (2024), for the distance between ethnomathematics studies that catalog mathematical concepts in cultural contexts and those that produce validated, classroom-ready instructional frameworks.

Ethnomathematics: The field founded by D'Ambrosio (1985) studying mathematical knowledge embedded in human cultural traditions and applying those findings to formal mathematics instruction.

Frozen Mathematics: Gerdes' (1988) term for mathematical knowledge embedded in cultural artifacts through accumulated practice, no longer consciously articulated by contemporary practitioners, with the bumbungan melentik construction technique as a clear instance.

Kurikulum Merdeka: Indonesian national curriculum reform under Permendikbudristek No. 56/M/2022 that organizes primary education into three developmental phases (A, B, C) and foregrounds contextual, project-based, culturally grounded learning.

Phase C: The Kurikulum Merdeka phase covering Grades 5 to 6, which is the target level of the pedagogical framework in this chapter.

Rumah Lontiok: Traditional elevated timber house of the Kampar Malay community in Riau Province, Indonesia, designated as National Intangible Cultural Heritage since 2017 and also known as Rumah Lancang or Rumah Pencalang.

Solid Geometry: The branch of geometry dealing with three-dimensional figures, including prisms, cubes, and composite solids, together with their faces, edges, vertices, volume, and surface area.

Tiang: Kampar Malay term for the timber pillars supporting Rumah Lontiok. At the primary research site, the building has 24 square-cross-section pillars arranged in a 4×6 grid, each measuring 0.20 × 0.20 × 1.75 m.

Van Hiele Theory: Developmental theory of geometric reasoning proposed by Van Hiele (1986) with five qualitative levels (Visualization, Analysis, Informal Deduction, Formal Deduction, Rigor) numbered 0 to 4, with Levels 0 to 2 being the target range of this framework.

Chapter 10
The Use of the Indigenous Game of Wera to Support the Acquisition of Number Sense in Namibian Primary Schools

Angelius Kanyanga Liveve
http://orcid.org/0009-0005-5669-3488
School of Primary, Secondary, and Postgraduate Education, International University of Management (IUM), Nkurenkuru, Namibia

ABSTRACT

This chapter examines how the indigenous game Wera can enhance the teaching and learning of number sense in primary mathematics in Namibia. Guided by Lev Vygotsky's social constructivist theory, the study views learning as a socially and culturally mediated process. A qualitative approach was employed using document analysis, lesson observations, semi-structured interviews, and stimulated recall with teachers, learners, and indigenous knowledge holders. Data were analysed thematically. The findings show that Wera strengthens counting skills and supports learners in connecting indigenous counting systems with the Hindu-Arabic numeral system. The intervention also resulted in a mean gain of 23.68 points across number-sense components, indicating improved mathematical understanding. The chapter recommends integrating indigenous games into mathematics teaching to promote culturally sustaining pedagogy and curriculum transformation in Namibia.

DOI: 10.4018/979-8-2600-3485-9.ch010

INTRODUCTION

This chapter provides strategies for using the Indigenous game of Wera to support the acquisition of number sense in Namibian primary schools. This is achieved by demonstrating our research on how engaging learners in the Wera game promotes play. This resulted from the need to develop a solution to the challenges learners face in constructing number sense concepts.

Background and Context

Mathematics may seem abstract without real-world connections (Roche, Gervasoni & Kalogeropoulos, 2023), but teaching should unify concepts with real-world patterns and relationships. Teachers develop learners' understanding through tasks and questions that challenge them and promote higher-order thinking, both inside and outside the classroom (Roche et al., 2023). Incorporating indigenous games linked to local cultures can help make math meaningful by engaging in activities such as counting, grouping, and spatial awareness, aligning with Vygotsky's (1978) theories of active, social learning. In Namibia, despite their value, math is often taught in a detached, teacher-centered way, hindering learners' understanding. Using indigenous games in math instruction offers a culturally relevant, learner-centered approach. Additionally, many learners struggle with math, which requires a better understanding of their difficulties. The EMU program aims to address this by using indigenous games such as Wera as effective interventions.

Problem Manifestation

As educators, I observe student teachers using content and pedagogical knowledge in schools. It was observed that learners' number sense was lacking; some couldn't count, but could when using their home counting system. The chapter explored why they could count at home but not use the Hindu-Arabic Numeral System. Findings showed that those with a base-ten system demonstrated better skills, while others had difficulty. In the chapter, it was also seen that children who played cultural games, such as swinging or see-sawing, understood the concepts of 'up' and 'down', which helped them grasp fractions. The chapter investigated whether the Wera game, common in Southern Africa and globally, embedded ethnomathematics to develop number sense and explained its theoretical basis.

The Purpose and Objective

The chapter promotes learning mathematics through the indigenous game of Wera, which integrates subject content into play, helping learners develop knowledge and skills beyond entertainment. It examines how indigenous games support the learning of the components of number sense at the primary level. Research shows that games can increase interest, reduce boredom, and improve engagement in math classrooms (Hui & Mahmud, 2023). Designed to develop skills through interactive experiences, games positively impact academic performance. Mathematics, essential globally and in Namibia, underpins fields like engineering and economics, fostering logical and critical thinking. Teaching math abstractly and disconnected from cultural contexts can reduce motivation. Despite efforts, many Namibian primary learners struggle with fundamental concepts. Indigenous games like Wera, involving mathematical reasoning, could help if integrated into classroom practice, utilizing culturally relevant resources. This chapter explores the integration of indigenous games into a problem-based learning framework to bridge the gap between school math language and home experiences. The key question is:

> To what extent does the indigenous game of Wera enable or constrain teaching and learning of number sense?

This question stems from observed gaps in primary learners' number sense.

LITERATURE REVIEW

Belonging as a Representation of Ethnomathematics

The first component is belonging or planting the seed (protection) to defend the stone representing cattle in the context of the Wera game. In ethnomathematics, Indigenous knowledge, practices, and teachings are centered. This means that learners first engage with an elder knowledge-keeper on a topic such as drum-making (Garcia-Olp et al., 2022). Once cultural relevance is established, learners learn how Indigenous knowledge informs the Westernized conceptualization of mathematics. Through this process, learners are reminded that they are here for a specific reason and, for that reason, they belong, thus creating a sense of community. When linking to Indigenous mathematics education, we must begin by having Indigenous elders facilitate the exploration of Indigenous mathematical concepts such as counting, the four operations, and algebra, since learners need to know how many stones are needed to reach our destination and/or the hole in the context of the chapter. Through this intentional process, Indigenous learners can begin to foster a sense of belonging and build relationships with both Indigenous and Westernized mathematical practices and teaching (Garcia-Olp et al., 2022).

Mastery Game for Ethnomathematics

The next component is mastery of winning or losing in the indigenous game. Through prediction and counting, one is reminded of the importance of fostering a sense of mastery, in which a person finds their foundation in a competency that enables their full expression and greatest potential, serving as the key to critical thinking (Garcia-Olp et al., 2022). To be clear, mastery within the decolonizing curriculum framework is not about oppressive power or domination. Rather, mastery enables our Indigenous learners to be confident in their understanding of aspects of Indigenous mathematical knowledge. Furthermore, by learning from Indigenous elders and community members, learners are modelling traditional learning modalities in which learning is grounded in Indigenous knowledge and, in the process of understanding mathematical concepts, provide learners with a purposeful drive to learn about both Indigenous and Western mathematics.

Independence for Indigenous Games

As mastery develops, there is a natural progression towards independence. This progression is cultivated during the stage of prosperity through the assistance of knowledgeable, independent individuals. At this stage of the learning process, the learner moves from completing a task with guidance (mastery) to completing it independently. For example, if a learner is working with the concept of drum-making in mathematics, the learner could not make a drum without guidance, even when engaging with Western concepts of geometry and engineering (Iyabode, 2024).

Generosity Towards Indigenous Games

The final component is generosity. Generosity is cultivated through giving and trust, as participants pass or donate stones to their peers without hesitation. This is developed by emphasizing the importance of becoming a holistic individual, applying acquired knowledge, and sharing this knowledge with the community (Garcia-Olp et al., 2022). In practice, this concluding step encourages learners to participate in future educational sessions and community events, as well as to support indigenous elders and community members in instructing a decolonized mathematics curriculum. Through this approach, learners practice generosity and reciprocate to fellow learners who are beginning their journey in this process (Garcia-Olp et al., 2022).

The pursuit of knowledge is fundamentally rooted in Indigenous methodologies of exploring mathematics education. Indigenous knowledge is regarded here as intrinsically linked to mathematics education. The chapter focused on a traditional game called Wera, an extension of Amagende Hadebe-Ndlovu and Manditereza (2025), a highly interactive game that connects learners to their culture and develops their mathematical literacy. Certainly, integrating Indigenous games into the curriculum provides a rich context for deep learning while covering essential foundational content areas in primary mathematics and beyond.

The Social Roles of Wera in the Language of Mathematics

The game of Wera, closely related to Mancala originating in southern Africa (Hadebe-Ndlovu & Manditereza, 2025), is a valuable indigenous African strategy game suited for formal classroom use, especially in Early Childhood Education (ECE) and primary math. It promotes cultural relevance and foundational mathematics skills like counting, addition, and subtraction, while fostering higher-order abilities such as pattern recognition, reasoning, spatial awareness, and probabilistic thinking, crucial from Grades 1 to 7. Integrating Wera aligns with UNESCO (2022)'s push to include indigenous knowledge in African Early Childhood Development (ECD), making learning more relevant and engaging by embedding mathematical concepts into familiar activities. Wera encourages deep learning through active interaction, reflection, and strategic thinking, providing experiential engagement with mathematical ideas rooted in learners' culture (Hadebe-Ndlovu & Manditereza, 2025). This process develops cognitive, social, and cultural skills, fostering critical thinking, problem-solving, and metacognition. The game also promotes social skills like fairness and cooperation, nurturing positive social interactions, cultural appreciation, and learner identity. Overall, Wera exemplifies harnessing indigenous practices to enrich math education, bridging concepts with cultural contexts to boost achievement and develop critical and social skills.

The Explanation of Playing Wera

This widespread African indigenous game is played in pairs, with both genders participating. It uses small objects, such as stones or seeds (called Nongombe in Rukwangali), as counters, with the number depending on the players and the game variation. The game is known by various names in Namibia, including ǁhus/xoros, onyune, wera, vuela, lochspiel, owela, and ogoro (Hadebe-Ndlovu & Manditereza, 2025). While there is no maximum number of players, at least two are needed. The game usually has two or four rows of holes, with the most common setup being four rows, each controlled by one or two players, using seeds or small stones about 19 mm in diameter that represent cattle heads, emphasising livestock's socio-economic importance.

The game involves moving cattle counterclockwise, with captured pieces redistributed across the board. The holes, representing kraals, ground the game in pastoral life (Hadebe-Ndlovu & Manditereza, 2025). Success is measured by capturing opponents' cattle. Cognitively, it's similar to chess, requiring strategic planning, anticipation, and the ability to predict moves, including identifying vulnerable cattle. The counting system is informal, based on local languages such as kashi (one), kali (two), and tatu (three), thereby supporting language and numeracy. Strategic decisions are evident in the placement of the final seed during a move called kurara (resting). Positional terms, such as mutwe (head) for the first hole, sikafi (neck) for the second, and yipepe (shoulder) for the third, incorporate embodied metaphors that enrich cultural meaning and aid spatial reasoning, memory, and strategies for winning more cattle (see Figure 1).

Figure 1. Outwit opponents to gain more cattle

Players must, in addition, outwit their opponents in subtraction, division, and multiplication; this enhances mental computation by requiring them to calculate the number of counters relative to the holes (Hadebe-Ndlovu & Manditereza, 2025). A player must have at least two seeds to initiate a move, distributing them sequentially across the holes within a row. If the final seed lands in a player's inner row and results in two or more "heads of cattle," the player may capture the opponent's corresponding pieces, provided that the opponent's inner row contains at least one seed. This capture may also extend to the outer row under specific conditions. However, if the opponent's inner hole is empty, no capture occurs, even if the outer row contains seeds. The game concludes when one player has successfully transferred all the opponent's seeds to their side or when the opponent is unable to make a valid move.

CONCEPTUAL FRAMEWORK

In this section, I begin with the concepts that underpin interaction in the indigenous game of Wera. The concept is now discussed as follows.

Zone of Proximal Development (ZPD)

Vygotsky (1978) defined the zone of proximal development (ZPD) in terms of actual and potential development. The characteristic of actual development is independence, that is, what the learner can do independently, using the number sense he has from the Wera game as a base to support the acquisition of number sense as viewed in the official curriculum, the target. Glukhova et al. (2022) consider Vygotsky's ZPD multidimensional model as a conceptual tool for promoting primary school mathematics learners' approaches to learning. The ZPD provides a conceptualization of developmental potential. Vygotsky (1978) defines the ZPD as:

> *The distance between the actual developmental level, as determined by independent problem solving, and the level of potential development, as determined through problem solving under adult guidance or in collaboration with more capable peers. (p. 86)*

The introduction of the zone of proximal development (ZPD) constitutes a foundational contribution to the elaboration of cultural-historical theory as advanced by Lev Vygotsky. This is so because the Wera game has a historical aspect and has existed throughout the players' history. Within this framework, the ZPD is conceptualised as the distance between a child's actual developmental level, determined through independent problem-solving, and the potential developmental level, which is realised through guided participation, whether facilitated by an adult or through collaboration with more capable peers (Piaget, 2003). This distinction underscores the inherently social nature of cognitive development and foregrounds the importance of mediation in learning processes.

Vygotsky's Theory of Play-Based Learning, as supported by Ndlovu et al. (2023), which posits that play is a primary driver of development during early childhood, is principal in our analysis. According to Vygotsky, play is not merely recreational but serves as a critical medium for the development of higher psychological processes, including abstract thinking, self-regulation, and problem-solving. Through imaginative and rule-based play, children are afforded opportunities to experiment with social roles, negotiate meaning, and practise emerging competencies in a context that is both engaging and cognitively demanding. Citing Vygotsky, (1978). Sakti, Endraswara, and Rohman (2024) emphasise that learning is fundamentally a social and cultural process, deeply embedded within the practices and values of a given community.

Scaffolding

Using scaffolding in math education presents a complex pedagogical challenge, requiring teachers to have strong subject knowledge, instructional judgment, and deliberate planning. Effective scaffolding goes beyond strategies; it involves dynamic support aligned with learners' evolving needs. Sarmiento-Campos et al. (2022) note that core features of collaboration, contingency, fading, and responsibility transfer must be intentionally enacted for meaningful learning. Collaboration emphasises interaction; contingency involves adapting support; and fading means gradually withdrawing help to foster learner autonomy, using Wera as a scaffolding tool.

Indigenous knowledge is often used synonymously with 'local knowledge,' 'ethnoscience,' or 'ethnomathematics,' all of which refer to knowledge specific to a culture (Philip, 2015). Indigenous knowledge encompasses a broad spectrum of understandings, practices, and beliefs that are often deeply connected to a particular community's land, culture, and social systems (Larasati, Pramasdyahsari & Harun, 2025). This knowledge is characterised by its holistic nature, emphasising relationships among all elements of fauna and flora. In contrast, scientific knowledge, while valuable and informative, often adopts a more reductionist approach, prioritising empirical data and systematic experimentation over contextual and experiential understanding (Lebaka, 2024). This knowledge is delineated as "unwritten, unsaid, just known" (Ayre & Mackenzie, 2013, p. 764). Scholars such as Garcia-Olp, Nelson and Saiz (2022) highlight the role of Indigenous knowledge by referring to the Tree of Life centres, which draw on learners' cultural and Indigenous knowledge to teach mathematical concepts in a healthy, heartfelt environment. These components are belonging, mastery, independence, and generosity (Garcia-Olp et al., 2022). Indigenous knowledges matter in this chapter because they might provide an additional lens and a novel angle for understanding mathematics, using indigenous games as a vehicle.

THEORETICAL FRAMEWORK

The theoretical framework for this is grounded in Vygotsky's (1978) Social Constructivism, with the key idea that learning occurs through social interaction and cultural tools. Wera, as a cultural game, is well-suited for social interaction. This framework is suitable because indigenous games serve as cultural tools that mediate mathematical thinking and deep learning (Iyabode, 2024). Besides mentioning social interaction, Vygotsky (1978) also views the use of language and its development as very important in developing one's cognitive system. This aligns well with the manifestation of the problem: learners who use their home language can easily learn to count and other number-sense concepts.

Also, in this theoretical framework, using a cultural practice to facilitate deep learning of number sense is sensible, as the theory advocates this. For this reason, the chapter adopts one strand of constructivism: cultural constructivism, as this framework values and embraces what culture brings forth. Finally, because the Wera game has been in use for a long time, I can conclude that it is a historical artefact in many cultures across this part of the earth. So, since the Wera game is found across diverse cultures, it is suitable for teaching and learning number sense, as it is not restricted to a single geographical locality. Historically, games have been designed and developed for both genders. Nevertheless, when playing the Wera game, boys and girls differ in their enjoyment of it (Hui & Mahmud, 2023). We investigate some of the concepts and constructs that will emerge. This will be as follows: (i) the focused conceptual framework. (ii) The social roles of Wera in the language of mathematics. Then, I focus on learning and teaching Mathematics in the subheadings of (iii) counting and strategy through Wera, and finally present the socially situated interactions with peers during gameplay, supported by concepts (iv) Zone of Proximal Development (ZPD), Vygotsky (1978) and scaffolding, Bruner (1978), which emphasise that mathematics emerges through culturally meaningful activities.

Indigenous knowledge is often used synonymously with 'local knowledge,' 'ethnoscience,' or 'ethnomathematics,' all of which refer to knowledge specific to a culture (Philip, 2015). Scholars such as Tassell-Matamua (2025) argued that Indigenous knowledges might be understood as a body of knowledge that has accumulated over time and has been informed by the worldviews of the communities that have generated, practised, refined, and modified them. The worldviews inherent in such knowledge tend to reflect common principles about the relationships among living beings (including humans) and their environment. These principles are notably inclusive of both tangible and intangible phenomena and extend from the micro- to the macro, and from the terrestrial to the celestial (Tassell-Matamua, 2025). Indigenous knowledge encompasses a broad spectrum of understandings, practices, and beliefs that are often deeply connected to a particular community's land, culture, and social systems (Larasati et al., 2025). Further elucidation: indigenous knowledge is deeply rooted in tradition, characterised by distinctive attributes that are tacit, unique, and exclusive to a particular culture or society (Madonsela & Machete, 2025). This knowledge is delineated as "unwritten, unsaid, just known" (Ayre & Mackenzie, 2013, p. 764). Consequently, Hudson et al. (2016) describe indigenous knowledge as the systems of understanding embedded within the cultural traditions of regional, Indigenous, or local communities.

In the context of the chapter, learners might benefit from Indigenous knowledge in mathematics, as part of everyday knowledge, which plays a significant role in developing mathematical concepts such as counting through critical thinking. Pedagogical practices engage not only human teachers but also modalities of teaching, such as drumming, to encourage a deeper connection to learning and living. Scholars such as Garcia-Olp, Nelson and Saiz (2022) highlight the role of Indigenous knowledge by referring to the Tree of Life centres, which draw on a student's cultural and Indigenous knowledge to teach mathematical concepts in a healthy, heartfelt environment. When working with Indigenous populations through a place-based mathematics education model, the learning environment is inherently connected to the land, the community, and that which sustains us. Furthermore, Indigenous people's knowledge is not consumed for individual learning. Rather, knowledge is rooted in the concept of belonging, where local elders and community members teach Indigenous youth to find a sense of belonging by centring Indigenous ways of exploring mathematics education. From the Tree of Life perspective, four components are identified to ensure that the application of the Tree of Life is reciprocal and that relationship-building is at the heart of the learning community. These components are belonging, mastery, independence, and generosity (Garcia-Olp et al., 2022). Indigenous game drives the intention of place-based and mathematics education, as well as our interactions with the community and the land where all that Indigenous people seek to protect exists, including the playing of the Wera game. Indigenous knowledges matter in this chapter because they might provide an additional lens and a novel angle for understanding mathematics, using indigenous games as a vehicle. The methodology confirming Wera's role in developing number sense is discussed below.

METHODOLOGY

For this empirical chapter, the methodology encompassed research design, participants, data collection techniques, data analysis, and ethical considerations.

Research Design

This chapter adopted a qualitative case study research design within an interpretivist paradigm to explore the use of the Indigenous game of Wera to support the acquisition of number sense in Namibian primary schools. Moloeng (2018) states that qualitative research seeks to understand phenomena related to human conditions, behaviours, perceptions, motivations, and actions. Furidha (2023) adds that it examines phenomena in natural settings using non-numerical data and interpretive techniques. In the context of the chapter, note that qualitative research is ideal for exploring complex social problems, especially in education, with multiple interacting variables.

Participants

The qualitative descriptive approach guided the selection of the population and sample, consisting of participants from four public primary schools in the Kavango West Region, Kandjimi Circuit. The sample comprised two purposively selected Grade 7 mathematics teachers from each school (a total of 8 teachers), learners, and a community member knowledgeable about indigenous games, particularly Wera, to facilitate their utilisation as educational tools (Crossman, 2020). Teachers with at least three years of experience teaching number sense were selected to ensure they had the pedagogical knowledge, classroom management skills, and understanding of learners' development needed to implement indigenous game-based instruction.

Data Collection Techniques

Data collection involved semi-structured interviews, lesson observations and document analysis of the pre- and post-tests administered to Grade 7 learners. The pre-test measures learners' number sense before they are exposed to the Wera game. The test assesses conceptual understanding rather than memorisation. The post-tests assessed the same number sense skills but used different questions. To align with the intervention, the post-test questions were framed around the Wera game while still measuring mathematical understanding. Importantly, this post-test mirrors the pre-test in difficulty and progression while embedding questions in authentic Wera game situations. Comparing scores can show if the Wera game improved number sense. For research, I add an observation checklist and learner interview questions to triangulate results.

Data Analysis Procedures

Data analysis combined quantitative and qualitative methods to assess the effectiveness of the Wera game in improving learners' number sense. Thematic analysis (Ahmed et al., 2025) involved coding, organising, and interpreting patterns related to research goals. Variations in performance within the Zone of Proximal Development (Bruner, 1978) were also examined. Quantitative data from pre- and post-tests were coded, entered into a spreadsheet, and summarised using descriptive statistics like frequencies, percentages, means, and standard deviations. Improvements were measured by comparing scores, and paired-samples t-tests were used to determine the statistical significance of differences.

Qualitative data were gathered through classroom observations, learners' activities during the Wera game, and semi-structured interviews with the teacher and selected learners. Audio recordings of interviews were transcribed accurately, and observation notes were organised immediately after each lesson. The data were analysed using Braun and Clarke's (2006) six-phase thematic analysis, which involved familiarising with the data through repeated reading of transcripts and notes; generating initial codes; identifying patterns among codes; reviewing and refining candidate themes; defining and naming themes; and developing a coherent interpretation. The themes were derived inductively from the data to reflect participants' experiences, perceptions, and observed changes in learners' number sense during the implementation of the Wera game, as reiterated by Ahmed et al. (2025).

Ethical Consideration

Ethical principles guided the research to protect participants' rights, dignity, and confidentiality. Ethical approval was secured from the relevant ethics committee and education authorities of the Kavango West Directorate of the Ministry of Education, Innovation, Youth, Sports, Arts and Culture (MEIYSAC). Participation was voluntary, with informed consent obtained from teachers and guardians and assent from learners under 18 (Meinck et al., 2025). Participants' identities were protected through pseudonyms, and data was securely stored. Respect for Indigenous Knowledge and community protocols was maintained, with voluntary participation and proper acknowledgement.

FINDINGS

In this section, the chapter presents data from pre- and post-tests administered to grade 7 learners by teachers both before and after the intervention of the indigenous games of Wera, with the aim of integrating these games into the teaching of mathematics in primary schools in Namibia, as well as from a community member who is the custodian of the Wera game at the study site. Data presentation is split into two forms. Data from document analysis and from semi-structured and stimulated recall interviews.

Data Emerging from Pre-Test

The pre-test, worth a total of 40 marks, was administered to 60 Grade 7 learners, 15 from four primary schools in the Kandjimi Circuit. The pre-test lasted for 35 minutes. The scores are collated and tallied.

Table 1. Scores of learners from four schools on the pre-test of number sense

Scores	Tally	Frequency	Percentage (%)
0-5	//// //// ///	13	22
6-10	//// //	7	12
11-15	//// //// ////	15	25
16-20	//// //// ////	14	23
21-25	////	4	7
26-30	//	2	3
31-35	///	3	5
36-40	//	2	3
Total		60	100

The data revealed that the participating learners from four primary schools demonstrated prior everyday knowledge of number sense before engaging in the Wera game intervention. Based on the outcomes, only 35 learners scored below average (59%), while 25 scored above average (41%). The tally showed that only 20 learners scored 10 or fewer marks, while 15 scored between 10 and 15 marks. Eventually, only two learners scored between 35 and 40 marks, representing 3%.

Data Emerged from the Post-Test

The post-test was administered after participants had engaged in the intervention using the Wera game. Teachers played the Wera game under the supervision of a knowledgeable community member and later acted as knowledge brokers, leading learners in playing it to advance number-sense acquisition. After learners played the Wera game, the researchers observed lessons to gain insight into learners' understanding of number sense.

Table 2. Scores of learners from four schools on the post-test of number sense

Scores	Tally	Frequency	Percentage (%)
0-5	////	5	8
6-10	////	5	8
11-15	//// /-	6	10
16-20	//// //	7	12
21-25	//// ////	10	17
26-30	//// ////	9	15
31-35	//// /////	10	17
36-40	//// ///	8	13
Total		60	100

The post-test results showed varying levels of performance among learners in primary mathematics number sense after the intervention. Twenty-three learners scored 20 or below, indicating many still struggled with the concepts. Eight learners scored 36 or higher, about 13% of the sample, demonstrating a solid understanding. The score distribution suggests that although some learners achieved high scores, many still need additional instruction to improve their number sense.

Comparing these assessments identified learning gains, changes in reasoning, and ongoing challenges. The findings show the intervention's effectiveness in developing number-sense understanding and in progressing learners through ethnomathematics. Table 5 presents the pre- and post-assessment results, detailing the distribution of learners across mathematics and changes in their number sense understanding.

Table 3. Quantitative results for pre- and post-test (n = 60)

Number sense components	No.	%	No.	%
Ordering numbers	18	30	29	48.3
Comparing numbers	28	46.6	45	75
Number patterns	29	48.3	42	70
Algebraic thinking	15	25	29	48.3
Operation: subtraction and addition	32	53.3	48	80

Inferential Analysis of Improvement and Average Percentage-Point Gain

$$Mean\ pre-test\ \% \ = \frac{30 + 46.6 + 48.3 + 25 + 53.3}{5} = 40.64$$

$$Mean\ post-test\ \% \frac{48.3 + 75 + 70 + 48.3 + 80}{5} = 64.32$$

Mean Gain = 64.32 – 40.64 = 23.68 percentage points

The findings show a significant mean gain from pre-test to post-test scores, indicating that learners' number sense improved after the Wera game intervention. In the pre-test, 30% of learners (30) answered questions on ordering numbers correctly, increasing to 48.3% (29 learners) post-intervention, an 18.3-point rise. This suggests the Wera game, with teacher scaffolding, boosted their ability to order numbers. The greatest improvement was in addition and subtraction, with a 26.7-point increase, showing the intervention strengthened computational skills. Overall, a mean gain of 23.68 points across number-sense components indicates that the intervention positively impacted learners' understanding and application.

Data Emerging from Document Analysis - Mathematics Curriculum

The data from document analysis were used to determine the curriculum expectations for community-based education. The rationale for Mathematics in Grades 4-7 in the senior phase of education emphasises that Mathematics involves observing, representing and investigating patterns and quantitative relationships in social and physical phenomena, as well as between mathematical objects themselves. Through these processes, new mathematical ideas and insights are generated. Mathematics uses its own specialised language, including notations and symbols for describing numerical, geometric and graphical relations. Mathematics is a purposeful activity in the context of social, political, technological and economic goals and constraints. It is a discipline that is not value-free or culturally neutral. This knowledge is part of culture, and we need it and these skills to handle many everyday tasks and fulfil personal interests and duties. It contributes to the development of logical and scientific thinking and provides pleasure and satisfaction when learners solve problems and enjoy number games (Ministry of Education, Arts and Culture, 2023). According to the Grade 4-7 syllabus articulated by the national institutions of education development (Ministry of Education, Arts and Culture, 2023), the following was observed.

The relationships between concepts and topics should continuously be developed throughout the phase. The learners' prior knowledge should always be elicited before commencing with a new topic. Where rote learning is not the most appropriate method for certain things, it should be done through interesting activities and games to motivate learners and enable them to learn by heart (p. 3)

Mathematics education promotes the development of functional numeracy and mathematical thinking. It enables learners to apply mathematics in everyday life. Table 3 shows the learning content.

Table 4. Mathematical learning content for grades 6 and 7 (Ministry of Education, Arts and Culture, 2023).

Pleasure and satisfaction when learners solve problems and enjoy number games		
Topic	**Grade 6**	**Grade 7**
Numbers	Order, compare, round, estimate and classify sets of numbers from 0 to 1 000 000. Select and apply appropriate strategies and methods for the four operations, including long multiplication and long division. Apply number properties and relationships. Solve two-step problems within the range 0 to 1,000,000. Find equivalent fractions by extending or reducing them. Compare and order common fractions with denominators up to 100. Perform simple addition, subtraction and multiplication with common fractions. Solve two-step problems involving common fractions.	Order, compare, round, estimate, and classify sets of numbers from 0 to 1 000 000. Select and apply appropriate strategies to the four operations. Apply number properties and relationships. Solve three-step problems within the range of 0 to 1 000 000. Order and compare tenths and hundredths. Add, subtract and multiply common fractions. Use percentage notation correctly.
Patterns, Functions and Algebra	Complete and describe number patterns. Construct and solve simple sentences.	Use algebraic notation to represent numbers with letters.

Numbers and algebra are inherently embedded in the indigenous game of Wera, making it a valuable pedagogical tool for developing mathematical understanding in a meaningful, culturally relevant context (Kancanadana et al., 2021). The relationship to Wera is outlined as follows.

Numbers in Wera

The concept of numbers is central to Wera's gameplay. Learners continuously engage in counting, sequencing, and quantifying as they distribute stones (often called counters) into holes. Each move requires players to count the stones in a selected hole, distribute them one by one into subsequent holes (one-to-one correspondence), and keep track of the total in each hole after redistribution, thereby facilitating mathematical operations (Kancanadana et al., 2021).

Through repeated play, learners develop strong number sense, including an understanding of magnitude, conservation of number, and the relationships between quantities. For example, when a learner picks up 5 stones and distributes them across 5 consecutive holes, they are implicitly reinforcing addition (adding one to each hole) and subtraction (removing stones from their palms).

Moreover, Wera supports the development of mental arithmetic skills, as players must quickly calculate outcomes of moves without external aids. This aligns with foundational mathematical competencies such as addition, subtraction, and even early multiplication through repeated distribution.

Algebraic Thinking in Wera

Although Wera is not explicitly designed as an algebraic activity, it promotes early algebraic thinking through pattern recognition, generalisation, and strategic reasoning. Algebra in this context is not about formal symbols but about identifying relationships and predicting outcomes. It encompasses pattern recognition, enabling players to notice recurring patterns in how stones move across the ground (Hadebe-Ndlovu & Manditereza, 2025). Moreover, it promotes functional thinking, in which learners begin to understand relationships between variables, such as the number of stones in a hole (input) and the resulting distribution across the pits (Philip, 1979). The game reflects the concept of a function, in which one quantity depends on another. For instance, players develop strategies based on general rules. If he or she begins with an even number of stones, they will end in this or that hole. This is a key aspect of algebraic reasoning. Wera requires players to anticipate future moves, predict opponents' actions, and plan strategically (Hadebe-Ndlovu & Manditereza, 2025). These processes involve logical reasoning and abstraction, which are foundational to algebra.

Integration of Numbers and Algebra

The interplay between numbers and algebra in Wera is evident in learners' progression from concrete counting to abstract reasoning. Initially, learners focus on counting stones and performing basic operations. As they gain experience, they recognise numerical patterns, formulate strategies, test hypotheses and make predictions during gameplay. This progression reflects a shift from arithmetic to algebraic thinking, consistent with constructivist learning principles (Vygotsky,

1978). Importantly, learners often do not perceive the activity as formal learning, which increases motivation and reduces anxiety associated with mathematics.

In summary, Wera provides a rich, informal environment in which both numerical skills and algebraic thinking are developed simultaneously. Through counting, distributing, pattern recognition, and strategic reasoning, learners build a strong mathematical foundation that bridges concrete experiences and abstract concepts (Philip, 1979).

DISCUSSION

This section covers data from teachers and learners involved in indigenous games and their integration into primary school maths across Namibia. It also includes insights from a community member who oversees the Wera game at the study site.

Data Emerged from Interviews with Teachers

The teacher used a social learning approach to teach maths, emphasizing learners' ability to count and relate learning to their lived experiences. This aligns with modern theories that see learning as socially mediated, especially in African communities where participation and cultural activities are central (Batiibwe, 2019). Wera enables learners to build mathematical knowledge through interaction with peers, teachers, and the community, reflecting Vygotsky's (1978) idea that social interaction fosters cognitive development. Linking counting to familiar objects and daily activities makes maths meaningful, engaging learners and connecting it to their cultural context, supporting culturally responsive pedagogy. Research by Piaget (2003) shows children learn best through concrete experiences, so Wera's use of tangible materials helps develop numeracy skills. Scholars like Rogoff (2025) emphasize that learning occurs via participation in community tasks, making Wera culturally grounded for acquiring maths skills. Scholars such as Maimela and Nedombeloni (2026) argue that African curricula often privilege Western knowledge, marginalising local culture. Training teachers to use indigenous games such as Wera helps validate local knowledge and foster an inclusive curriculum that reflects learners' identities and cultural heritage (Sakti et al., 2024). This boosts participation, confidence, and a sense of belonging.

Overall, teacher training equips educators to incorporate indigenous games into math instruction, aligning with culturally relevant pedagogy (Ladson-Billings, 1995), sociocultural learning (Vygotsky, 1978), and indigenous knowledge (da Silva, Pereira & Amorim, 2024).

Data Emerged from Semi-Structured and Stimulated Recall Interviews

In this section, the participating teachers are identified as T1–T8, and a community member as "Tate Lyambombola". Findings from semi structured interviews and stimulated recall interviews were coded into four emergent themes: (1) teachers' use of Wera in teaching mathematics; (2) training and qualifications for using indigenous games to teach mathematics in primary school; (3) the rewards and challenges of using indigenous games to teach mathematics in primary education; (4) teachers' experience of academic and support-curricular activities.

Teachers' Use of Wera in Teaching Math

Teachers use indigenous Wera games as culturally grounded learning activities, reporting on learners' mathematical reasoning, the mathematical components embedded in Wera, and the impact of integrating these games. Through interviews, teachers were able to interpret Wera and its role in their mathematics lessons. T4 expressed that:

If I allow learners to play Wera before teaching the operation, most of them will develop critical thinking skills through interactions. Learners will be able to add and subtract stones as part of the learning and teaching material.

The excerpt resonates with Vygotsky (1978), who postulated that learning occurs when learners interact with peers rather than in isolation. The Wera game supports the development of number sense by engaging players in grouping, counting, and practising operations such as addition, subtraction, and division through the distribution of stones. The game reflects knowledge of number-sense concepts, serving as a foundational domain that aligns with the official curriculum and provides scaffolding for learning. Therefore, using Wera to teach number sense poses no challenges, as summarised in the table below, confirming that it embeds numerous relevant concepts.

Table 5. Components of mathematics embedded in the indigenous game -Wera

Indigenous mathematics and mathematical concepts		
The core mathematical components	**İndigenous game's expectation**	**Sample statement from the curriculum**
Arithmetic	Players or learners continuously practice by adding stones to holes and removing them, fostering basic mental arithmetic skills.	Addition and Subtraction
Counting	The game requires constant counting of seeds or stones, both when placing them and when moving them around the board and/or ground.	Solve two-step problems within the range 0 to 1,000,000
Spatial Awareness	The game develops spatial recognition as players understand the positions and relationships among objects (holes/seeds) relative to the ground and/or a grid-like board.	Mapping, drawing (distance, size, etc.
Patterns and Sequencing	Players must understand sequences and patterns to plan moves, such as predicting where the last seed will fall (resting).	Ordering, ascending, descending, etc
Logical Thinking and Problem Solving	Requires strategic planning, decision-making, and critical thinking to manage resources and predict opponent moves.	Probabilities
Multiplication and Division (Grouping	The movement of stones, such as doubling or halving groups of stones during play	Basic operations
Basic Number Sense	The game strengthens the ability to visualise and manipulate numbers up to a certain limit	algebra

Table 5 provides a self-explanation of the zone of proximal development and scaffolding (Bruner, 1978). Teachers play a significant role in guiding learners until the level of such assistance is withdrawn. Meanwhile, T2 said as follows:

I will share the knowledge gained from the game Wera, prepared by Tate Lyambombola. I use this knowledge to guide my learners until they master number sense and probability skills.

This type of guiding and mediating of learning, whether curricular or otherwise, is what Bruner (1978) referred to as scaffolding, and the area in which learners perform a certain task with the assistance of a knowledgeable person was referred to as the zone of proximal development (Vygotsky, 1978). Teachers learned to play the Wera game from Tate Lyambombola, the game's custodian. Teachers then became knowledge brokers because knowledge brokers require high standards of trans- and interdisciplinary (Bortolus Hewitt, Mitova, Schwindt, Sogbanmu, Chukwu, & Yu, 2025).

Training and Qualifications on the Use of Indigenous Games To Teach Mathematics

The second theme highlighted teachers' training for teaching primary school mathematics through indigenous games. Participants knew about skills practically and academically, but some were unclear about what an indigenous game is and how to integrate it into math teaching. All could explain community-based education in Namibia. T3 and T8 agreed that understanding one another (ubuntu) supports social learning. Play, discovery, and games foster learning as teachers use play-based strategies. T5 noted that training requires commitment and collaboration. Respondents recognised that learners learn through play, supporting Tassell-Matamua (2025), who states that facilitating play helps learners construct knowledge in an enjoyable way. They see learners learn through games and singing. One respondent said, "I will incorporate games to deepen understanding and make learning enjoyable, less pressured, and encourage collaboration and creativity, avoiding reliance on structured themes" (T7). Indigenous knowledge encompasses practices and beliefs related to the environment, as exemplified by the Wera game.

Rewards and Challenges of Using Indigenous Games

In the third theme of this chapter, teachers revealed that indigenous games have a strong contribution to the learners' outcomes. Learners lay a strong foundation for inquiry-based learning and promote collaboration as they team up to defeat opponents. Moreover, the most interesting responses came from the study participants.

> *"I can remember our time at school. My classmate and I enjoyed the practical activities conducted in the physics laboratory at Leevi Hakusembe Senior Secondary School in early 1999. The activities enabled most learners in our science class to perform exceptionally well compared with those in the social science class. I do not doubt that imparting the knowledge learnt today will also benefit most primary mathematics learners, and even those who transition to secondary school will not have a learning gap in mathematics" (T1).*

This excerpt discusses active learning in classrooms (Roche et al., 2023) that cater to different thinking styles, challenge learners with multiple solutions, and emphasise hands-on activities that motivate learning and collaboration in mathematics. Teachers' understanding of autonomy affects how they deliver content, shaping their pedagogy. Their knowledge of planning and teaching primary mathematics, including the use of Indigenous games like Wera, reflects their efforts to improve learner performance. However, integrating indigenous games into math instruction remains underexplored (Hadebe-Ndlovu & Manditereza, 2025) and requires careful planning and training. No research has yet examined Wera as an educational resource in primary math, despite global efforts to preserve Indigenous knowledge in Namibia.

Academic Games as Support For Extracurricular Activities

The chapter's final theme, emerging from findings, emphasises the value of indigenous games in primary mathematics teaching. Teachers noted their usefulness, especially Wera, which benefits learners by developing skills like coordination, concentration, and critical thinking. These games help connect classroom concepts with out-of-school experiences, enhancing understanding and engagement. Incorporating Wera can improve learning outcomes, boost self-confidence, and strengthen links between school and life outside. Participants in interviews suggested that:

> *"I desire that the learners in my classes collaborate effectively and be encouraged to explore a diverse range of materials both within and outside the classroom setting. Additionally, I wish for learners to have opportunities to experiment with and familiarise themselves with these materials before their use. I plan to introduce learners to this type of activity, such as the game Wera, which I believe is familiar to my learners' immediate environment.*

The excerpt above resonates with the chapter's notion, confirming that Indigenous games such as Wera can effectively promote the teaching and learning of mathematics by instilling the principles of ubuntu in primary grades. This approach stimulates critical thinking, creativity, and the assimilation of collective moral values among learners by incorporating culturally specific teaching and learning activities, as alluded to by Nxumalo and Mncube (2019). In most cases, teachers are forced by their tight schedules to lean toward academic teaching, inadvertently neglecting other important aspects of supportive curriculum activities.

Data emerged from lesson observations

Two of the eight Mathematics teachers, T2 and T4, were observed teaching lessons on the indigenous Wera game, each lasting 40 minutes with learner activities. T2 taught about ordering numbers, asking about the minimum number of cattle in Wera to move and about ascending/descending arrangements. T2 distributed number cards and asked groups to send members forward with cards. T2 drew a circle representing the cattle in the kraal, a source of wealth. Each learner displayed the number on the card, and it was represented by dots in a circle, as shown in Table 6.

Table 6. Activity on ordering numbers during lesson observation

Dots in pit	••• •	•••• ••• ••	••••• •••• •••	•• •	••• ••
Frequency	4	8	12	3	5

Learners order quantities with dots, from smallest to largest (3, 4, 5, 8, 12) and from largest to smallest (12, 8, 5, 4, 3). They observed learning as fun, with a game integrated into ordering numbers. Some individual learners struggled to grasp the ascending and descending arrangements

T4 taught comparing numbers and introduced the symbols using cards for greater than (>), less than (<), and equal to (=). For instance, T4 divide learners into pairs and allows them to play one round of Wera on the wooden board. At the end of the game, each learner counts the stones they have collected and records on the chalkboard the stones picked by four learners from the pits.

Table 7. Activity on comparing numbers during lesson observation

Player	**Stones picked from the pit**
1	5
2	6
3	8
4	5

T4 inquiries of learners: Who achieved the highest score? Who attained the lowest? Which player gathered more stones, 1 or 3? Identify the two players with identical scores. Determine which score is higher: 6 or 8? Which is lower: 5 or 6? Learners are instructed to use comparison symbols.

Write the comparable expressions on the board and request learners to complete them: $8 > 5$, $5 < 6$, $5 = 5$, $8 > 5$. It was observed that T4 encouraged learners to articulate each statement aloud, for example:

"Eighteen is greater than five; five is less than six; five is equal to five."

Analysis of lesson observations demonstrates that learners acquire knowledge effectively through game-based activities, as reaffirmed by Yonwilad et al. (2025).

Data Emerged from Interviews with Learners

A sample of 6 learners from the population of 60 was interviewed to find out how Wera helped them and how the Wera game can be improved so that every learner in the region can benefit from learning number sense. These learners are identified by pseudonyms as Tulinendi, Kim, Sapira, Tumona, Ndambu and Kashati. Learners demonstrated limited knowledge of all assessed number-sense components in the pre-test. Meanwhile, the post-test revealed significant improvement as learners learned number sense components through social interaction, in which knowledge is constructed (Murat & Demirci, 2026). In the pre- and post-tests, these six learners have demonstrated their knowledge on all five assessed components of number sense, as shown in Table 8.

Table 8. Six learners' pre-test and post-test performance on number sense components

Name of learner (Pseudonym)	Pre-test Marks 40	Percentage (%)	Post-test Marks: 40	Percentage (%)	Variation in (%)
Tulinendi	22	55	33	82.5	27.5
Kim	19	47.5	25	62.5	15
Sapira	11	27.5	19	47.5	20
Tumona	08	20	17	42.5	22.5
Ndambu	18	45	28	70	25
Kashati	19	47.5	29	72.5	25

According to the data in Table 8, the pre-test results showed that Sapira and Tumona achieved the lowest scores among the selected learners. However, both learners demonstrated notable improvement following the Wera game intervention, with gains of 20 and 22.5 percentage points, respectively, in the post-test. The remaining four learners, including Tulinendi, Kim, Ndambu, and Kashati, performed better on the pre-test, suggesting they drew on prior mathematical knowledge acquired in previous grades. Among these learners, Tulinendi exhibited the greatest improvement, with a gain of 27.5 percentage points between the pre-test and post-test scores. These findings suggest that although learners entered the intervention with varying levels of prior knowledge, all demonstrated measurable improvement in their number sense following participation in the Wera game activities. Tumona has echoed the following during interviews:

> *The assessment was tough and left me with 08 points. However, I performed much better on the second test, scoring 17 out of 40. I am very happy that the Wera game is being played at home, and I did not realise it had mathematics related to the class.*

The excerpt aligns with ethnomathematics (Batiibwe, 2025), which views it as vital for connecting learners' experiences with the formal curriculum. Using cultural practices, indigenous knowledge, and community activities, ethnomathematics helps learners relate abstract math to real life, making learning more meaningful. This approach improves understanding, engagement, and appreciation of cultural knowledge in math. Integrating ethnomathematics bridges learners' everyday experiences and school math, boosting understanding and outcomes.

In the interviews, Tulinendi commends the following:

"After participating in the Wera game, I felt supported by my fellow learners and learned through competing with my opponents. The game allowed me to apply and improve my mathematics skills in an enjoyable way. Although Wera is a familiar game in our community, it requires guidance from teachers or someone who understands the rules and how to play it effectively for learning."

The excerpt supports learning through games, as emphasised by Vygotsky (1978) and sociocultural theory (learning through interaction and scaffolding), in which learners acquire knowledge and skills through active participation, interaction, and problem-solving. This excerpt also aligns with Yonwilad et al. (2025), who noted that learning by games provides meaningful, enjoyable engagement with mathematical concepts and fosters collaboration, competition, and peer support. In the Wera game intervention, learners applied their skills in practice, received peer support, and deepened their understanding through repeated play. This indicates that game-based learning effectively links abstract concepts to real-world experiences.

These findings highlight the importance of culturally embedded games like Wera across all levels of mathematics education, from primary to tertiary, in enhancing mathematical learning. Wera connected ethnomathematical practices with the formal curriculum by linking learners' cultural experiences to school-based concepts, fostering deeper understanding beyond rote memorisation. Kheirollahi Meidani et al. (2025) stated that integrating indigenous games can promote meaningful learning by engaging learners in familiar cultural settings. At primary levels, it supports foundational number sense and problem-solving; at secondary levels, it enhances mathematical reasoning in real-world contexts; and at tertiary levels, it informs teacher training in adopting culturally responsive pedagogies (Fitrianto & Farisi, 2025). Overall, Wera shows how ethnomathematical knowledge enriches formal curriculum, encouraging inclusive, relevant, and conceptually rich mathematics education.

IMPLICATION

The chapter has significant educational implications, as it demonstrates the potential of the indigenous game of Wera to serve as a meaningful link between ethnomathematics and the formal mathematics curriculum. By integrating culturally grounded games into mathematics teaching and learning, teachers can create opportunities for learners to connect their everyday experiences, indigenous knowledge, and cultural practices with abstract mathematical concepts taught in schools

(Yonwilad et al., 2025). This alignment promotes culturally responsive mathematics education by making learning more relevant, accessible, and meaningful to learners.

Furthermore, the use of Wera highlights the importance of recognising indigenous knowledge as a valuable resource for enhancing mathematical understanding, as reiterated by Fitrianto and Farisi (2025). The findings suggest that indigenous games can support the development of conceptual understanding, problem-solving abilities, collaboration, and positive attitudes towards mathematics. Therefore, mathematics teachers, curriculum developers, and teacher education institutions can consider incorporating indigenous games as complementary pedagogical tools to enrich mathematics instruction and bridge the gap between community-based mathematical knowledge and formal school mathematics.

CONCLUSION

This chapter demonstrates that indigenous games like Wera enhance native learners' critical reasoning and support culturally connected math education. Wera teaches counting and geometric concepts with simple materials, fostering understanding of mathematical processes. When tied to culture and language, these games promote learning through interaction, boosting performance, retention, motivation, and engagement. Indigenous games embed concepts like counting, sequencing, and measurement in real contexts, supported by scaffolding within the Zone of Proximal Development. They also preserve culture and aid holistic cognitive growth. Challenges include limited teacher training and inconsistent community engagement, with socio-economic differences across Namibia urging caution in generalizing findings. Learners showed a 23.68-point gain in number sense, indicating positive impact. The chapter recommends integrating indigenous games into primary math lessons, emphasizing the need for teacher training, flexible, play-based methods, and recognition of these games as valuable pedagogical tools. Digital platforms can help preserve their cultural and mathematical importance.

REFERENCES

Ahmed, S. K., Mohammed, R. A., Nashwan, A. J., Ibrahim, R. H., Abdalla, A. Q., Ameen, B. M. M., & Khdhir, R. M. (2025). Using thematic analysis in qualitative research. *Journal of Medicine, Surgery, and Public Health*, *6*, 100198. DOI: 10.1016/j.glmedi.2025.100198

Ayre, M., & Mackenzie, J. (2013). "Unwritten, unsaid, just known": *The role of Indigenous knowledge (s) in water planning in Australia. Local Environment*, *18*(7), 753–768. DOI: 10.1080/13549839.2012.665864

Batiibwe, M. S. K. (2019). Using Cultural Historical Activity Theory to understand how emerging technologies can mediate teaching and learning in a mathematics classroom: A review of literature. *Research and Practice in Technology Enhanced Learning*, *14*(1), 12. DOI: 10.1186/s41039-019-0110-7

Batiibwe, M. S. K. (2025). Ethnomathematics as a pedagogical tool for mathematics education: Opportunities and challenges. *SN Social Sciences*, *5*(12), 221. DOI: 10.1007/s43545-025-01260-0

Bortolus, A., Hewitt, C. L., Mitova, V., Schwindt, E., Sogbanmu, T. O., Chukwu, E. E., Heesen, R., Kaufer, R., Rubin, H., Schneider, M. D., Schwenkenbecher, A., Slanickova, H., Woolaston, K., & Yu, L. A. (2025). Knowledge brokers at the science–policy interface: Insights from biosecurity and environmental management. *Ambio*, *54*(12), 1–13. DOI: 10.1007/s13280-025-02208-7 PMID: 40536587

Braun, V., & Clarke, V. (2006). Using thematic analysis in psychology. *Qualitative Research in Psychology*, *3*(2), 77–101. DOI: 10.1191/1478088706qp063oa

Bruner, J. (1978). The role of dialogue in language acquisition. In Sinclair, A., Jarvella, R., & Levelt, W. J. M. (Eds.), *The Child's Conception of Language* (pp. 241–256).

Crossman, A. (2020). Purposive Sampling - Definition and Types. In Thoughtco.

da Silva, C., Pereira, F., & Amorim, J. P. (2024). The integration of indigenous knowledge in school: A systematic review. *Compare: A Journal of Comparative Education*, *54*(7), 1210–1228. DOI: 10.1080/03057925.2023.2184200

Fitrianto, I., & Farisi, M. (2025). Integrating local wisdom into 21st century skills: A contextual framework for culturally relevant pedagogy in rural classrooms. *International Journal of Post Axial: Futuristic Teaching and Learning*, 109-121. https://doi.org/DOI: 10.59944/postaxial.v3i2.444

Furidha, B. W. (2023). Comprehension of the descriptive qualitative research method: A critical assessment of the literature. *Acitya Wisesa: Journal Of Multidisciplinary Research*, 1- 8.

Garcia-Olp, M., Nelson, C., & Saiz, L. (2022). Decolonizing mathematics curriculum and pedagogy: *Indigenous knowledge has always been mathematics education. Educational Studies (Ames)*, *58*(1), 1–16. DOI: 10.1080/00131946.2021.2010079

Glukhova, O. V., Volikova, S. V., Zaretskii, Y. V., & Zaretskii, V. K. (2022). The results of a longitudinal diagnostic study on the project «chess for overall development». *И ПСИХОТЕРАПИЯ*, *51*. Advance online publication. DOI: 10.24234/miopap.v2i10.7

Hadebe-Ndlovu, B., & Manditereza, B. (2025). Using indigenous games as a resource for teaching mathematics. [IJEMST]. *International Journal of Education in Mathematics, Science, and Technology*, *13*(5), 1144–1158. DOI: 10.46328/ijemst.5703

Hudson, M., Milne, M., Russell, K., Smith, B., Reynolds, P., & Atatoa-Carr, P. (2016). The development of guidelines for indigenous research ethics in Aotearoa/New Zealand.

Hui, H. B., & Mahmud, M. S. (2023). Influence of game-based learning in mathematics education on the learners' cognitive and affective domain: A systematic review. *Frontiers in Psychology*, *14*, 1105806. DOI: 10.3389/fpsyg.2023.1105806 PMID: 37057144

Iyabode, A. J. A. (2024). Influence of indigenous games on the mathematics foundation knowledge for young children in Afijio local government. *World of Science: Journal of Modern Research Technologies*, *3*(3), 139–149.

Kancanadana, G., Saputri, O., & Tristiana, V. (2021). The existence of traditional games as a learning media in Elementary School. In *International Conference on Early and Elementary Education (ICEEE)* (Vol. 4, pp. 31-39).

Kheirollahi Meidani, H., Sefidgar, A., Aali, S., Imani, F., & Ashrafi, N. (2025). The Effectiveness of Teaching Indigenous and Local Games on The Development of Cultural and Sports Dimensions in Secondary Schools. *Research on Educational Sport*, *13*(41), e4741. DOI: 10.22089/res.2025.17701.2612

Ladson-Billings, G. (1995b). Toward a theory of culturally relevant pedagogy. *American Educational Research Journal*, *32*(3), 465–491. DOI: 10.3102/00028312032003465

Larasati, I., Pramasdyahsari, A. S., & Harun, L. (2025). Enhancing elementary student's numeracy skills through ethnomathematics-based learning: An analysis of minimum competency assessment results. *Journal of Honai Math*, *8*(1), 71–88. DOI: 10.30862/jhm.v8i1.861

Lebaka, M. E. K. (2024). The Understanding of Indigenous Knowledge as Indigenous Practice, Skill and Know-How: The Case of Communal Music-Making in Bapedi Culture as a Social Practice. *European Journal of Social Science Education and Research*, *11*(2), 1–14. DOI: 10.26417/vtwts124

Madonsela, B. S., & Machete, M. (2025). A systematic review of Indigenous knowledge in global contexts. *Diaspora, Indigenous, and Minority Education*, *19*(2), 71–77. DOI: 10.1080/15595692.2023.2249565

Maimela, G., & Nedombeloni, R. (2026). Decolonising the mainstream education and preserving Indigenous knowledge systems: learning through African Indigenous games. *AlterNative: An International Journal of Indigenous Peoples*, https://doi.org/ .DOI: 10.11771801251403201

Meinck, F., Franchino-Olsen, H., Hafstad, G. S., Jud, A., Liel, C., Nikolaidis, G., & Græsholt-Knudsen, T. (2025). Who gets to decide when children participate? Considering the complexities of consent and gatekeeping in violence research. *Child Abuse & Neglect*, *169*, 107679. DOI: 10.1016/j.chiabu.2025.107679 PMID: 40976033

Ministry of Education. Arts and Culture. (2023). *Mathematics syllabus Grade 4–7*. National Institute for Educational Development (NIED). Windhoek, Namibia

Moleong, L. C. (2018). Pengaruh Real Interest Rate dan Leverage Terhadap Financial Distress. *Modus*, *30*(1), 71–86.

Murat, M., & Demirci, Ö. (2026). Improving Primary School Learners' Number Sense Levels Through Scratch-Supported and Game-Based Activities. *Erzincan Üniversitesi Eğitim Fakültesi Dergisi*, (2026), 1-16. https://doi.org/DOI: 10.17556/erziefd.1801045

Nxumalo, S. A., & Mncube, D. W. (2019). Using Indigenous games and knowledge to decolonise the school curriculum: Ubuntu perspectives. https://core.ac.uk/download/480900151.pdf

Philip, K. S. (2015). Indigenous knowledge: Science and technology studies. In Wright, J. D. (Ed.), *International encyclopedia of the social & behavioral sciences* (2nd ed., pp. 113–114). Elsevier., DOI: 10.1016/B978-0-08-097086-8.85012-6

Philip, T. (1979). Mankala in Eastern and Southern Africa: A Distributional Analysis. *Azania*, *14*(1), 109–138. DOI: 10.1080/00672707909511266

Piaget, J. (2003). Part I: Cognitive Development in Children—Piaget Development and Learning. *Journal of Research in Science Teaching*, •••, 40.

Roche, A., Gervasoni, A., & Kalogeropoulos, P. (2023). Factors that promote interest and engagement in learning mathematics for low-achieving primary learners across three learning settings. *Mathematics Education Research Journal, 35*(3), 525–556. DOI: 10.1007/s13394-021-00402-w

Rogoff, B. (2025). Transforming by participating. *Pillars of developmental psychology: Recollections and reflections*, 278-291.

Sakti, S. A., Endraswara, S., & Rohman, A. (2024). Revitalizing local wisdom within character education through ethnopedagogy apporach: A case study on a preschool in Yogyakarta. *Heliyon, 10*(10), e31370. DOI: 10.1016/j.heliyon.2024.e31370 PMID: 38803959

Sarmiento-Campos, N. V., Lázaro-Guillermo, J. C., Silvera-Alarcón, E. N., Cuellar-Quispe, S., Huamán-Romaní, Y. L., Apaza, O. A., & Sorkheh, A. (2022). A look at Vygotsky's sociocultural theory (SCT): The effectiveness of scaffolding method on EFL learners' speaking achievement. *Education Research International, 2022*(1), 3514892. DOI: 10.1155/2022/3514892

Tassell-Matamua, N. (2025). Indigenous knowledges. What they are and why they matter. *explore, 21*(3), 103144.https://doi.org/DOI: 10.1016/j.explore.2025.103144

Vygotsky, L. S. (1978). *Mind in society: The development of higher psychological processes*. Harvard University Press.

Yonwilad, W., Nongharnpituk, P., Khansila, P., & Pongwiritthon, K. (2025). Enhancing Teamwork Skills and Mathematical Conceptual Connections through Board Game Development in a Cooperative Learning Environment. *Educational Process: International Journal, 17*(1), e2025361. DOI: 10.22521/edupij.2025.17.361

Compilation of References

Abd-El-Khalick, F. (2003). Socioscientific issues in pre-college science classrooms. *The role of moral reasoning on socioscientific issues and discourse in science education*, 41–61. DOI: 10.1007/1-4020-4996-X_3

Abdioğlu, C., Yılmaz, E., & Çevik, M. (2020). 8. sınıf öğrencilerine yönelik fen-matematik temalı bilim kampının değerlendirilmesi: "Gelin tanış olalım; fen ve matematiği eğlenceli kılalım!" projesi. [Evaluation of science-math themed science camp for 8th grade students: The project "let's meet; let's make science and mathematics fun!]. *OPUS International Journal of Society Researches*, *15*(22), 1031–1058. DOI: 10.26466/opus.635705

Adam, S. (2004). Ethnomathematical ideas in the curriculum. *Mathematics Education Research Journal*, *16*(2), 49–68. DOI: 10.1007/BF03217395

Adleff, A.-K., Ross, N., König, J., & Kaiser, G. (2023). Types of mathematical tasks in lower secondary classrooms in Germany: Statistical findings from a latent class analysis based on general mathematical competencies. *Educational Studies in Mathematics*, *114*(3), 371–392. DOI: 10.1007/s10649-023-10254-9 PMID: 37273842

Agbozo, K. K., Agyei, E., & Arthur, Y. D. (2024). Socio-cultural foundations in teacher education: Fostering egalitarianism among preservice teachers in the mathematics classroom. *Golden Ratio of Social Science and Education*, *4*(2), 136–146. DOI: 10.52970/grsse.v4i2.339

Aguirre, J. M., Mayfield-Ingram, K., & Martin, D. B. (2013). *The impact of identity in K–8 mathematics: Rethinking equity-based practices*. National Council of Teachers of Mathematics.

Aguirre, J. M., & Zavala, M. R. (2013). Making culturally responsive mathematics teaching explicit: A lesson analysis tool. *Pedagogies*, *8*(2), 163–190. DOI: 10.1080/1554480X.2013.768518

Ahmed, S. K., Mohammed, R. A., Nashwan, A. J., Ibrahim, R. H., Abdalla, A. Q., Ameen, B. M. M., & Khdhir, R. M. (2025). Using thematic analysis in qualitative research. *Journal of Medicine, Surgery, and Public Health, 6*, 100198. DOI: 10.1016/j.glmedi.2025.100198

Aiken, L. R. (1985). Three coefficients for analyzing the reliability and validity of ratings. *Educational and Psychological Measurement, 45*(1), 131–142. DOI: 10.1177/0013164485451012

Akdemir, M., & Şengül Akdemir, T. (2024). Kültürel matematiksel bilginin eğitime entegrasyonu: İlköğretim matematik öğretmen adaylarının oluşturdukları etnomatematiksel etkinlikler [Integration of cultural mathematical knowledge into education: Ethnomathematical activities created by middle school mathematics teacher candidates]. *International Journal of New Trends in Arts, Sports & Science Education, 13*(4), 218–226.

Aktuna, H. E. (2013). *Sixth grade students' perceptions of and engagement in ethnomathematical tasks in the area measurement concept* [Unpublished master's thesis]. Middle East Technical University.

Aktuna, H. E. (2013). *Sixth grade students' perceptions of and engagement in ethnomathematical tasks in the area measurement concept (Unpublished master's thesis).* Ortadoğu Teknik Üniversitesi.

Al Husaini, M. A., Hidayat, W., Mashuri, Djuniati, S., & Nawawi, A. (2023). Study of existing conditions as preservation Rumah Lontiok of Kendil Kemilau Emas Foundation. [MARCOPOLO]. *Indonesian Journal of Interdisciplinary Research in Science and Technology, 1*(8), 719–730. DOI: 10.55927/marcopolo.v1i8.6255

Alghar, M. Z., & Jamaluddin. (2024). Ethnomodelling: Fractal geometry on the door ornament of the Sumenep palace using the Lindenmayer system. *Euclid, 11*(1), 1–16. DOI: 10.33603/x5gk7n46

Alghar, M. Z., & Marhayati, M. (2023). Ethnomathematics: Exploration of fractal geometry in gate ornaments of the Sumenep Jamik Mosque using the Lindenmayer system. *Indonesian Journal of Science and Mathematics Education, 6*(3), 311–329. DOI: 10.24042/ijsme.v6i3.18219

Alghar, M. Z., & Radjak, D. S. (2024). Systematic literature review: Implementation of ethnomodelling in mathematics learning. Union. *Jurnal Ilmiah Pendidikan Matematika, 12*(1), 67–81. DOI: 10.30738/union.v12i1.17004

Alkan, S. H., & Ipek, D. (2023). Addressing Language Diversity in Early Years Mathematics: Proposed Classroom Practices through a Live Brief Assessment. *Education Sciences, 13*(10), 1025. DOI: 10.3390/educsci13101025

Anderson, D., Lawson, B., & Mayer-Smith, J. (2006). Investigating the impact of a practicum experience in an aquarium on pre-service teachers. *Teaching Education, 17*(4), 341–353. DOI: 10.1080/10476210601017527

Anderson, D., Piscitelli, B., Weier, K., Everett, M., & Tayler, C. (2002). Children's museum experiences: Identifying powerful mediators of learning. *Curator (New York, N.Y.), 45*(3), 213–231. DOI: 10.1111/j.2151-6952.2002.tb00057.x

Anderson, T., & Shattuck, J. (2012). Design-based research: A decade of progress in education research? *Educational Researcher, 41*(1), 16–25. DOI: 10.3102/0013189X11428813

Anhalt, C. O., & Cortez, R. (2016). Developing understanding of mathematical modeling in secondary teacher preparation. *Journal of Mathematics Teacher Education, 19*(6), 523–545. DOI: 10.1007/s10857-015-9309-8

Appelbaum, P., & Stathopoulou, C. (2023). History of Ethnomathematics: Recent developments. In S. R. Sanchez, A. S. Bayés, P. Appelbaum, & G. Aldon (Eds.). (2023). *The role of the history of mathematics in the teaching/learning process: A CIEAEM sourcebook* (pp. 189-207). Springer Nature. DOI: 10.1007/978-3-031-29900-1_8

Apple, M. (2011). *Education and power*. Routledge., DOI: 10.4324/9780203143124

Arcavi, A. (2002). The everyday and the academic in mathematics. *Journal for Research in Mathematics Education Monograph, 11*, 12–29. DOI: 10.2307/749962

Arı, A. A. (2022). Ethnomodeling reflections from Turkish culture. *The Journal of Kesit Academy, 8*(33), 200–223. DOI: 10.29228/kesit.66661

Arı, A. A., & Demir, B. (2022). Matematiksel modellemenin kültürel perspektifi: Etnomodelleme [Cultural perspective of mathematical modeling: Ethnomodeling]. *Kültür Araştırmaları Dergisi, 12*(12), 253–270. DOI: 10.46250/kulturder.1067167

Arnesen, A.-L., Hadjitheodoulou-Loizidou, P., Allan, J., Trasberg, K., Chavdarova-Kostova, S., Furch, E., Vallianatos, A., Dumont, B., & Qiriazi, V. (2010). *Policies and practices for teaching socio-cultural diversity: A framework of teacher competences for engaging with diversity*. Council of Europe Publishing.

Asare, B. (2026). Influence of ethnomathematics-based instruction on students' attitudes, participation, and cultural connections in mathematics learning in Ghana. *Cogent Education, 13*(1), 2612395. Advance online publication. DOI: 10.1080/2331186X.2025.2612395

Ascher, M. (1991). *Ethnomathematics: A multicultural view of mathematical ideas.* Routledge.

Atit, K., Power, J. R., Pigott, T., Lee, J., Geer, E. A., Uttal, D. H., Ganley, C. M., & Sorby, S. A. (2022). Examining the relations between spatial skills and mathematical performance: A meta-analysis. *Psychonomic Bulletin & Review, 29*(3), 699–720. DOI: 10.3758/s13423-021-02012-w PMID: 34799844

Atweh, B., Forgasz, H., & Nebres, B. (2001). *Sociocultural research on mathematics education: An international perspective.* Psychology Press., DOI: 10.4324/9781410600042

Atweh, B., Meaney, T., McMurchy-Pilkington, C., Neyland, J., & Trinick, T. (2004). Social justice and socio-cultural perspectives in mathematics education. In Perry, B., Anthony, G., & Diezmann, C. (Eds.), *Research in mathematics education in Australasia 2000–2003* (pp. 29–52). Post Pressed.

Aydemir, H., & Kubanç, Y. (2014). Problem çözme sürecinde üst bilişsel davranışların incelenmesi. *Journal of Turkish Studies, 9*(2), 203–219. DOI: 10.7827/Turkish-Studies.6555

Aydın, C. S., & Ata, R. (2024). Use of digital games in mathematics education in Turkey: A systematic review study. *Inonu University Journal of the Faculty of Education, 25*(2), 570–596. DOI: 10.17679/inuefd.1415803

Aydoğdu, A. S. E., Aydoğdu, M., & Aktaş, V. (2022). Matematik dersinde bir eğitim aracı olarak sanal müze kullanımı. [Virtual museum use as an educational tool in math class]. *International Journal of Social Science Research, 11*(1), 51–70. https://dergipark.org.tr/tr/pub/ijssresearch

Ayre, M., & Mackenzie, J. (2013). ?oUnwritten, unsaid, just known ??: *The role of Indigenous knowledge (s) in water planning in Australia. Local Environment, 18*(7), 753 ?"768. DOI: 10.1080/13549839.2012.665864

Bahadır, E. (2021). Göçmen öğrencilerin matematik eğitiminde etnomatematik yaklaşımının kullanılması [Ethnomathematics approach in mathematics education for migrant students]. *Milli Eğitim Dergisi, 50*(1), 577–594. DOI: 10.37669/milliegitim.959829

Bakhtin, M. M. (1981). *The dialogic imagination: Four essays (M. Holquist* (Emerson, C., & Holquist, M. Eds. & Trans.). University of Texas Press.

Bakker, A. (2018). *Design research in education: a practical guide for early career researchers*. Routledge. DOI: 10.4324/9780203701010

Barbosa, J. C. (2006). Mathematical modelling in classroom: A socio-critical and discursive perspective. *ZDM Mathematics Education, 38*(3), 293–301. DOI: 10.1007/BF02652812

Barton, B. (1996). *Ethnomathematics: Exploring cultural diversity in mathematics* [Doctoral dissertation, University of Auckland]. University of Auckland ResearchSpace.

Barton, W. D. (1996). *Ethnomathematics: Exploring cultural diversity in mathematics* [Unpublished doctoral dissertation]. University of Auckland.

Bassanezi, R. C. (2002). *Ensino-aprendizagem com modelagem matemática*. Editora Contexto.

Bassarear, T. (1995). *Mathematics for elementary school teachers*. Houghton Mifflin Company.

Batiibwe, M. S. K. (2019). Using Cultural Historical Activity Theory to understand how emerging technologies can mediate teaching and learning in a mathematics classroom: A review of literature. *Research and Practice in Technology Enhanced Learning, 14*(1), 12. DOI: 10.1186/s41039-019-0110-7

Batiibwe, M. S. K. (2024). The role of ethnomathematics in mathematics education: A literature review. *Asian Journal for Mathematics Education, 3*(4), 383–405. DOI: 10.1177/27527263241300400

Batiibwe, M. S. K. (2025). Ethnomathematics as a pedagogical tool for mathematics education: Opportunities and challenges. *SN Social Sciences, 5*(221), 1–38. DOI: 10.1007/s43545-025-01260-0

Bellone, T., Fiermonte, F., & Mussio, L. (2017). The common evolution of geometry and architecture from a geodetic point of view. *International Archives of the Photogrammetry, Remote Sensing and Spatial Information Sciences - ISPRS Archives, XLII-5(W1)*, 623–630. DOI: 10.5194/isprs-archives-XLII-5-W1-623-2017

Berkowitz, M. W., & Simmons, P. (2003). Integrating science education and character education. *The role of moral reasoning on socioscientific issues and discourse in science education*, 117–138. DOI: 10.1007/1-4020-4996-X_7

Bingölbali, E., Arslan, S., & Zembat, İ. Ö. (Eds.). (2016). *Matematik eğitiminde teoriler*. Pegem Akademi.

Binns, C. (2015). What can 'social practice' theory and 'socio-cultural' theory contribute to our understanding of the processes of module design? *Journal of Further and Higher Education, 39*(5), 758–775. DOI: 10.1080/0309877X.2014.953462

Bishop, A. J. (1994). Cultural Conflicts in Mathematics Education: Developing a Research Agenda. *For the Learning of Mathematics, 14*(2), 15–18. JSTOR. https://www.jstor.org/stable/40248109

Bishop, A. J. (1997, August). The relationship between mathematics education and culture. Opening address delivered at the Iranian Mathematics Education Conference in Kermanshah, Iran.

Bishop, A. (2020). Values in mathematics education. In *Encyclopedia of mathematics education* (pp. 893–896). Springer International Publishing., DOI: 10.1007/978-3-030-15789-0_160

Bishop, A. J. (1988). *Mathematical enculturation: a cultural perspective on mathematics education*. Kluwer Academic Publishers. DOI: 10.1007/978-94-009-2657-8

Bishop, A. J. (1990). Western mathematics: The secret weapon of cultural imperialism. *Race & Class, 32*(2), 51–65. DOI: 10.1177/030639689003200204

Bishop, A. J. (1991). *Mathematical enculturation: A cultural perspective on mathematics education*. Kluwer Academic Publishers.

Bishop, A. J. (1994). Cultural conflicts in mathematics education: Developing a research agenda. *For the Learning of Mathematics, 14*(2), 15–18. https://www.jstor.org/stable/40248109

Blomhøj, M. (2009). Different perspectives in research on the teaching and learning mathematical modelling. *Mathematical applications and modelling in the teaching and learning of mathematics*, 1.

Blum, W., & Ferri, R. (2009). Mathematical modelling: Can it be taught and learnt? 1, 45–58.

Blum, W., & Leiss, D. (2007). How do students and teachers deal with modelling problems. *Mathematical modelling (ICTMA 12): Education, engineering and economics*, 222-231. DOI: 10.1533/9780857099419.5.221

Boaler, J. (1993). Encouraging the transfer of 'school' mathematics to the 'real world' through the integration of process and content, context and culture. *Educational Studies in Mathematics, 25*(4), 341–373. DOI: 10.1007/BF01273906

Boaler, J. (1993). The role of contexts in the mathematics classroom: Do they make mathematics more "real"? *For the Learning of Mathematics, 13*(2), 12–17. https://www.jstor.org/stable/40248079

Boaler, J. (2002). Exploring the nature of mathematical activity: Using theory, research and working hypotheses to broaden conceptions of mathematics knowing. *Educational Studies in Mathematics, 51*(1-2), 3–21. DOI: 10.1023/A:1022468022549

Bortolus, A., Hewitt, C. L., Mitova, V., Schwindt, E., Sogbanmu, T. O., Chukwu, E. E., Heesen, R., Kaufer, R., Rubin, H., Schneider, M. D., Schwenkenbecher, A., Slanickova, H., Woolaston, K., & Yu, L. A. (2025). Knowledge brokers at the science ?"policy interface: Insights from biosecurity and environmental management. *Ambio, 54*(12), 1 ?"13. DOI: 10.1007/s13280-025-02208-7 PMID: 40536587

Bosch, M., & Gascón, J. (2014). Introduction to the Anthropological Theory of the Didactic (ATD). In Bikner-Ahsbahs, A., & Prediger, S. (Eds.), *Networking of Theories as a Research Practice in Mathematics Education* (pp. 67–83). Springer International Publishing., DOI: 10.1007/978-3-319-05389-9_5

Bowen, G. A. (2009). Document analysis as a qualitative research method. *Qualitative Research Journal, 9*(2), 27–40. DOI: 10.3316/QRJ0902027

Brandt, A. L., & Chernoff, E. J. (2015). The importance of ethnomathematics in the math class. *Ohio Journal of School Mathematics, 71*, 31–36.

Braun, V., & Clarke, V. (2006). Using thematic analysis in psychology. *Qualitative Research in Psychology, 3*(2), 77 ?"101. DOI: 10.1191/1478088706qp063oa

Brennan, K., & Resnick, M. (2012-April). New frameworks for studying and assessing the development of computational thinking. *Proceedings of the 2012 Annual Meeting of the American Educational Research Association,* Vol. 1, Vancouver, 13-17 April 2012, 25 p. https://scratched.gse.harvard.edu/ct/files/AERA2012.pdf

Britton, J., & Seymour, D. (1989). *Introduction to tessellations*. Dale Seymour Publications.

Brown, A. L. (1992). Design experiments: Theoretical and methodological challenges in creating complex interventions in classroom settings. *Journal of the Learning Sciences, 2*(2), 141–178. DOI: 10.1207/s15327809jls0202_2

Brown, M. W. (2009). The teacher-tool relationship: Theorizing the design and use of curriculum materials. In Remillard, J. T., Herbel-Eisenmann, B. A., & Lloyd, G. M. (Eds.), *Mathematics teachers at work: Connecting curriculum materials and classroom instruction* (pp. 17–36). Routledge.

Bruner, J. (1978). The role of dialogue in language acquisition. In Sinclair, A., Jarvella, R., & Levelt, W. J. M. (Eds.), *The Child ?(tm)s Conception of Language* (pp. 241 ?"256).

Bukova Güzel, E. (2021). *Matematik eğitiminde matematiksel modelleme.* Pegem Akademi.

Büyüköztürk, Ş., Kılıç Çakmak, E., Akgün, Ö. E., Karadeniz, Ş., & Demirel, F. (2019). *Bilimsel araştırma yöntemleri* (26. bs.). Pegem Akademi.

Cadena, J. R., Machado, J. L., & Muso, M. A. (2024). The formation of critical thinking in mathematics learning through a model of contextualization with the social, political, and historical environment in Ecuadorian education. In C. Nicol, G. Knijnik, A, Peng, M. Cherinda, A. Bose (Eds.), *Ethnomathematics and mathematics education international perspectives in times of local and global change* (pp. 185-212). Springer. DOI: 10.1007/978-3-031-60680-9_9

Çağlar, S., Ünal, Y., Çalışkan, B., Gürel, R., & Durmaz, B. (2018). İnformal öğrenme ortamlarının ortaokul öğrencilerinin matematik tutumuna etkisi. [The effect of the mathematıcal learnıng areas to the attıtudes towards mathematıcs]. *Mehmet Akif Ersoy Üniversitesi Sosyal Bilimler Enstitüsü Dergisi*, *10*(23), 11–26. DOI: 10.20875/makusobed.357694

Cahyani, A. L., Putra, Z. H., Witri, G., & Ahmad, S. (2025). Exploring ethnomathematics in a Malay traditional house and architecture on Meranti Island as a geometry learning resource for elementary school. *Journal of Teaching and Learning in Elementary Education*, *8*(2), 105–117. DOI: 10.33578/jtlee.08.02.105.117

Caldeira, A. D. (2009). Mathematical modeling and environmental education. *Mathematical applications and modelling in the teaching and learning of mathematics*, 145.

Caniglia, J. (2003). Math in the city: Experiencing mathematics through visiting black historic sites. *Journal of Experiential Education*, *26*(2), 70–74. DOI: 10.1177/105382590302600204

Carroll, J. B. (1963). A model of school learning. *Teachers College Record*, *64*(8), 723–733. DOI: 10.1177/016146816306400801

Casi, R., & Sabena, C. (2024). Mathematics in art and history museums: An informal mathematics education case for teachers' in-service training. *Education Sciences*, *14*(5), 489. DOI: 10.3390/educsci14050489

Çelebi, M. (2022). *Nitel araştırma yöntemleri*. Pegem Akademi.

Çenberci, S., & Horzum, T. (2023). Matematik öğretmeni adaylarının etnomatematik farkındalıklarının incelenmesi. [Investigation of ethnomathematics awareness of preservice mathematics teachers]. *Ahmet Keleşoğlu Eğitim Fakültesi Dergisi, 5*(3), 713-732. https://dergipark.org.tr/tr/download/article-file/3187289

Çenberci, S., & Horzum, T. (2023). Matematik öğretmeni adaylarının etnomatematik farkındalıklarının incelenmesi [Investigation of ethnomathematics awareness of preservice mathematics teachers]. *Ahmet Keleşoğlu Eğitim Fakültesi Dergisi, 5*(3), 713–732. DOI: 10.38151/akef.2023.80

Chaudhary, P. R., Panthi, D., & Bhatta, C. R. (2023). Ethnomathematics: Geometry and architecture of the tharu's traditional houses at Chakhoura museum. *Journal of Mathematical Problems. Equations and Statistics*, *4*(2), 9–13. DOI: 10.22271/math.2023.v4.i2a.96

Cherinda, M. (2024). From ethnomathematical reasoning to mathematical research. In C. Nicol, G. Knijnik, A, Peng, M. Cherinda, A. Bose (Eds.), *Ethnomathematics and mathematics education international perspectives in times of local and global change* (pp. 101-134). Springer. DOI: 10.1007/978-3-031-60680-9_6

Chevallard, Y. (2006). Steps towards a new epistemology in mathematics education. In *Proceedings of the IV Congress of the European Society for Research in Mathematics Education, 1, 21–30.*

Chou, P.-I., Su, M.-H., & Wang, Y.-T. (2018). Transforming teacher preparation for culturally responsive teaching in Taiwan. *Teaching and Teacher Education*, *75*, 116–127. DOI: 10.1016/j.tate.2018.06.013

Cirade, G., & Crumière, A. (2019). The study of teachers' mathematical and didactic praxeologies as a tool for teacher education. In M. Bosch, Y. Chevallard, F. J. García, & J. Monaghan (Eds.), *Working with the Anthropological Theory of the Didactic in Mathematics Education* (1st ed., pp. 159–168). Routledge. DOI: 10.4324/9780429198168-10

Clements, D. H., & Battista, M. T. (1992). Geometry and spatial reasoning. In Grouws, D. A. (Ed.), *Handbook of research on mathematics teaching and learning* (pp. 420–464). Macmillan.

Cobb, P., Confrey, J., diSessa, A., Lehrer, R., & Schauble, L. (2003). Design experiments in educational research. *Educational Researcher*, *32*(1), 9–13. DOI: 10.3102/0013189X032001009

Cobb, P., Gresalfi, M., & Hodge, L. L. (2009). An interpretive scheme for analyzing the identities that students develop in mathematics classrooms. *Journal for Research in Mathematics Education, 40*(1), 40–68. DOI: 10.5951/jresematheduc.40.1.0040

Cohen, J. (1960). A coefficient of agreement for nominal scales. *Educational and Psychological Measurement, 20*(1), 37–46. DOI: 10.1177/001316446002000104

Conway, J. H., Burgiel, H., & Strauss, C. (2008). *The symmetries of things*. A K Peters.

Cooper, S. (2011). An exploration of the potential for mathematical experiences in informal learning environments. *Visitor Studies, 14*(1), 48–65. DOI: 10.1080/10645578.2011.557628

Cortés, D. P. O., & Orey, D. C. (2020). Connecting ethnomathematics and modelling: A mixed methods study to understand the dialogic approach of ethnomodelling. *REVEMOP, 2*, 1–25. DOI: 10.33532/revemop.e202011

Cotton, T., & Hardy, T. (2004). Problematising culture and discourse for mathematics education research. *Researching the socio-political dimensions of mathematics education*, 85–103.

Council of Higher Education (CoHE). (2018). *İlköğretim matematik öğretmenliği lisans programı* [Elementary mathematics education undergraduate program] https://www.yok.gov.tr/Documents/Kurumsal/egitim_ogretim_dairesi/Yeni-Ogretmen-Yetistirme-Lisans Programlari/Ilkogretim_Matematik_Lisans_Programi.pdf

Creswell, J. W. (2015). *Educational research: Planning, conducting, and evaluating quantitative and qualitative research* (5th ed.). Pearson.

Crossman, A. (2020). Purposive Sampling - Definition and Types. In Thoughtco.

Crowley, M. L. (1987). The van Hiele model of the development of geometric thought. In Lindquist, M. M., & Shulte, A. P. (Eds.), *Learning and teaching geometry, K–12* (pp. 1–16). National Council of Teachers of Mathematics.

Csikszentmihalyi, M. (1990). *Flow: The psychology of optimal experience*. Harper & Row.

D'Ambrosio, U. (1985). Ethnomathematics and Its Place in the History and Pedagogy of Mathematics. *For the Learning of Mathematics, 5*(1), 44–48. JSTOR. https://www.jstor.org/stable/40247876

D'Ambrosio, U. (1993). Etnomatemática: Um programa. *A Educação Matemática em Revista, 1*(1), 5–11.

D'Ambrosio, U. (1985). Ethnomathematics and its place in the history and pedagogy of mathematics. *For the Learning of Mathematics*, *5*(1), 44–48.

D'Ambrosio, U. (1990). *Etnomatemática*. Editora Ática.

D'Ambrosio, U. (1997). Ethnomathematics and its place in the history and pedagogy of mathematics. In Powell, A., & Frankenstein, M. (Eds.), *Ethnomathematics: Challenging eurocentrism in mathematics education* (pp. 13–24). SUNY Press.

D'Ambrosio, U. (2001). *Etnomatemática: Elo entre as tradições e a modernidade*. Autêntica.

D'Ambrosio, U. (2001). General remarks on ethnomathematics. *Zentralblatt für Didaktik der Mathematik, ZDM*, *33*(3), 67–69. DOI: 10.1007/BF02655696

D'Ambrosio, U. (2006). *Ethnomathematics: Link between traditions and modernity*. Sense Publishers. DOI: 10.1163/9789460911415

D'Ambrosio, U. (2006). The program ethnomathematics and the challenges of globalization. *Circumscribere International Journal for the History of Science*, *1*(1), 74–82.

D'Ambrosio, U., & D'Ambrosio, B. S. (2013). The role of ethnomathematics in curricular leadership in mathematicseducation. *Journal of Mathematics Education at Teachers College*, *4*(1), 19–25. DOI: 10.7916/jmetc.v4i1.767

da Silva, C., Pereira, F., & Amorim, J. P. (2024). The integration of indigenous knowledge in school: A systematic review. *Compare: A Journal of Comparative Education*, *54*(7), 1210 ?"1228. DOI: 10.1080/03057925.2023.2184200

Darragh, L. (2016). Identity research in mathematics education. *Educational Studies in Mathematics*, *93*(1), 19–33. DOI: 10.1007/s10649-016-9696-5

Darvas G. (2007). *Symmetry: Cultural-historical and ontological aspects of science-arts relations; the natural and man-made world in an interdisciplinary approach*. Basel.

Davis, E. A., & Krajcik, J. S. (2005). Designing educative curriculum materials to promote teacher learning. *Educational Researcher*, *34*(3), 3–14. DOI: 10.3102/0013189X034003003

de Abreu, G. (2000). Relationships between macro and micro socio-cultural contexts: Implications for the study of interactions in the mathematics classroom. *Educational Studies in Mathematics*, *41*(1), 1–29. DOI: 10.1023/A:1003875728720 de Castro, R. S. (2024). Weaving the harmony of cultures into the infinite tapestry of mathematics. *Aracê*, *6*(1), 219–230. DOI: 10.56238/arev6n1-013

de Oliveira Vidal, E., de Sousa, V. F., Alves, L. C. D. S. D., de Costa, D., Lopes, R. L. F., de Pontes, J. S., & de Luna, J. L. (2025). The socio-critical perspective of mathematical modeling: A proposal on prostate cancer. *Journal of Education and Learning*, *14*(6), 57–72. DOI: 10.5539/jel.v14n6p57

Demir, G., & Seçkin Kapucu, M. (2023). Ethnomathematics approaches at middle school textbooks. *Osmangazi Journal of Educational Research*, *10*(1), 120–141. DOI: 10.59409/ojer.1269449

Desai, S., Kurtz, B., & Safi, F. (2021). Mathematics heritage project: An exploration empowering students' mathematical identities. *Journal of Humanistic Mathematics*, *11*(2), 106–122. DOI: 10.5642/jhummath.202102.05

Desai, S., Safi, F., Bush, S. B., Wilkerson, T., Andreasen, J., & Orey, D. C. (2022). Ethnomodeling: Extending mathematical modeling research in teacher education. *Investigations in Mathematics Learning*, *14*(4), 305–319. DOI: 10.1080/19477503.2022.2139092

Design-Based Research Collective. (2003). Design-Based research: An emerging paradigm for educational inquiry. *Educational Researcher*, *32*(1), 5–8. DOI: 10.3102/0013189X032001005

Di Martino, P., Gregorio, F., & Iannone, P. (2023). The transition from school to university in mathematics education research: New trends and ideas from a systematic literature review. *Educational Studies in Mathematics*, *113*(1), 7–34. DOI: 10.1007/s10649-022-10194-w PMID: 36408299

Dictado, M., & Dagdag, J. D. (2025). Barriers to Achieving Mathematics Proficiency: Synthesis of Research. *Journal of Interdisciplinary Perspectives*. https://api.semanticscholar.org/CorpusID:280731060

Dierking, L. D., Falk, J. H., Rennie, L., Anderson, D., & Ellenbogen, K. (2003). Policy statement of the "informal science education" ad hoc committee. *Journal of Research in Science Teaching*, *40*(2), 108–111. DOI: 10.1002/tea.10066

Doerr, H. M. (2007). What knowledge do teachers need for teaching mathematics through applications and modelling? In *Modelling and applications in mathematics education: The 14th ICMI study* (pp. 69–78). Springer US., DOI: 10.1007/978-0-387-29822-1_5

Doerr, H. M., Ärlebäck, J. B., & Misfeldt, M. (2017). Representations of modelling in mathematics education. In *Mathematical modelling and applications: Crossing and researching boundaries in mathematics education* (pp. 71–81). Springer International Publishing., DOI: 10.1007/978-3-319-62968-1_6

Duatepe-Paksu, A., Kazak, S., & Çontay, E. G. (2022). Okul dışı ortamlarda gerçekleştirilen matematik etkinliklerinin değerlendirilmesi: "Her Yer Matematik Projesi". [Evaluation of mathematics activities performed in out-of-school environments: "Mathematics Everywhere" project]. *Muğla Sıtkı Koçman Üniversitesi Eğitim Fakültesi Dergisi*, *9*(2), 541–558. DOI: 10.21666/muefd.1094581

Dumlu, B. Ö., & Alkaş Ulusoy, Ç. (2025). Using an ethnomathematical approach in teaching transformational geometry: the kilim activity. *Çukurova Üniversitesi Eğitim Fakültesi Dergisi, 54*(2), 670-714. DOI: 10.14812/cuefd.1525565

Ebby, C. B., Hess, B., Pecora, L., & Valerio, J. (2026). Facilitating collaborative inquiry into practice around artifacts of mathematics teaching. *Journal of Mathematics Teacher Education, 29*(1), 59–82. DOI: 10.1007/s10857-024-09649-z

Eğe, E. F., Temel, H., & Üzel, D. (2025). Trends in research on out-of-school learning environments in mathematics education: A systematic review. *Balıkesir Üniversitesi Fen Bilimleri Enstitüsü Dergisi*, *27*(2), 802–823. DOI: 10.25092/baunfbed.1659445

Eka, W. D., Tola, T. T., & Negasi, R. D. (2025). Effects of students' mathematics learning strategy and perceived task difficulty on their achievement of mathematics. *PLoS One*, *20*(11), e0337115. DOI: 10.1371/journal.pone.0337115 PMID: 41270042

El Bedewy, S., Lavicza, Z., & Lyublinskaya, I. (2024). STEAM practices connecting mathematics, arts, architecture, culture and history in a non-formal learning environment of a museum. *Journal of Mathematics and the Arts*, *18*(1-2), 101–134. DOI: 10.1080/17513472.2024.2321563

Ergene, Ö., Çaylan Ergene, B., & Yazıcı, E. Z. (2020). Ethnomathematics activities: Reflections from the design and implementation process. *Turkish Journal of Computer and Mathematics Education, 11*(2), 402–437. DOI: 10.16949/turkbilmat.688780

Ernest, P., Sriraman, B., & Ernest, N. (2016). *Critical mathematics education: Theory, praxis, and reality*. IAP., DOI: 10.1108/978-1-68123-261-4

Eş, H., & Öztürk, N. (2021). An activity for transferring the multidimensional structure of SSI to middle school science courses: I discover myself in the decision-making process with SEE-STEP! *Research in Science Education, 51*(3), 889–910. DOI: 10.1007/s11165-019-09865-1

Eş, H., & Varol, V. (2019). Fen bilgisi öğretmenliği ve ilahiyat öğrencilerinin nükleer santral sosyo-bilimsel konusuyla ilgili informal argümanları [Informal arguments of pre-service science teachers and theology students regarding the socioscientific issue of nuclear power plants]. *Mersin Üniversitesi Eğitim Fakültesi Dergisi, 15*(2), 437–454. DOI: 10.17860/mersinefd.533013

Eshach, H. (2007). Bridging in-school and out-of-school learning: Formal, non-formal, and informal education. *Journal of Science Education and Technology, 16*(2), 171–190. DOI: 10.1007/s10956-006-9027-1

Faisal, G. (2019). Arsitektur Melayu: Identifikasi Rumah Melayu Lontiak Suku Majo Kampar. *Langkau Betang: Jurnal Arsitektur, 6*(1), 1–12. DOI: 10.26418/lantang.v6i1.31007

Falk, J. H., & Dierking, L. D. (2000). *Learning from museums: Visitor experiences and the making of meaning*. AltaMira Press.

Falk, J. H., & Dierking, L. D. (2016). *The museum experience revisited*. Routledge., DOI: 10.4324/9781315417851

Fatani, K., Mohamed, M., & Al-Khateeb, S. (2017). Sustainable socio-cultural guidelines for neighborhood design in Jeddah. *Procedia Environmental Sciences, 37*, 584–593. DOI: 10.1016/j.proenv.2017.03.045

Fatkhurohman, F., Ayuningtyas, A. D., Noto, M. S., & Widodo, S. A. (2021). Etnomathematics: Exploration of Geblek Renteng Batik in transformation geometry. *Numerical: Jurnal Matematika dan Pendidikan Matematika, 5*(2), 79-90. DOI: 10.25217/numerical.v5i2.1506

Ferri, R. (2007). Modelling from a cognitive perspective: Individual modelling routes of pupils. In C. Haines, P. Galbraith, W. Blum, & S. Khan (Eds.), *Mathematical modelling. Education, engineering, and economics* (pp. 260–270). Horwood Publishing.DOI: 10.1533/9780857099419.5.260

Ferri, R. (2006). Theoretical and empirical differentiations of phases in the modelling process. *ZDM Mathematics Education, 38*(2), 86–95. DOI: 10.1007/BF02655883

Fitrianto, I., & Farisi, M. (2025). Integrating local wisdom into 21st century skills: A contextual framework for culturally relevant pedagogy in rural classrooms. *International Journal of Post Axial: Futuristic Teaching and Learning*, 109-121. https://doi.org/DOI: 10.59944/postaxial.v3i2.444

Fouze, A. Q., & Amit, M. (2018). Development of mathematical thinking through integration of ethnomathematic folklore game in math instruction. *Eurasia Journal of Mathematics, Science and Technology Education, 14*(2), 617–630. DOI: 10.12973/ejmste/80626

Fouze, A. Q., & Amit, M. (2018). On the importance of an ethnomathematical curriculum in mathematics education. *Eurasia Journal of Mathematics, Science and Technology Education, 14*(2), 561–567. DOI: 10.12973/ejmste/76956

Fouze, A. Q., & Amit, M. (2023). The importance of ethnomathematics education. *Creative Education, 14*(4), 729–740. DOI: 10.4236/ce.2023.144048

Franita, Y., Hendrastuti, Z. R., Pamungkas, M. D., Chasanah, A. N., Rahmawati, F., 'Adna, S. F., & Wicaksono, A. B. (2024). Development of mathematics teaching materials for junior high school based on ethnomathematics. *Formatif: Jurnal Ilmiah Pendidikan MIPA, 14*(2), 413–418. DOI: 10.30998/formatif.v14i2.28002

Frankenstein, M. (1983). Critical mathematics education: An application of Paulo Freire's epistemology. *Journal of Education, 165*(4), 315–339. DOI: 10.1177/002205748316500403

Freire, P. (2014). *Pedagogy of the oppressed* (30th anniversary ed.). Bloomsbury Academic.

Freire, P. (1970). *Pedagogia do oprimido*. Paz e Terra.

Fuentes-Cabrera, A., Parra-González, M. E., Lopez-Belmonte, J., & Segura-Robles, A. (2020). Learning mathematics with emerging methodologies-The escape room as a case study. *Mathematics, 8*(9), 1586. DOI: 10.3390/math8091586

Furidha, B. W. (2023). Comprehension of the descriptive qualitative research method: A critical assessment of the literature. *Acitya Wisesa: Journal Of Multidisciplinary Research*, 1- 8.

Garcia-Olp, M., Nelson, C., & Saiz, L. (2022). Decolonizing mathematics curriculum and pedagogy: *Indigenous knowledge has always been mathematics education. Educational Studies (Ames), 58*(1), 1 ?"16. DOI: 10.1080/00131946.2021.2010079

Gay, G. (2018). *Culturally responsive teaching: Theory, research, and practice* (3rd ed.). Teachers College Press.

Gerber, B. L., Marek, E. A., & Cavallo, A. M. L. (2001). Development of an informal learning opportunities assay. *International Journal of Science Education, 23*(6), 569–583. DOI: 10.1080/09500690116959

Gerdes, P. (1988). On culture, geometrical thinking and mathematics education. *Educational Studies in Mathematics, 19*(2), 137–162. DOI: 10.1007/BF00751229

Gerdes, P. (1994). Reflections on ethnomathematics. *For the Learning of Mathematics, 14*(2), 19–22. https://www.jstor.org/stable/40248110

Gerdes, P. (1998). On culture and mathematics teacher education. *Journal of Mathematics Teacher Education, 1*(1), 33–53. DOI: 10.1023/A:1009955031429

Gibbs, A. M. (2019). *Socio-critical mathematical modeling and the role of mathematics in society* (Doctoral dissertation, Florida Institute of Technology).

Gibbs, A. M., & Park, J. Y. (2022). Unboxing mathematics: Creating a culture of modeling as critic. *Educational Studies in Mathematics, 110*(1), 167–192. DOI: 10.1007/s10649-021-10119-z

Gilligan-Lee, K. A., Hawes, Z. C. K., & Mix, K. S. (2022). Spatial thinking as the missing piece in mathematics curricula. *npj Science of Learning, 7*(1), 10. DOI: 10.1038/s41539-022-00128-9

Glukhova, O. V., Volikova, S. V., Zaretskii, Y. V., & Zaretskii, V. K. (2022). The results of a longitudinal diagnostic study on the project A"chess for overall developmentA". *D~ DYCD~XOTEPADYD~D_, 51*. Advance online publication. DOI: 10.24234/miopap.v2i10.7

Godino, J. D., Batanero, C., Burgos, M., & Wilhelmi, M. R. (2024). Understanding the onto-semiotic approach in mathematics education through the lens of the cultural historical activity theory. *ZDM Mathematics Education, 56*(6), 1331–1344. DOI: 10.1007/s11858-024-01590-y

Greer, B., & Mukhopadhyay, S. (2015). Honoring diversity in intercultural mathematics education. *Intercultural Education, 26*(4), 261–265. DOI: 10.1080/14675986.2015.1073876

Grothérus, A., & Fägerstam, E. (2017). Impact of long-term regular outdoor learning in mathematics: The case of John. In T. Dooley & G. Gueudet (Eds.), *Proceedings of the tenth congress of the European Society for Research in Mathematics Education (CERME10)* (pp. 1074–1081). Dublin City University and ERME. https://hal.science/hal-01936003

Günay, K. (2023). *Etnomatematik uygulamalarının kullanıldığı öğretimin öğrencilerin problem çözme becerileri ve duyuşsal davranışları üzerindeki etkisinin incelenmesi [Examining the effects of instruction using ethnomathematics applications on students' problem-solving skills and affective behaviors]* [Unpublished master thesis]. Sakarya University.

Günay, K., & Takunyacı, M. (2023). Examining the effect of teaching with ethnomathematics on students' problem-solving skills on transformation geometry. *Journal of Interdisciplinary Education: Theory and Practice*, *5*(2), 86–106. DOI: 10.47157/jietp.1384160

Gürefe, N. (2025). Investigation of the effect of geometry teaching with ethnomodeling on students' academic achievement in area measurement. *Ordu Üniversitesi Sosyal Bilimler Enstitüsü Sosyal Bilimler Araştırmaları Dergisi, 15*(ISRIS 2025), 508-535. DOI: 10.48146/odusobiad.1728168

Gürefe, N. (2025). Investigation of the effect of geometry teaching with ethnomodeling on students' academic achievement in area measurement. *Ordu Üniversitesi Sosyal Bilimler Enstitüsü Sosyal Bilimler Araştırmaları Dergisi*, *15*, 508-535.

Güreş, H., & Ada, T. (2025). Revealing and examining students' cultural mathematical knowledge and thoughts through an ethnomathematics activity. *Journal of Mathematics and the Arts*, *19*(3-4), 136–159. DOI: 10.1080/17513472.2025.2576831

Gutiérrez, R. (2008). A "gap-gazing" fetish in mathematics education? Problematizing research on the achievement gap. *Journal for Research in Mathematics Education*, *39*(4), 357–364. https://www.jstor.org/stable/40539302. DOI: 10.5951/jresematheduc.39.4.0357

Gutiérrez, R. (2013). The sociopolitical turn in mathematics education. *Journal for Research in Mathematics Education*, *44*(1), 37–68. DOI: 10.5951/jresematheduc.44.1.0037

Gutstein, E. (2003). Teaching and learning mathematics for social justice in an urban, Latino school. *Journal for Research in Mathematics Education*, *34*(1), 37–73. DOI: 10.2307/30034699

Gutstein, E. (2006). *Reading and writing the world with mathematics: Toward a pedagogy for social justice*. Routledge.

Gutstein, E., & Peterson, B. (2005). Rethinking mathematics: Teaching social justice by the numbers. *Rethinking Schools*.

Hadebe-Ndlovu, B., & Manditereza, B. (2025). Using indigenous games as a resource for teaching mathematics. [IJEMST]. *International Journal of Education in Mathematics, Science, and Technology*, *13*(5), 1144 ?"1158. DOI: 10.46328/ijemst.5703

Haggarty, L., & Pepin, B. (2002). An investigation of mathematics textbooks and their use in English, French and German classrooms: Who gets an opportunity to learn what? *British Educational Research Journal*, *28*(4), 567–590. DOI: 10.1080/0141192022000005832

Hajazizi, F., & Gooya, Z. (2024). Internalized mathematics as intuitive as breathing: an ethnomathematics study of traditional architecture in Iran. In C. Nicol, G. Knijnik, A, Peng, M. Cherinda, A. Bose (Eds.), *Ethnomathematics and mathematics education international perspectives in times of local and global change* (pp. 135-164). Springer. DOI: 10.1007/978-3-031-60680-9_7

Hall, A. (2007). Vygotsky goes online: Learning design from a socio-cultural perspective. *Learning and Socio-cultural Theory: Exploring Modern Vygotskian Perspectives International Workshop, 1*(1). http://ro.uow.edu.au/llrg/vol1/iss1/6(open in a new window)

Handelzalts, A. (2019). Collaborative Curriculum Development in Teacher Design Teams. In Pieters, J., Voogt, J., & Pareja Roblin, N. (Eds.), *Collaborative Curriculum Design for Sustainable Innovation and Teacher Learning* (pp. 159–173). Springer International Publishing., DOI: 10.1007/978-3-030-20062-6_9

Hart, L. C. (1993). Some factors that impede or enhance performance in mathematical problem solving. *Journal for Research in Mathematics Education, 24*(2), 167–171.

Hawes, Z. C. K., Gilligan-Lee, K. A., & Mix, K. S. (2022). Effects of spatial training on mathematics performance: A meta-analysis. *Developmental Psychology, 58*(1), 112–137. DOI: 10.1037/dev0001281 PMID: 35073120

Heimlich, J. E. (1993). *Nonformal environmental education: Toward a working definition* (ERIC/CSMEE Informational Bulletin No. 2). ERIC Clearinghouse for Science, Mathematics, and Environmental Education. https://files.eric.ed.gov/fulltext/ED360154.pdf

Hendriyanto, A., Priatna, N., Juandi, D., Dahlan, J. A., Hidayat, R., Sahara, S., & Muhaimin, L. H. (2023). Learning mathematics using an ethnomathematics approach: A systematic literature review. *Journal of Higher Education Theory and Practice, 23*(7), 59–74. DOI: 10.33423/jhetp.v23i7.6012

Hernandez, C. M., Morales, A. R., & Shroyer, M. G. (2013). The development of a model of culturally responsive science and mathematics teaching. *Cultural Studies of Science Education, 8*(4), 803–820. DOI: 10.1007/s11422-013-9544-1

Herzig, A., & Steinthorsdottir, O. B. (2020). Cultural influences in mathematics education. In *Encyclopedia of Mathematics Education* (pp. 168–172). Springer International Publishing., DOI: 10.1007/978-3-030-15789-0_38

Hobbs, L. (2018). *Developing culturally responsive mathematics teachers: Understanding equity and access in math education* [Doctoral thesis, University of Houston]. University of Houston Institutional Repository.

Högström, P., Gericke, N., Wallin, J., & Bergman, E. (2025). Teaching Socioscientific Issues: A Systematic Review. *Science & Education, 34*(5), 3079–3122. https://doi .org/. DOI: 10.1007/s11191-024-00542-y

Hsiao, Y. J. (2015). The culturally responsive teacher preparedness scale: An exploratory study. *Contemporary Issues in Education Research, 8*(4), 241–250. DOI: 10.19030/cier.v8i4.9432

Huang, S.-Y., Kuo, Y.-H., & Chen, H.-C. (2020). Applying digital escape rooms infused with science teaching in elementary school: Learning performance, learning motivation, and problem-solving ability. *Thinking Skills and Creativity, 37*, 100681. DOI: 10.1016/j.tsc.2020.100681

Hudson, M., Milne, M., Russell, K., Smith, B., Reynolds, P., & Atatoa-Carr, P. (2016). The development of guidelines for indigenous research ethics in Aotearoa/ New Zealand.

Hui, H. B., & Mahmud, M. S. (2023). Influence of game-based learning in mathematics education on the learners ?(tm) cognitive and affective domain: A systematic review. *Frontiers in Psychology, 14*, 1105806. DOI: 10.3389/fpsyg.2023.1105806 PMID: 37057144

Husén, T. (Ed.). (1967). *International study of achievement in mathematics: A comparison of twelve countries* (Vols. 1–2). Almqvist & Wiksell; John Wiley.

Isa, B. (2017). *Museum pedagogy and learning experiences: An investigation into museum education from institutional perspectives* [Doctoral dissertation, RMIT University]. RMIT Research Repository.

Işık, C., & Kar, T. (2012). Sınıf öğretmeni adaylarının sözel problemlere yönelik kurdukları problemlerin analizi. *Kuram ve Uygulamada Eğitim Bilimleri, 12*(1), 201–221.

Işık, H. K., & Liman Kaban, A. (2025). Escaping traditional learning: The role of educational digital escape games on achievement and motivation in primary math education. *Journal of Educational Technology and Online Learning, 8*(1), 136–151. DOI: 10.31681/jetol.1603015

Iyabode, A. J. A. (2024). Influence of indigenous games on the mathematics foundation knowledge for young children in Afijio local government. *World of Science: Journal of Modern Research Technologies, 3*(3), 139 ?"149.

Janiola, F. R. (2024). The dynamic components of Janiola's ethnolearning framework. In C. Nicol, G. Knijnik, A, Peng, M. Cherinda, A. Bose (Eds.), *Ethnomathematics and mathematics education international perspectives in times of local and global change* (pp. 235-256). Springer. DOI: 10.1007/978-3-031-60680-9_11

Johnson, D. T., & Chandler, F. J. (2009). Pre-service teachers' fieldtrip to the battleship: Teaching and learning mathematics through an ınformal learning experience. *Issues in the Undergraduate Mathematics Preparation of School Teachers, 2*, 1-9. https://eric.ed.gov/?id=EJ859285

Jorgensen, R., Gates, P., & Roper, V. (2014). Structural exclusion through school mathematics: Using Bourdieu to understand mathematics as a social practice. *Educational Studies in Mathematics*, *87*(2), 221–239. DOI: 10.1007/s10649-013-9468-4

Junger, A. P., Jacomossi, R. R., Oliveira, V. I. D., Araújo, M. M. P., & Galoni, N. C. (2026). *Digital Technologies And The Overcoming Of Difficulties In Mathematical Learning*. DOI: 10.9790/7388-1601026064

Kahila, J., Parkki, T., Gröhn, A., Karvinen, A., Telimaa, E., Riikonen, P., Tiitta, R., Haantio, P., Keinänen, A., Kerkkänen, T., Jormanainen, I., Penttinen, S., & Tedre, M. (2020-November). Escape room game for CT learning activities in the primary school. *Koli Calling 20: Proceedings of the 20th Koli Calling International Conference on Computing Education Research*. 9, 1-5. Association for Computing Machinery. DOI: 10.1145/3428029.3428063

Kaiser, G., Blomhøj, M., & Sriraman, B. (2006). Towards a didactical theory for mathematical modelling. *ZDM Mathematics Education*, *38*(2), 82–85. DOI: 10.1007/BF02655882

Kaiser, G., & Sriraman, B. (2006). A global survey of international perspectives on modelling in mathematics education. *ZDM Mathematics Education*, *38*(3), 302–310. DOI: 10.1007/BF02652813

Kalinec-Craig, C., Prasad, P. V., & Luna, C. (2019). Geometric transformations and Talavera tiles: A culturally responsive approach to teacher professional development and mathematics teaching. *Journal of Mathematics and the Arts, 13*(1–2), 72–90. DOI: 10.1080/17513472.2018.1504491

Kancanadana, G., Saputri, O., & Tristiana, V. (2021). The existence of traditional games as a learning media in Elementary School. In *International Conference on Early and Elementary Education (ICEEE)* (Vol. 4, pp. 31-39).

Karadag, Z., & Karagöz Akar, G. (2020). El sanatları ve matematik: Cebir dünyasına bakış. *Boğaziçi University Journal of Education, 37*(2), 123–146. https://izlik.org/JA92TY58KB

Karademir, A., & Girgin, D. (2024). Eğitimde araştırma ve uygulamanın buluşması: Tasarım tabanlı araştırmaya kuramsal bir bakış [Integration of research and practice in education: A theoretical perspective on design based research]. *Journal of Computer and Education Research, 12*(24), 780–799. DOI: 10.18009/jcer.1522744

Karina, A., Faizah, H., Elmustian, & Syafrial. (2022). Filosofi rumah adat Lontiok khas Desa Wisata Pulau Belimbing Kuok Kabupaten Kampar Provinsi Riau. *Jurnal Pendidikan dan Konseling, 4*(6), 10188–10202. DOI: 10.31004/jpdk.v4i6.10007

Kastamonu Governorship Provincial Directorate of Culture and Tourism. (n.d.). *Kastamonu e-broşür* [Kastamonu e-brochure] [Brochure]. https://kastamonu.ktb.gov.tr/Eklenti/112738,kastamonu-e-brosurpdf.pdf?0

Kayhan Altay, M., & Yetkin Özdemir, E. (2023). The use of museum resources in mathematics education: A study with preservice middle-school mathematics teachers. *Journal of Education for Teaching, 49*(4), 616–629. DOI: 10.1080/02607476.2022.2150534

Kelton, M. L. (2015). Math on the move: a video-based study of school field trips to a mathematics exhibition. (Publication No. 3712516) [Doctoral dissertation, San Diego State University]. ProQuest Dissertations & Theses Global.

Kelton, M. L. (2021). Mathematics learning pathways on a school fieldtrip: Interactional practices linking school and museum activity. *Visitor Studies, 24*(2), 220–242. DOI: 10.1080/10645578.2021.1939984

Kemendikbudristek. (2022). *Permendikbudristek No. 56/M/2022 tentang pedoman penerapan kurikulum dalam rangka pemulihan pembelajaran.* Kementerian Pendidikan, Kebudayaan, Riset, dan Teknologi Republik Indonesia.

Kementerian Pendidikan dan Kebudayaan. (2017). *Penetapan warisan budaya takbenda Indonesia: Rumah Lontiok.* Kementerian Pendidikan dan Kebudayaan Republik Indonesia.

Ketenci Tona, F. M., & Akman, B. (2026). Mathematics from a cultural perspective: Systematic review article. *Bayburt Journal of Education Faculty, 21*(49), 424–458. DOI: 10.35675/befdergi.1677972

Khalil, I. A. (2023). Mathematics teachers' attitudes towards integrating ethnic mathematics in teaching mathematical topics: A mixed methods study. *International Journal of Instruction, 16*(4), 1061–1080. DOI: 10.29333/iji.2023.16458a

Khasanah, B. A., Prahmana, R. C. I., Adiputra, S., & Arnal-Palacián, M. (2025). The beauty of mathematics in Indonesian culture: An impactful and meaningful context in number patterns learning. *JRAMathEdu (Journal of Research and Advances in Mathematics Education)*. DOI: 10.23917/jramathedu.v10i2.10441

Kheirollahi Meidani, H., Sefidgar, A., Aali, S., Imani, F., & Ashrafi, N. (2025). The Effectiveness of Teaching Indigenous and Local Games on The Development of Cultural and Sports Dimensions in Secondary Schools. *Research on Educational Sport, 13*(41), e4741. DOI: 10.22089/res.2025.17701.2612

Kieran, C., Forman, E., & Sfard, A. (2002). *Learning discourse: Discursive approaches to research in mathematics education.* DOI: 10.1007/0-306-48085-9

Kim, C., Na, H., Zhang, N., & Bai, C. (2024). Escape rooms for education: A meta-analysis. *International Journal of Instruction, 17*(4), 219–234. DOI: 10.29333/iji.2024.17413a

Kinuthia, W. (2012). Reported challenges for socio-cultural curricula in instructional design. *Multicultural Education & Technology Journal, 6*(2), 87–99. DOI: 10.1108/17504971211236281

Kırıkçılar, R. G. (2023). *Matematik merkezli etnosteam ders planlarinin geliştirme ve uygulama süreçlerinin incelenmesi [Examining the development and implementation processes of mathematics-centered ethnosteam lesson plans]* [Unpublished doctoral dissertation]. Marmara University.

Kisiel, J. (2013). Introducing future teachers to science beyond the classroom. *Journal of Science Teacher Education, 24*(1), 67–91. DOI: 10.1007/s10972-012-9288-x

Knijnik, G. (2012). Differentially positioned language games: Ethnomathematics from a philosophical perspective. *Educational Studies in Mathematics, 80*(1–2), 87–100. DOI: 10.1007/s10649-012-9396-8

Knuchel, C. (2004). Teaching symmetry in the elementary curriculum. *The Montana Math Enthusiast, 1*(1), 3–8. DOI: 10.54870/1551-3440.1001

Kong, J. E., Arizmendi, G. D., & Doabler, C. T. (2023). Implementing the science of math in a culturally sustainable framework for students with and at risk for math learning disabilities. *Teaching Exceptional Children, 56*(1), 44–51. DOI: 10.1177/00400599221127385

Krippendorff, K. (2018). *Content analysis: An introduction to its methodology* (4th ed.). SAGE Publications.

Küçük, A. (2014). Ethnomathematics in Anatolia-Turkey: Mathematical thoughts in multiculturalism. *Revista Latinoamericana de Etnomatemática Perspectivas Socioculturales de la Educación Matemática, 7*(1), 171-184. https://www.redalyc.org/pdf/2740/274030901008.pdf

Küçük, A. (2013). Ethnomathematics in Anatolia (in Turkey): Mathematical thoughts in multiculturalism. *Revista Latinoamericana de Etnomatemática, 7*(1), 171–184.

Küçük, A. (2013). Ethnomathematics in Anatolia-I (in Turkey): Geometry perception in multiculturalism. *Educational Research Review, 8*(16), 1360. DOI: 10.5897/ERR012.231

Küçük, A. (2014). Ethnomathematics in Anatolia-Turkey: Mathematical thoughts in multiculturalism. *Revista Latinoamericana de Etnomatemática, 7*(1), 171–184.

Küçükoğlu, A., & Kaya, H. İ. (2020). *Kuramdan uygulamaya okul dışı öğrenme ortamları* [Out-of-school learning environments: From theory to practice]. Pegem Akademi.

Kusi, P., & Bonyah, E. (2025). A systematic review and bibliometric report: Ethnomathematics and culturally responsive pedagogy in mathematics education. *European Journal of Education Studies, 12*(9), 113–140. DOI: 10.46827/ejes.v12i9.6187

Kuş, M. (2024). Conceptualization of out-of-school mathematics education. *Yaşadıkça Eğitim, 38*(1), 245–257. DOI: 10.33308/26674874.2024381644

Kyeremeh, P., Awuah, F. K., & Dorwu, E. (2023). Integration of ethnomathematics in teaching geometry: A systematic review and bibliometric report. *Journal of Urban Mathematics Education, 16*(2), 68–89. https://journals.tdl.org/jume/index.php/JUME. DOI: 10.21423/jume-v16i2a519

Ladson-Billings, G. (1995). Toward a theory of culturally relevant pedagogy. *American Educational Research Journal, 32*(3), 465–491. DOI: 10.3102/00028312032003465

Landis, J. R., & Koch, G. G. (1977). The measurement of observer agreement for categorical data. *Biometrics, 33*(1), 159–174. DOI: 10.2307/2529310 PMID: 843571

Larasati, I., Pramasdyahsari, A. S., & Harun, L. (2025). Enhancing elementary student ?(tm)s numeracy skills through ethnomathematics-based learning: An analysis of minimum competency assessment results. *Journal of Honai Math, 8*(1), 71 ?"88. DOI: 10.30862/jhm.v8i1.861

Lave, J., & Wenger, E. (1991). *Situated learning: Legitimate peripheral participation.* Cambridge University Press. DOI: 10.1017/CBO9780511815355 le Roux, K., & Swanson, D. (2021). Toward a reflexive mathematics education within local and global relations: Thinking from critical scholarship on mathematics education within the sociopolitical, global citizenship education and decoloniality. *Research in Mathematics Education, 23*(3), 323–337. DOI: 10.1080/14794802.2021.1993978

Lebaka, M. E. K. (2024). The Understanding of Indigenous Knowledge as Indigenous Practice, Skill and Know-How: The Case of Communal Music-Making in Bapedi Culture as a Social Practice. *European Journal of Social Science Education and Research, 11*(2), 1 ?"14. DOI: 10.26417/vtwts124

Lerman, S. (1998). Research on socio-cultural perspectives of mathematics teaching and learning. In Kilpatrick, J., & Sierpinska, A. (Eds.), *Mathematics education as a research domain: A search for identity* (pp. 333–350). Kluwer Academic Publisher., DOI: 10.1007/978-94-011-5196-2_6

Lerman, S. (2001). Cultural, discursive psychology: A sociocultural approach to studying the teaching and learning of mathematics. *Educational Studies in Mathematics, 46*(1), 87–113. DOI: 10.1023/A:1014031004832

Lerman, S. (2006). Socio-cultural research in PME. In *Handbook of research on the psychology of mathematics education* (pp. 347–366). Brill., DOI: 10.1163/9789087901127_013

Louie, N., & Zhan, W. (2022). A socio-ecological framework for research in mathematics education. *Journal for Research in Mathematics Education, 53*(5), 365–371. Advance online publication. DOI: 10.5951/jresematheduc-2022-0003

Machaba, F., & Dhlamini, J. (2021). Ethnomathematics as a fundamental teaching approach. In Luneta, K. (Ed.), *Mathematics teaching and professional learning in sub-Sahara Africa* (pp. 59–76). Springer International Publishing., https://link.springer.com/book/10.1007/978-3-030-82723-6 DOI: 10.1007/978-3-030-82723-6_5

Madonsela, B. S., & Machete, M. (2025). A systematic review of Indigenous knowledge in global contexts. *Diaspora, Indigenous, and Minority Education, 19*(2), 71 ?"77. DOI: 10.1080/15595692.2023.2249565

Maimela, G., & Nedombeloni, R. (2026). Decolonising the mainstream education and preserving Indigenous knowledge systems: learning through African Indigenous games. *AlterNative: An International Journal of Indigenous Peoples*, https://doi.org/ .DOI: 10.11771801251403201

Mainali, B. (2021). Representation in teaching and learning mathematics. *International Journal of Education in Mathematics. Science and Technology, 9*(1), 1–21. DOI: 10.46328/ijemst.1111

Makramalla, M. (2022). Commentary: Mathematics education in a globally precarious reality–A simulated dialogue between two key players. *Research in Mathematics Education, 24*(2), 249–261. DOI: 10.1080/14794802.2022.2090423

Mania, S., & Alam, S. (2021). Teachers' perception toward the use of ethnomathematics approach in teaching math. [IJEMST]. *International Journal of Education in Mathematics, Science, and Technology, 9*(2), 282–298. DOI: 10.46328/ijemst.1551

Marbun, L. R. P., Putra, Z. H., Noviana, E., Gunawan, Y., Afrillia, Y. M., Anriana, R., & Ahmad, S. (2026). Ethnomathematics exploration of tandok of the Batak Toba tribe as geometry learning resources for elementary school. *Ethnography and Education*, •••, 1–24. DOI: 10.1080/17457823.2026.2641176

Mariamah, M., Muslim, M., Amrullah, A., Prayitno, S., & Kusuma, A. B. (2021). Geometry from Lambitu tribe ethnomathematics. *Journal of Physics: Conference Series, 1778*(1), 012015. DOI: 10.1088/1742-6596/1778/1/012015

Marini, M., Tohir, M., Muhasshanah, M., & Anam, A. C. (2025). Mathematical Calculations in Determining the Amount of Zakat Mal and Zakat Fitrah as an Effort to Empower the Community Economy. In Alfarisi, R., Mason, J., Abdullah, N. B., Margunayasa, I. G., & Chen, H. H. (Eds.), *Proceedings of International Conference on Neuroscience and Learning Technology (ICONSATIN 2025)* (Vol. 980, pp. 355–375). Atlantis Press SARL. DOI: 10.2991/978-2-38476-525-6_37

Martin, D. (2019). Equity, inclusion, and antiblackness in mathematics education. *Race, Ethnicity and Education, 22*(4), 459–478. DOI: 10.1080/13613324.2019.1592833

Mason, M. (2002). The van Hiele levels of geometric understanding. In Jane, L. (Ed.), *Professional handbook for teachers, geometry: Explorations and applications* (pp. 4–8). McDougal Littell.

McKenney, S., & Reeves, T. C. (2012). *Conducting educational design research*. Routledge.

Medina, M. A., Gerofsky, S., & Nicol, C. (2024). Weaving indigenous mathematics: Bringing indigenous ways and stories into conversation with ethnomathematics. In C. Nicol, G. Knijnik, A, Peng, M. Cherinda, A. Bose (Eds.), *Ethnomathematics and mathematics education international perspectives in times of local and global change* (pp. 77-100). Springer.

Meeran, S., & Niranjan, C. (2026). The game of Kgati as an implementation tool in teaching intermediate phase mathematics: A case of two schools in rural Kwazulu-natal, South Africa. *Education Sciences, 16*(2), 294. DOI: 10.3390/educsci16020294

Meinck, F., Franchino-Olsen, H., Hafstad, G. S., Jud, A., Liel, C., Nikolaidis, G., & GrA sholt-Knudsen, T. (2025). Who gets to decide when children participate? Considering the complexities of consent and gatekeeping in violence research. *Child Abuse & Neglect, 169*, 107679. DOI: 10.1016/j.chiabu.2025.107679 PMID: 40976033

Melber, L. H., & Abraham, L. M. (1999). Beyond the classroom: Linking with informal education. *Science Activities, 36*(1), 3–4. DOI: 10.1080/00368129909601027

Mendoza, J. E. D., & Deri, R. A. (2025). Real-World Context in Mathematics Teaching: Teachers' Insights. *International Journal of Innovative Science and Research Technology*. https://api.semanticscholar.org/CorpusID:277686467

Ministry of Education. Arts and Culture. (2023). *Mathematics syllabus Grade 4 ?"7*. National Institute for Educational Development (NIED). Windhoek, Namibia

Ministry of National Education (MoNE). (2018). 2023 vizyon belgesi. [2023 vision document]. https://baklan.meb.gov.tr/www/meb-2023-vizyon-belgesi/icerik/330

Ministry of National Education (MoNE). (2024). *Ortaokul matematik dersi öğretim programı (5, 6, 7 ve 8. sınıflar)* [Middle school mathematics curriculum (Grades 5, 6, 7, and 8)] https://tymm.meb.gov.tr/upload/program/2024programmat5678Onayli.pdf

Ministry of National Education [MoNE]. (2024). *Ortaokul matematik dersi öğretim program (5, 6, 7 ve 8. sınıflar) [Middle school mathematics curriculum (grades 5, 6, 7, and 8)]*. https://tymm.meb.gov.tr/upload/program/2024programmat5678Onayli.pdf

Ministry of National Education [MoNE]. (2025). *Türkiye yüzyılı maarif modeli öğretim programları ortak metni [Joint text of the Turkish century education model curriculum]*. https://tymm.meb.gov.tr/upload/brosur/ortak_metin.pdf

Ministry of National Education. (2024a). *Türkiye Yüzyılı Maarif Modeli: Ortak metinler*. MEB.

Ministry of National Education. (2024b). *Matematik dersi öğretim programı (1–8. sınıflar)*. MEB.

Ministry of National Education. (2024c). *Matematik ders kitabı 5. sınıf, 1. kitap*. MEB Yayınları.

Ministry of National Education. (2024d). *Matematik ders kitabı 5. sınıf, 2. kitap*. MEB Yayınları.

Ministry of National Education. (2025a). *Matematik ders kitabı 6. sınıf, 1. kitap*. MEB Yayınları.

Ministry of National Education. (2025b). *Matematik ders kitabı 6. sınıf, 2. kitap*. MEB Yayınları.

Ministry of National Education. (2025c). *Ders kitabı dağıtım bilgileri*. MEB TYMM Portalı. Retrieved March 2025, from https://tymm.meb.gov.tr

Mishra, N. R. (2023). Constructivist Approach to Learning: An Analysis of Pedagogical Models of Social Constructivist Learning Theory. *Journal of Research and Development (Srinagar)*, *6*(01), 22–29. DOI: 10.3126/jrdn.v6i01.55227

Moalosi, R. (2007). *The impact of socio-cultural factors upon human-centred design in Botswana* (Doctoral dissertation, Queensland University of Technology).

Moalosi, R., Popovic, V., & Hudson, A. (2004) Socio-cultural factors that impact upon human-centred design in Botswana. In *Proceedings Design Research Society International Conference* 1–11. https://eprints.qut.edu.au/2740/

Moleong, L. C. (2018). Pengaruh Real Interest Rate dan Leverage Terhadap Financial Distress. *Modus*, *30*(1), 71 ?"86.

Morgan, C. (2014). Social theory in mathematics education: Guest editorial. *Educational Studies in Mathematics*, *87*(2), 123–128. DOI: 10.1007/s10649-014-9572-0

Moschkovich, J. N. (2010). *Language and mathematics education: Multiple perspectives and directions for research*. Information Age Publishing. DOI: 10.1108/978-1-61735-161-7

Motseki, P. D., Jojo, Z. M. M., & Gumbo, M. T. (2026). Transforming mathematics teaching through Ndebele art: An ethnomathematics perspective. *Diaspora, Indigenous, and Minority Education*, *20*(1), 1–7. DOI: 10.1080/15595692.2025.2451239

Muhtadi, D., Sukirwan, S., Warsito, W., & Prahmana, R. C. I. (2017). Sundanese ethnomathematics: Mathematical activities in estimating, measuring, and making patterns. *Journal on Mathematics Education*, *8*(2), 185–198. DOI: 10.22342/jme.8.2.4055.185-198

Müller, D. (2025). Towards a critical pragmatic philosophy of sustainable mathematics education. arXiv preprint arXiv:2504.17149.

Munjiatun, K., O., Guslinda, Zufriady, Santika, Y., & Rahayu, S. K. (2022). Typology of Rumah Lontiok architecture in Kuok Kampar Riau. In *Proceedings of the 5th International Conference on Language, Literature, and Education (ICLLE-5 2022)* (pp. 420–425). Atlantis Press. DOI: 10.2991/978-2-494069-85-5_45

Murat, M., & Demirci, A-. (2026). Improving Primary School Learners' Number Sense Levels Through Scratch-Supported and Game-Based Activities. *Erzincan Aoniversitesi E Yitim FakA1/4ltesi Dergisi*, (2026), 1-16. https://doi.org/DOI: 10.17556/erziefd.1801045

Naresh, N., & Kasmer, L. (2018). Using ethnomathematics perspective to widen the vision of mathematics teacher education curriculum. In Bartell, T. G. (Ed.), *Toward equity and social justice in mathematics education* (pp. 309–328). Springer. DOI: 10.1007/978-3-319-92907-1_18

Nasir, N. I. S. (2002). Identity, goals, and learning: Mathematics in cultural practice. *Mathematical Thinking and Learning*, *4*(2-3), 213–247. DOI: 10.1207/S15327833MTL04023_6

Nasir, N. I. S., Hand, V., & Taylor, E. V. (2008). Culture and mathematics in school: Boundaries between "cultural" and "domain" knowledge in the mathematics classroom and beyond. *Review of Research in Education*, *32*(1), 187–240. DOI: 10.3102/0091732X07308962

Nasrum, A., Salido, A., & Chairuddin, C. (2025). Unveiling emerging trends and potential research themes in future ethnomathematics studies: A global bibliometric analysis (from inception to 2024). *International Journal of Learning, Teaching and Educational Research*, *24*(2), 206–226. DOI: 10.26803/ijlter.24.2.11

National Council of Teachers of Mathematics. (2000). *Principles and standards for school mathematics*. NCTM.

National Council of Teachers of Mathematics. (2014). *Principles to actions: Ensuring mathematical success for all*. Author.

National Research Council. (1996). *National science education standards*. National Academic Press.

National Research Council. (2001). *Adding it up: Helping children learn mathematics*. National Academies Press.

Nemirovsky, R., Kelton, M. L., & Civil, M. (2017). Toward a vibrant and socially significant informal mathematics education. In Cai, J. (Ed.), *Compendium for research in mathematics education* (pp. 968–979). National Council of Teachers of Mathematics.

Nicol, C. (2002). Where's the math? Prospective teachers visit the workplace. *Educational Studies in Mathematics*, *50*(3), 289–309. DOI: 10.1023/A:1021211207232

Nicol, C., Knijnik, G., Peng, A., Cherinda, M., & Bose, A. (Eds.). (2024). *Ethnomathematics and mathematics education: International perspectives in times of local and global change*. Springer. DOI: 10.1007/978-3-031-60680-9

Nolan, K. T., & Xenofontos, C. (2023). The role of insights in becoming a culturally responsive mathematics teacher. *Education Sciences, 13*(10), 1028. DOI: 10.3390/educsci13101028

Novak, J. D., & Cañas, A. J. (2008). *The theory underlying concept maps and how to construct and use them.* Technical Report IHMC CmapTools 2006-01 Rev 01-2008, Florida Institute for Human and Machine Cognition, Pensacola. https://cmap.ihmc.us/Publications/ResearchPapers/TheoryUnderlyingConceptMaps.pdf

Nxumalo, S. A., & Mncube, D. W. (2019). Using Indigenous games and knowledge to decolonise the school curriculum: Ubuntu perspectives. https://core.ac.uk/download/480900151.pdf

Olson, J. K., Cox-Petersen, A. M., & McComas, W. F. (2001). The inclusion of informal environments in science teacher preparation. *Journal of Science Teacher Education, 12*(3), 155–173. DOI: 10.1023/A:1016715127697

Orey, D. C., & Rosa, M. (2010). Ethnomodeling: A pedagogical action for uncovering ethnomathematical practices. *Journal of Mathematical Modelling and Application, 1*(3), 58–67.

Orey, D. C., & Rosa, M. (2024). Ethnomodelling research: glocalizing educational systems from exclusion to inclusion at local and global levels. In C. Nicol, G. Knijnik, A, Peng, M. Cherinda, A. Bose (Eds.), *Ethnomathematics and mathematics education international perspectives in times of local and global change* (pp. 213-234). Springer. DOI: 10.1007/978-3-031-60680-9_10

Orey, D. C., & Rosa, M. (2010). Ethnomodeling: A pedagogical action for uncovering ethnomathematical practices. *Journal of Mathematical Modelling and Application, 1*(3), 58–67.

Otte, C. R., Bølling, M., Elsborg, P., Nielsen, G., & Bentsen, P. (2019). Teaching maths outside the classroom: Does it make a difference? *Educational Research, 61*(1), 38–52. DOI: 10.1080/00131881.2019.1567270

Padilla, M. O., Velasco, N. Y. G., & Ortiz, S. A. (2025). Analysis of pedagogical content knowledge for teaching mathematics in early childhood education. *Journal of Infrastructure Policy and Development, 9*(1), 10135. DOI: 10.24294/jipd10135

Pais, A., & Mesquita, M. (2013). Ethnomathematics in non-formal educational settings: The urban boundaries project. *Revista Latinoamericana de Etnomatemática*, *6*(3), 134–144.

Pais, S., Sousa, A., & Pires, A. (2023-October). Using an escape room activity to enhance the motivation of undergraduate life science students in mathematics classes: A case study. *Proceedings of the 17th European Conference on Games Based Learning, 17* (1), 490-496. DOI: 10.34190/ecgbl.17.1.1431

Pansell, A. (2023). Mathematical knowledge for teaching as a didactic praxeology. *Frontiers in Education, Volume 8-2023*. DOI: 10.3389/feduc.2023.1165977

Paris, D. (2012). Culturally sustaining pedagogy: A needed change in stance, terminology, and practice. *Educational Researcher, 41*(3), 93–97. DOI: 10.3102/0013189X12441244

Paris, D., & Alim, H. S. (2017). *Culturally sustaining pedagogies: Teaching and learning for justice in a changing world*. Teachers College Press.

Pattison, S., Rubin, A., & Wright, T. (2017). *Mathematics in informal learning environments: A summary of the literature* [Literature review report]. Center for Advancement of Informal Science Education (CAISE). https://www.informalscience.org/mathematics-informal-learning-environments-summary-literature

Payadnya, I. P. A. A., Wulandari, I. G. A. P. A., Puspadewi, K. R., & Saelee, S. (2024). Student responses to culture-based mathematics learning in the Indonesian and Thailand education curricula. *Jurnal Pendidikan Matematika, 18*(1), 113–128. DOI: 10.22342/jpm.v18i1.pp113-128

Pepin, B., & Haggarty, L. (2001). Mathematics textbooks and their use in English, French and German classrooms: A way to understand teaching and learning cultures. *ZDM Mathematics Education*, *33*(5), 158–175. DOI: 10.1007/BF02656616

Philip, K. S. (2015). Indigenous knowledge: Science and technology studies. In Wright, J. D. (Ed.), *International encyclopedia of the social & behavioral sciences* (2nd ed., pp. 113 ?"114). Elsevier., DOI: 10.1016/B978-0-08-097086-8.85012-6

Philip, T. (1979). Mankala in Eastern and Southern Africa: A Distributional Analysis. *Azania, 14*(1), 109 ?"138. DOI: 10.1080/00672707909511266

Piaget, J. (2003). Part I: Cognitive Development in Children ?"Piaget Development and Learning. *Journal of Research in Science Teaching, ? ? ?*, 40.

Pinero Charlo, J. C. (2020). Educational escape rooms as a tool for horizontal mathematization: Learning process evidence. *Education Sciences*, *10*(9), 213. DOI: 10.3390/educsci10090213

Pischetola, M., Møller, J. K., & Malmborg, L. (2023). Enhancing teacher collaboration in higher education: The potential of activity-oriented design for professional development. *Education and Information Technologies*, *28*(6), 7571–7600. DOI: 10.1007/s10639-022-11490-x PMID: 36532797

Pittalis, M., Mousoulides, N., & Christou, C. (2010). Students' 3D geometry thinking profiles. In V. Durand-Guerrier, S. Soury-Lavergne, & F. Arzarello (Eds.), *Proceedings of the Sixth Congress of the European Society for Research in Mathematics Education* (pp. 816–825). Institut National de Recherche Pédagogique.

Pittalis, M., & Christou, C. (2010). Types of reasoning in 3D geometry thinking and their relation with spatial ability. *Educational Studies in Mathematics*, *75*(2), 191–212. DOI: 10.1007/s10649-010-9251-8

Plomp, T., & Nieveen, N. (2013). *Educational design research: An introduction*. Netherlands Institute for Curriculum Development (SLO).

Pöhler, B., & Prediger, S. (2015). Intertwining lexical and conceptual learning trajectories. *Eurasia Journal of Mathematics, Science and Technology Education*, *11*(6), 1697–1716. DOI: 10.12973/eurasia.2015.1497a

Pradana, K. C., Putra, R. A., & Rahmawati, Y. (2022). Ethnomathematics on traditional culture: A bibliometric mapping analysis and systematic review on database scopus. *International Journal Corner of Educational Research*, *1*(1), 1–8. DOI: 10.54012/ijcer.v1i1.61

Prahmana, R. C. I., & D'Ambrosio, U. (2020). Learning geometry and values from patterns: Ethnomathematics on the batik patterns of Yogyakarta, Indonesia. *Journal on Mathematics Education*, *11*(3), 439–456. DOI: 10.22342/jme.11.3.12949.439-456

Prasad, R. (2017). Trainee Teachers' Feedback and Reflection on Math Lesson Study. *International Journal of Advanced Research*, *5*(12), 237–242. 10.21474/IJAR01/5959

Prasad, D. B. (2008). Content analysis: A method of social science research. In Das, D. K. L., & Bhaskaran, V. (Eds.), *Research methods for social work* (pp. 174–193). Rawat.

Pumpian, I., Fisher, D., & Wachowiak, S. (Eds.). (2006). *Challenging the classroom standard through museum-based education: School in the park*. Lawrence Erlbaum Associates. DOI: 10.4324/9781410617279

Purniati, T., Turmudi, T., Juandi, D., & Suhaedi, D. (2021). Ethnomathematics exploration of the masjid raya bandung ornaments in transformation geometry materials. *Journal of Medives: Journal of Mathematics Education IKIP Veteran Semarang*, *5*(2), 235–243. DOI: 10.31331/medivesveteran.v5i2.1639

Pusat Asesmen Pendidikan. (2023). *Laporan hasil Asesmen Nasional 2022: Rapor Pendidikan Indonesia*. Kementerian Pendidikan, Kebudayaan, Riset, dan Teknologi Republik Indonesia.

Putra, E. C. S., & Mahmudah, F. N. (2021). The implementation of ethnomathematics based-learning for students. *SJME (Supremum Journal of Mathematics Education,* *5*(2), 162–169.

Putra, Z. H. (2017). Pre-service elementary teachers' knowledge of comparing decimals based on the anthropological theory of the didactic. *HAL Open Science*, 01948872.

Putra, Z. H., Ramiati, Zufriady, Hidayat, R., Jismulatif, Hermita, N., & Sulistiyo, U. (2024). Development of computational thinking tasks based on Riau Malay culture: A study of fifth-grade public school students in Pekanbaru, Indonesia. *Education 3-13, 52*(8), 1387–1397. DOI: 10.1080/03004279.2022.2150063

Putra, Z. H., Alim, J. A., Melihayatri, N., Dahnilsyah, , Afrillia, Y. M., & Aljarrah, A. (2026). Project-based learning in teacher education: A reflection on analysing prospective elementary teachers' knowledge of math, didactics, and technology. *Journal on Mathematics Education*, *17*(1), 179–198. DOI: 10.22342/jme.v17i1.pp179-198

Putra, Z. H., Hermita, N., Alim, J. A., Dahnilsyah, D., & Gunawan, Y. (2025). The role of project-based ethno-didactics of mathematics learning toward pre-service elementary teachers' perspectives, mathematical, and didactic knowledge. *Frontiers in Education*, *10*, 1568366. DOI: 10.3389/feduc.2025.1568366

Putri, R. A., & Faradiba, S. S. (2025). *Kajian Literatur Integrasi Aritmetika Sosial dengan Nilai-Nilai Fiqih dalam Zakat pada Pembelajaran Matematika*. Mandalika Mathematics and Educations Journal., DOI: 10.29303/jm.v7i3.9900

Rachman, K., Mas'ud, A. I., Vesya, S. R., Arbiyanti, A. P., & Rasyid, I. (2025). Development of digital puzzle based on Indonesian traditional houses as virtual manipulative learning media for plane geometry. *Unnes Journal of Mathematics Education*, *14*(2), 167–178. DOI: 10.15294/ujme.v14i2.29951

Radford, L. (1997). On psychology, historical epistemology, and the teaching of mathematics: Towards a socio-cultural history of mathematics. *For the Learning of Mathematics*, *17*(1), 26–33.

Ratinho, E., Martins, C., & Machado, J. (2023). The role of gamified learning strategies in student's motivation in high school and higher education: A systematic review. *Heliyon*, *9*(8), e19033. DOI: 10.1016/j.heliyon.2023.e19033

Remillard, J. T. (2005). Examining key concepts in research on teachers' use of mathematics curricula. *Review of Educational Research*, *75*(2), 211–246. DOI: 10.3102/00346543075002211

Remillard, J. T., & Heck, D. J. (2014). Conceptualizing the curriculum enactment process in mathematics education. *ZDM Mathematics Education*, *46*(5), 705–718. DOI: 10.1007/s11858-014-0600-4

Rezat, S. (2024). Research on curriculum resources in mathematics education: A survey of the field. *ZDM Mathematics Education*, *56*(2), 223–237. DOI: 10.1007/s11858-024-01559-x

Rizki, L. M., & Priatna, N. (2019). Mathematical literacy as the 21st century skill. *Journal of Physics: Conference Series*, *1157*, 042088. DOI: 10.1088/1742-6596/1157/4/042088

Roche, A., Gervasoni, A., & Kalogeropoulos, P. (2023). Factors that promote interest and engagement in learning mathematics for low-achieving primary learners across three learning settings. *Mathematics Education Research Journal*, *35*(3), 525 ?"556. DOI: 10.1007/s13394-021-00402-w

Rogoff, B. (2025). Transforming by participating. *Pillars of developmental psychology: Recollections and reflections*, 278-291.

Rosa, M. (2000). From reality to mathematical modelling: A proposal for using ethnomathematical knowledge (Unpublished master's thesis). California State University, Sacramento.

Rosa, M., & Orey, D. C. (2013). Ethnomodelling as a research lens on ethnomathematics and modelling. In G. A. Stillman, G. Kaiser, W. Blum, & J. P. Brown (Eds.), *Teaching mathematical modelling: Connecting to research and practice* (pp. 117–127). Springer. DOI: 10.1007/978-94-007-6540-5_10

Rosa, M., & Orey, D. C. (2015). Social-critical dimension of mathematical modelling. In G. A. Stillman, W. Blum, & M. S. Biembengut (Eds.), *Mathematical modelling in education research and practice* (pp. 385–395). Springer. https://doi.org/DOI: 10.1007/978-3-319-18272-8_32

Rosa, M., & Orey, D. (2025). *A didactic approach of sociocultural perspectives of mathematical modelling for classrooms: An ethnomodelling perspective*. REMATEC., DOI: 10.37084/REMATEC.1980-3141.2025.n54.e2025005.id772

Rosa, M., & Orey, D. C. (2006). Abordagens atuais do programa etnomatemática: Delineando-se um caminho para a ação pedagógica. *Bolema, 19*(26), 19–48.

Rosa, M., & Orey, D. C. (2010b). Ethnomodeling as a pedagogical tool for the ethnomathematics program. *Revista Latinoamericana de Etnomatemática, 3*(2), 14–23.

Rosa, M., & Orey, D. C. (2011). Ethnomathematics: The cultural aspects of mathematics. *Revista Latinoamericana de Etnomatemática, 4*(2), 32–54.

Rosa, M., & Orey, D. C. (2013). Ethnomodeling as a research theoretical framework on ethnomathematics and mathematical modeling. *Journal of Urban Mathematics Education, 6*(2), 62–80. DOI: 10.21423/jume-v6i2a195

Rosa, M., & Orey, D. C. (2015). A trivium curriculum for mathematics based on literacy, matheracy, and technoracy: An ethnomathematics perspective. *ZDM Mathematics Education, 47*(4), 587–598. DOI: 10.1007/s11858-015-0688-1

Rosa, M., & Orey, D. C. (2016). Humanizing mathematics through ethnomodelling. *Journal of Humanistic Mathematics, 6*(2), 3–22. DOI: 10.5642/jhummath.201602.03

Rosa, M., & Orey, D. C. (2016). State of the Art in Ethnomathematics. In Rosa, M., D'Ambrosio, U., Orey, D. C., Shirley, L., Alangui, W. V., Palhares, P., & Gavarrete, M. E. (Eds.), *Current and Future Perspectives of Ethnomathematics as a Program* (pp. 11–37). Springer International Publishing., DOI: 10.1007/978-3-319-30120-4_3

Rosa, M., & Orey, D. C. (2016). State of the art in ethnomodelling: Research on local mathematical knowledge and perspectives. In Rosa, M., D'Ambrosio, U., Orey, D. C., Shirley, L., Alangui, W. V., Palhares, P., & Gavarrete, M. E. (Eds.), *Current and future perspectives of ethnomathematics as a program* (pp. 71–80). Springer. DOI: 10.1007/978-3-319-30120-4

Rosa, M., & Orey, D. C. (2018). The anthropological dimension on ethnomodelling research based on ethnomathematics and modelling. *Open Access Journal of Archaeology & Anthropology, 1*(1), 1–8. DOI: 10.33552/OAJAA.2018.01.000503

Rosa, M., & Orey, D. C. (2019). Ethnomodelling as the translation of diverse cultural mathematical practices. In *Handbook of the Mathematics of the Arts and Sciences*. Springer., DOI: 10.1007/978-3-319-70658-0_70-1

Rosa, M., & Orey, D. C. (2020). Principles of culturally relevant education in an ethnomathematical perspective. *Revista de Educação Matemática*, (17), 1–24. DOI: 10.37001/remat25269062v17id306

Rosa, M., & Orey, D. C. (2021). An ethnomathematical perspective of STEM education in a glocalized world. *Bolema: Boletim de Educação Matemática, 35*(70), 840–876. https://doi.org/DOI: 10.1590/1980-4415v35n70a14

Rosa, M., & Orey, D. C. (2021). Ethnomodelling as the translation of diverse cultural mathematical practices. In Sriraman, B. (Ed.), *Handbook of the mathematics of the arts and sciences* (pp. 2607–2635). Springer. DOI: 10.1007/978-3-319-57072-3_70

Rosa, M., & Orey, D. C. (2021a). Applying ethnomodelling to explore glocal mathematical knowledge systems. *Acta Scientiae*, *23*(1), 199–232. DOI: 10.17648/acta.scientiae.6241

Rosa, M., & Orey, D. C. (2021b). Ethnomodelling as a globalization process of mathematical practices through cultural dynamism. *The Montana Math Enthusiast*, *18*(3), 439–468. DOI: 10.54870/1551-3440.1533

Rosa, M., Orey, D. C., & Cortes, D. P. D. O. (2020). Resignifying function concept: A mixed methods study to understand the contributions of the dialogic approach of ethnomodelling. In *Mathematical modelling education and sense-making* (pp. 107–116). Springer., DOI: 10.1007/978-3-030-37673-4_10

Rosenman, M. A., & Gero, J. S. (1998). Purpose and function in design: From the socio-cultural to the techno-physical. *Design Studies*, *19*(2), 161–186. DOI: 10.1016/S0142-694X(97)00033-1

Ruiz, J. J. R., Sanchez, A. D. V., & Figueredo, O. R. B. (2024). Impact of gamification on school engagement: A systematic review. *Frontiers in Education*, *9*, 1466926. DOI: 10.3389/feduc.2024.1466926

Sadler, T. D. (2004). Informal reasoning regarding socioscientific issues: A critical review of research. *Journal of Research in Science Teaching*, *41*(5), 513–536. DOI: 10.1002/tea.20009

Sadler, T. D. (2009). Situated learning in science education: Socio-scientific issues as contexts for practice. *Studies in Science Education*, *45*(1), 1–42. DOI: 10.1080/03057260802681839

Sadler, T. D., & Zeidler, D. L. (2004). The morality of socioscientific issues: Construal and resolution of genetic engineering dilemmas. *Science Education*, *88*(1), 4–27. DOI: 10.1002/sce.10101

Sadler, T. D., & Zeidler, D. L. (2005). Patterns of informal reasoning in the context of socioscientific decision making. *Journal of Research in Science Teaching, 42*(1), 112–138. DOI: 10.1002/tea.20042

Sagala, S. A., & Hasanah, R. U. (2023). Ethnomathematics exploration at the state museum of North Sumatra. *Mathline: Jurnal Matematika dan Pendidikan Matematika, 8*(1), 123-136. DOI: 10.31943/mathline.v8i1.364

Şahin, H. B. (2025). Türk-İslam kültür ve mimarisi bağlamında geometri öğretimi: Mimar Sinan'ın eserlerindeki geometrik desenler [Teaching geometry within the context of Turkish-Islamic culture and architecture: Geometric patterns in the works of Architect Sinan]. (Publication No. 933490). [Doctoral dissertation, Bursa Uludağ University]. YÖK Tez Merkezi.

Sakti, S. A., Endraswara, S., & Rohman, A. (2024). Revitalizing local wisdom within character education through ethnopedagogy apporach: A case study on a preschool in Yogyakarta. *Heliyon, 10*(10), e31370. DOI: 10.1016/j.heliyon.2024.e31370 PMID: 38803959

Sarmiento-Campos, N. V., LA zaro-Guillermo, J. C., Silvera-AlarcA3n, E. N., Cuellar-Quispe, S., HuamA n-RomanA-, Y. L., Apaza, O. A., & Sorkheh, A. (2022). A look at Vygotsky ?(tm)s sociocultural theory (SCT): The effectiveness of scaffolding method on EFL learners ?(tm) speaking achievement. *Education Research International, 2022*(1), 3514892. DOI: 10.1155/2022/3514892

Schattschneider, D. (2004). *Escher: Visions of symmetry* (2nd ed.). Harry N. Abrams.

Schmidt, W. H., McKnight, C. C., & Raizen, S. A. (1997). *A splintered vision: An investigation of U.S. science and mathematics education*. Kluwer Academic Publishers.

Schoenfeld, A. H. (2016). Learning to think mathematically: Problem solving, metacognition, and sense making in mathematics. *Journal of Education, 196*(2), 1–38. DOI: 10.1177/002205741619600202

Schoevers, E. M., Leseman, P. P. M., & Kroesbergen, E. H. (2020). Enriching mathematics education with visual arts: Effects on elementary school students' ability in geometry and visual arts. *International Journal of Science and Mathematics Education, 18*(8), 1613–1634. DOI: 10.1007/s10763-019-10018-z

Seligmann, T. (2014). Learning museum: A meeting place for pre-service teachers and museums. *Journal of Museum Education, 39*(1), 42–53. DOI: 10.1080/10598650.2014.11510794

Şen, A. İ. (2019). *Okul dışı öğrenme ortamları* [Out-of-school learning environments]. Pegem Akademi. DOI: 10.14527/9786052419700

Setiaputra, F. I., Subanti, S., Usodo, B., Triyanto, T., Fitriana, L., & Hendriyanto, A. (2025). A systematic review of ethnomathematics research (2019–2023): Cultural integration in mathematics teaching and learning. *Jurnal Pendidikan MIPA*, *26*(1), 195–211. DOI: 10.23960/jpmipa.v26i1.pp195-211

Sevgi, S. (2019). Dönüşüm geometrisi etkinliklerine kültürel bağlamın dahil edilmesi sonucu oluşan öğrenci yaklaşımları (Master's thesis, Dokuz Eylul Universitesi (Turkey)).

Sevgi, S. (2019). Dönüşüm geometrisi etkinliklerine kültürel bağlamın dahil edilmesi sonucu oluşan öğrenci yaklaşımları [Student approaches resulting from the inclusion of cultural context in transformation geometry activities]. (Publication No. 593584) [Master's thesis, Dokuz Eylül University]. YÖK Tez Merkezi.

Sfard, A. (2008). *Thinking as communicating: Human development, the growth of discourses, and mathematizing*. Cambridge university press., DOI: 10.1017/CBO9780511499944

Sharma, T., & Orey, D. C. (2017). Meaningful Mathematics Through the Use of Cultural Artifacts. In Rosa, M., Shirley, L., Gavarrete, M. E., & Alangui, W. V. (Eds.), *Ethnomathematics and its diverse approaches for mathematics education* (pp. 153–182). Springer. DOI: 10.1007/978-3-319-59220-6_7

Shield, M., & Dole, S. (2013). Assessing the potential of mathematics textbooks to promote deep learning. *Educational Studies in Mathematics*, *82*(2), 183–199. DOI: 10.1007/s10649-012-9415-9

Siliunas, V., Wan, G., & Edejer, E. (2024). Teacher preparation for culturally responsive teaching: Implications for teacher education. *Action in Teacher Education*, *46*(4), 313–331. DOI: 10.1080/01626620.2024.2358912

Simamora, R. E., Saragih, S., & Hasratuddin, H. (2019). Improving students' mathematical problem solving ability and self-efficacy through guided discovery learning in local culture context. *International Electronic Journal of Mathematics Education*, *14*(1), 61–72. DOI: 10.12973/iejme/3966

Simonneaux, L. (2007). Argumentation in socio-scientific contexts. In S. Erduran & M. P. JiménezAleixandre (Eds.), *Argumentation in science education: Perspectives from classroom-based research* (pp. 179–199). Springer. https://doi.org/10.1007/978-1-4020-6670-2_9.

Sinclair, N., & Bruce, C. D. (2015). New opportunities in geometry education at the primary school. *ZDM Mathematics Education, 47*(3), 319–329. DOI: 10.1007/s11858-015-0693-4

Skovsmose, O. (2005). *Travelling through education: Uncertainty, mathematics, responsibility*. Brill., DOI: 10.1163/9789087903626

Skovsmose, O. (2013). *Towards a philosophy of critical mathematics education* (Vol. 15). Springer Science & Business Media., DOI: 10.1007/978-94-017-3556-8

Skovsmose, O. (2014). *Critique as uncertainty*. IAP., DOI: 10.1108/978-1-62396-755-0

Skovsmose, O. (2022). Concerns of critical mathematics education and of ethnomathematics. *Revista Colombiana de Educación, 86*(86), 365–382. DOI: 10.17227/rce.num86-13713

Sleeter, C. E., & Delgado Bernal, D. (2004). Critical pedagogy, critical race theory, and antiracist education: Implications for multicultural education. In Banks, J. A., & Banks, C. A. M. (Eds.), *Handbook of research on multicultural education* (2nd ed., pp. 240–258). Jossey-Bass.

Solares-Rojas, A., Arellano-Aguilar, O., García González, M. M., López-Vargas, M. D. R., Coles, A., & Méndez Serrano, A. (2022). Mathematics education and social-environmental crises: An interdisciplinary proposal for didactic innovation with rural communities in Mexico. *Research in Mathematics Education, 24*(2), 202–223. DOI: 10.1080/14794802.2022.2062781

Song, K. H., & Coppersmith, S. A. (2020). Working toward linguistically and culturally responsive math teaching through a year-long urban teacher training program for English learners. *Journal of Urban Mathematics Education, 13*(2), 60–86. DOI: 10.21423/jume-v13i2a409

Steen, L. A., Turner, R., & Burkhardt, H. (2007). Developing Mathematical Literacy. In Blum, W., Galbraith, P. L., Henn, H.-W., & Niss, M. (Eds.), *Modelling and Applications in Mathematics Education: The 14th ICMI Study* (pp. 285–294). Springer US., DOI: 10.1007/978-0-387-29822-1_30

Steffensen, L., & Kasari, G. (2023). Integrating societal issues with mathematical modelling in pre-service teacher education. *Education Sciences, 13*(7), 721. DOI: 10.3390/educsci13070721

Stein, M. K., Remillard, J. T., & Smith, M. S. (2007). How curriculum influences student learning. In Lester, F. K. (Ed.), *Second handbook of research on mathematics teaching and learning* (pp. 319–370). Information Age Publishing. DOI: 10.1108/978-1-28141-200-320251012

Stein, M. K., Smith, M. S., Henningsen, M. A., & Silver, E. A. (2000). *Implementing standards-based mathematics instruction: A casebook for professional development*. Teachers College Press.

Štemberger, T., & Cencič, M. (2016). Design based research: The way of developing and ımplementing educational ınnovation. *World Journal on Educational Technology: Current Issues*, *8*(3), 180–189. DOI: 10.18844/wjet.v8i3.621

Stinson, D. W., & Bullock, E. C. (2012). Critical postmodern theory in mathematics education research: A praxis of uncertainty. *Educational Studies in Mathematics*, *80*(1), 41–55. DOI: 10.1007/s10649-012-9386-x

Sunzuma, G., & Umbara, U. (2025). Ethnomathematics-based technology in Indonesia: A systematic review. *Asian Journal for Mathematics Education*, *4*(1), 129–153. DOI: 10.1177/27527263241305812

Supriyadi, E., Turmudi, T., Dahlan, J. A., & Juandi, D. (2024). Development of Sundanese gamelan ethnomathematics e-module for junior high school mathematics learning. *Malaysian Journal of Learning and Instruction*, *21*(2), 139–178. DOI: 10.32890/mjli2024.21.2.6

Suryaningsih, Y., Fajriah, N., & Kamid, K. (2023). Geometry exploration for the development of ethnomathematics worksheet based on the ornament of Jingah River Jami Mosque. *Jurnal Teori dan. Aplikace Matematiky*, *7*(2), 298–309. DOI: 10.31764/jtam.v7i2.12366

Sweller, J., Ayres, P., & Kalyuga, S. (2011). *Cognitive load theory*. Springer. DOI: 10.1007/978-1-4419-8126-4

Taraldsen, L. H., Haara, F. O., Lysne, M. S., Jensen, P. R., & Jenssen, E. S. (2022). A review on use of escape rooms in education – touching the void. *Education Inquiry*, *13*(2), 169–184. DOI: 10.1080/20004508.2020.1860284

Tarr, J. E., Reys, R. E., Reys, B. J., Chávez, Ó., Shih, J., & Osterlind, S. J. (2008). The impact of middle-grades mathematics curricula and the classroom learning environment on student achievement. *Journal for Research in Mathematics Education*, *39*(3), 247–280. https://www.jstor.org/stable/30034970

Tassell-Matamua, N. (2025). Indigenous knowledges. What they are and why they matter. *explore*, *21*(3), 103144.https://doi.org/DOI: 10.1016/j.explore.2025.103144

Topan, B., & Bozkus, F. (2021). Türk mimari eserlerindeki geometrik desenlerin geometri öğretiminde kullanılabilirliğine ilişkin öğretmen görüşleri. *Mehmet Akif Ersoy Üniversitesi Eğitim Fakültesi Dergisi, 58*, 362–389. https://izlik.org/JA38WP94TX

Ugo, C. C., López-Pernas, S., Gordillo, A., & Tukiainen, M. (2025). Escape rooms for learning programming: A systematic literature review. *Review of Education, 13*(3), 1-26, e70123. DOI: 10.1002/rev3.70123

Umar, I. O., Tudunkaya, M. S., & Muawiya, H. U. (2019). Effect of ethnomathematics teaching approach on performance and retention in trigonometry among secondary school students in Zaria Local Government Area Kaduna State, Nigeria. *Abacus (Mathematics Education Series), 44*(1), 104–112.

Umbara, U., Wahyudin, W., & Prabawanto, S. (2021). Exploring ethnomathematics with ethnomodelling methodological approach: How does Cigugur indigenous people use calculations to determine good days to build houses? *Eurasia Journal of Mathematics, Science and Technology Education, 17*(2), 1–19. DOI: 10.29333/ejmste/9673

Utari, R. S., Putri, R. I. I., Zulkardi, , & Hapizah, . (2025). Seeing data differently: Developing a local instructional theory for culturally relevant statistical literacy in Indonesian teacher education. *Journal on Mathematics Education, 16*(3), 915–936. DOI: 10.22342/jme.v16i3.pp915-936

Valero, P. (2004). Socio-political perspectives on mathematics education. In Valero, P., & Zevenbergen, R. (Eds.), *Researching the socio-political dimensions of mathematics education: Issues of power in theory and methodology* (pp. 5–23). Springer., DOI: 10.1007/1-4020-7914-1_2

Valverde, G. A., Bianchi, L. J., Wolfe, R. G., Schmidt, W. H., & Houang, R. T. (2002). *According to the book: Using TIMSS to investigate the translation of policy into practice through the world of textbooks*. Springer. DOI: 10.1007/978-94-007-0844-0

Van Hiele, P. M. (1986). *Structure and insight: A theory of mathematics education*. Academic Press.

Verner, I., Massarwe, K., & Bshouty, D. (2019). Development of competencies for teaching geometry through an ethnomathematical approach. *The Journal of Mathematical Behavior, 56*, 1–14. DOI: 10.1016/j.jmathb.2019.05.002

Vidergor, H. E. (2021). Effects of digital escape room on gameful experience, collaboration, and motivation of elementary school students. *Computers & Education, 166*, 104156. DOI: 10.1016/j.compedu.2021.104156

Vygotsky, L. S. (1978). *Mind in society: The development of higher psychological processes*. Harvard University Press.

Wahyudi, H., Widodo, S. A., Setiana, D. S., & Irfan, M. (2021). Etnomathematics: Batik activities in tancep batik. *Journal of Medives: Journal of Mathematics Education IKIP Veteran Semarang*, *5*(2), 305–315. DOI: 10.31331/medivesveteran.v5i2.1699

Wenger, E. (1998). *Communities of practice: Learning, meaning, and identity*. Cambridge University Press. DOI: 10.1017/CBO9780511803932

Wertsch, J. V. (1998). *Mind as action*. Oxford University Press. DOI: 10.1093/acprof:oso/9780195117530.001.0001

Whitcombe, A. (1988). Mathematics creativity, imagination, beauty. *Mathematics in School*, *17*(2), 13–15.

Widada, W., Herawaty, D., & Lubis, A. N. M. T. (2018). Realistic mathematics learning based on ethnomathematics in Bengkulu to improve students' cognitive level. *Journal of Physics: Conference Series*, *1088*(1), 012028. DOI: 10.1088/1742-6596/1088/1/012028

Wildfeuer, S. (2022). Ethnomathematics and cultural identity to promote culturally responsive pedagogical choices for teachers in early childhood and elementary education. In Danesi, M. (Ed.), *Handbook of cognitive mathematics* (pp. 1–15). Springer. DOI: 10.1007/978-3-030-44982-7_1-1

Williams-Pierce, C., Simpson, A., & Eronen, L. (2024). Editorial: Mathematical thinking, practices, and processes in non-formal learning environments. *Frontiers in Education*, *9*, 1498414. DOI: 10.3389/feduc.2024.1498414

Wing, J. M. (2006). Computational thinking. *Communications of the ACM*, *49*(3), 33–35. DOI: 10.1145/1118178.1118215

Wulandari, D. U., Mariana, N., Wiryanto, W., & Amien, M. S. (2024). Integration of ethnomathematics teaching materials in mathematics learning in elementary school. *International Journal of Recent Educational Research*, *5*(1), 1–15. DOI: 10.46245/ijorer.v5i1.542

Xenofontos, C., & Andrews, P. (2014). Defining mathematics teaching: A comparative study of Cypriot and English primary teachers' beliefs. *Teaching and Teacher Education*, *37*, 1–10. DOI: 10.1016/j.tate.2013.08.006

Xenofontos, C., Fraser, S., Priestley, A., & Priestley, M. (2021). Mathematics teachers and social justice: A systematic review of empirical studies. *Oxford Review of Education*, *47*(2), 135–151. DOI: 10.1080/03054985.2020.1807314

Yazıcı, E. Z. (2021). *Etnomatematik uygulamalarında kültür ve matematiğin gerçekçi matematik eğitimi çerçevesinde incelenmesi [Examining culture and mathematics within the framework of realistic mathematics education in ethnomathematics applications]* [Unpublished doctoral dissertation]. Marmara University.

Yıldırım, A., & Şimşek, H. (2021). *Sosyal bilimlerde nitel araştırma yöntemleri* (12. bs.). Seçkin Yayıncılık.

Yin, R. K. (2014). *Case study research: Design and methods* (5th ed.). Sage Publications.

Yllana-Prieto, F., González-Gómez, D., & Jeong, J. S. (2025). The escape room and breakout as an aid to learning STEM contents in primary schools: an examination of the development of pre-service teachers in Spain. *Education 3-13*, 53(1), 15-31. DOI: 10.1080/03004279.2022.2163183

Yonwilad, W., Nongharnpituk, P., Khansila, P., & Pongwiritthon, K. (2025). Enhancing Teamwork Skills and Mathematical Conceptual Connections through Board Game Development in a Cooperative Learning Environment. *Educational Process: International Journal, 17*(1), e2025361. DOI: 10.22521/edupij.2025.17.361

Zahedi, M., Tessier, V., & Hawey, D. (2017). Understanding Collaborative Design Through Activity Theory. *The Design Journal, 20*(sup1), S4611–S4620. DOI: 10.1080/14606925.2017.1352958

Zapata, M., Ramos-Galarza, C., Valencia-Aragón, K., & Guachi, L. (2024). Enhancing mathematics learning with 3D augmented reality escape room. *International Journal of Educational Research Open*, *7*, 1–11, 100389. DOI: 10.1016/j.ijedro.2024.100389

Zeidler, D. L. (2003). The role of moral reasoning on socioscientific issues and discourse in science education. In Zeidler, D. L. (Ed.), *The role of moral reasoning on socioscientific issues and discourse in science education* (pp. 201–216). Kluwer Academic Publishers., DOI: 10.1007/1-4020-4996-X

Zeidler, D. L. (2014). Socioscientific issues as a curriculum emphasis: Theory, research, and practice. In Lederman, N. G., & Abell, S. K. (Eds.), *Handbook of research on science education* (Vol. 2, pp. 697–726). Routledge.

Zeidler, D. L., & Nichols, B. H. (2009). Socioscientific issues: Theory and practice. *Journal of Elementary Science Education*, *21*(2), 49–58. DOI: 10.1007/BF03173684

Zeidler, D. L., & Sadler, T. D. (2008). Social and ethical issues in science education: A prelude to action. *Science & Education*, *17*(8–9), 799–803. DOI: 10.1007/s11191-007-9130-6

Zeidler, D. L., Sadler, T. D., Simmons, M. L., & Howes, E. V. (2005). Beyond STS: A research-based framework for socioscientific issues education. *Science Education, 89*(3), 357–377. DOI: 10.1002/sce.20048

Zhu, J., Huang, X., & Trouche, L. (2025). The development of didactic praxeologies in lesson design through social resource interactions: Evidence from two novice mathematics teachers. *Educational Studies in Mathematics, 119*(3), 487–513. DOI: 10.1007/s10649-025-10403-2

About the Contributors

Lutfi Incikabi is a Professor of Mathematics Education at Kastamonu University. He earned an Ed.D. in Mathematics Education from Teachers College, Columbia University in 2011 and subsequently served as a Teaching Assistant at Mersin University (2011–2012). Since joining Kastamonu University in 2012, his scholarship has focused on elementary mathematics education, gender equity and stereotypes in mathematics, authentic mathematics teaching, comparative education, and technology-integrated instruction. His specific interests include the design of mathematics courses and professional learning grounded in the Technological Pedagogical Content Knowledge (TPACK) framework; national and international comparative analyses of assessments, curricula, and textbooks; and research on mathematical creativity, authentic problem development, and the influence of gender stereotypes on students' learning and participation. He contributes to teacher education through course design, practicum mentorship, and school–university partnerships that advance equity-oriented mathematics instruction, and he has collaborated on curriculum and assessment initiatives at multiple levels.

Hartono Tjoe is an Associate Professor of Mathematics Education at The Pennsylvania State University. He earned a Ph.D. in Mathematics Education from Teachers College, Columbia University. His research examines the cognitive role of mathematical aesthetics in students' reasoning; how learners make strategic choices among solution methods; and the relational roles of item discrimination and difficulty when evaluating performance and sequencing instruction. He designs and studies tasks that foreground elegance, structure, and efficiency in problem solving and investigates how such criteria shape students' preferences and affective engagement with mathematics. Tjoe teaches and advises on integrating problem-solving heuristics, psychometric evidence, and task design in teacher education, and he contributes to program development in mathematics education. His scholarship bridges psychometrics, problem-solving research, and teacher professional learning to inform evidence-based instructional decision-making and assessment.

Fadime Ulusoy is an Associate Professor in the Department of Mathematics and Science Education at Kastamonu University, Türkiye. She received her Ph.D. in Mathematics Education from Middle East Technical University (METU) in 2016 and was awarded the METU 2017 Thesis Award for her dissertation, "Developing Prospective Mathematics Teachers' Knowledge for Teaching Quadrilaterals Through a Video Case-Based Learning Environment." Her research spans elementary mathematics education, video-case methodologies in teacher preparation, active learning techniques, covariational reasoning, the teaching and learning of geometry, and multiple representations. She also investigates teachers' subject-matter knowledge and pedagogical content knowledge (PCK), teacher noticing, and technology integration in mathematics classrooms. Ulusoy designs and studies video-rich learning environments that help prospective teachers analyze student thinking and strengthen instructional decision-making, and she collaborates with schools on professional development that links discourse, representation use, and formative assessment.

***

Ahmet Oğuz Akçay is an Associate Professor Ahmet Oğuz Akçay is a faculty member at Eskişehir Osmangazi University, Türkiye. He earned his doctorate in Educational Technology from Duquesne University (USA) and his M.A. in Curriculum and Instruction from The University of Texas at San Antonio (USA). His doctoral research examined pre-service mathematics teachers' integration of technology into instructional activities through the lens of cognitive demand and levels of technology implementation. Dr. Akçay's research focuses on technology-enhanced mathematics teaching and learning, STEM education, mathematical literacy, teacher education, digital learning environments, and artificial intelligence in education. He has published extensively in leading international journals and has contributed numerous book chapters and edited volumes on mathematics education, STEM, educational technology, and refugee education. His recent work explores the integration of AI, digital games, metaverse applications, and innovative instructional approaches into mathematics and primary education. He has led and participated in various national and international research projects and actively contributes to the development of future teachers through research, teaching, and academic supervision.

Feyza Aliustaoğlu is an Associate Professor of Matmematics Education at Kastamonu University, Türkiye. She graduated from the primary school mathematics teacher training program in 2011. Between 2012 and 2014, she worked as a mathematics teacher. In 2014, she began working at the Department of Mathematics Education, Faculty of Education, Kastamonu University. She completed her master's degree in mathematics education in 2015 and her doctorate in 2019. She currently continues to work as a faculty member in the same department. Her research areas include teacher training, pedagogical content knowledge, problem solving and posing, out-of-school learning, and game-based mathematics learning.

Arzu Aydoğan Yenmez is a Professor of Mathematics Education at Niğde Ömer Halisdemir University, Türkiye. She received her BA, MA, and PhD in mathematics education from Middle East Technical University. Her research interests include mathematics teacher education, mathematical modelling, mathematical reasoning, sociomathematical norms, and technology-enhanced mathematics instruction. She has an extensive record of publications in national and international journals, has contributed to academic books and edited collections, and has supervised numerous graduate theses in the field of mathematics education.

Emel Çilingir Altıner is an Associate Professor in the Department of Elementary Education at Çukurova University, Türkiye. She received her B.A. in Primary Education from Gazi University, her M.A. in Primary Education from Çukurova University, and her Ph.D. in Primary Education from Marmara University. Her research focuses on primary mathematics education, with particular emphasis on spatial reasoning, computational thinking, and the integration of digital technologies into mathematics teaching. She has conducted numerous studies on escape-room-based instructional design and technology-supported mathematics education in primary classrooms. Dr. Çilingir Altıner has published in national and international journals and has contributed to several academic book chapters on mathematics education and digital pedagogy. Her recent work explores participation-rich mathematics learning environments that support students' mathematical reasoning.

Hermandra is an academic specializing in semantics and culture. He began his studies at the University of Riau (Bachelor's degree, S.Pd.), continued at the National University of Malaysia (Master's degree, M.A.), and completed his studies at Padjadjaran University (Doctoral degree, Dr.). His research has been published regionally, nationally, and internationally.

Semahat İncikabı, Associate Professor in the Department of Primary Education, Mathematics Teaching at Sinop University, Türkiye, holds a PhD in Mathematics Education. Her research focuses on middle school mathematics textbooks, multiple representations, mathematical modeling, socio-critical mathematical modeling, the role of social and cultural contexts in mathematics learning, and international assessments such as PISA and TIMSS. She examines intersections between mathematics education and social sciences, including gender stereotypes, social roles, creative problem solving, and equity in curricula. Dr. İncikabı has published in national and international journals and continues research aimed at supporting students' mathematical literacy and critical thinking in socio-cultural contexts. Research Areas / Keywords: Mathematics Education; Textbook Analysis and Multiple Representations; Mathematical Modeling and Socio-Critical Modeling; Gender and Social Roles in Education; Middle School Mathematics; PISA and TIMSS; Creativity in Mathematical Problem Solving; Equity and Sociocultural Perspectives in Education

Fatih Kaleci is an Assistant Professor of Mathematics Education at Necmettin Erbakan University, Türkiye. His primary expertise includes mathematical reasoning, modeling, problem solving, teacher education, teacher agency, assessment validity, culturally responsive mathematics education, ethnomathematics, the history of mathematics, and Maker-STEAM learning. His interdisciplinary work focuses on human-centred AI, epistemic governance, responsible AI assurance, research integrity, publication ethics, scholarly communication, intelligent knowledge discovery, higher education leadership, quality assurance, institutional capacity, research performance, and university rankings. His methodological expertise spans mixed-methods research, big data and learning analytics, bibliometrics, scientometrics, network analysis, longitudinal research mapping, qualitative inquiry, survey research, and advanced statistical modeling.

Fuji Lestari is a Doctor of Education candidate in the Educational Sciences program at Universitas Riau, Indonesia. She holds a Master's degree in Elementary Education and currently serves as an Elementary School Teacher in Duri, Bengkalis Regency, Riau Province, Indonesia. Her research focuses on the development of Malay Ethnomathematics e-Modules, integrating local cultural values into digital learning materials to enhance mathematics education at the elementary level. Her academic and professional work reflects a deep commitment to innovative, culturally responsive teaching and the advancement of educational quality in Indonesia.

Angelius Kanyanga Liveve holds a PhD and has a background in Natural Science, Mathematics, and research. Since 2018, he has taught natural science, trends in Mathematics and Science Education, inquiry-based learning, practical skills, and STEM at International University of Management (IUM), where he supervises master's and PhD students. His research focuses on science curriculum, transformative professional development, and Indigenous Knowledge Systems (IKS). He has a Basic Education Teacher Diploma from Rundu College, a bachelor's in educational management from Tswane University, an Honours in Natural Science from Unisa, and a Master's and PhD from Rhodes University. Though relatively new to publishing, with few articles and book chapters, he has presented 8 conference papers. He serves as a peer reviewer for SAARMMSTE, AASIKS and the Namibia Journal of Managerial Science (NJMS)

Muryanti is a lecturer at the Faculty of Teacher Training and education (FKIP), Universitas Riau. She holds a master's degree in Education (Policy and International Development) from the University of Bristol. Her research interests include inclusive education, Education for Sustainability, and education policy.

Mustafa Serkan Pelen is a mathematics teacher at the Ministry of National Education in Türkiye for 16 years. He received his PhD in Mathematics Education in 2023. Dr. Pelen's research investigates students' mathematical thinking, problem-solving strategies, proportional reasoning, and the integration of digital technologies. His current work extends into culturally responsive pedagogy, utilizing Design-Based Research (DBR) and ethnomodeling to bridge students' informal cultural knowledge with formal academic mathematics.

Zetra Hainul Putra is a professor in the didactics of mathematics at the Faculty of Teacher Training and Education, Universitas Riau, Pekanbaru, Indonesia, where he currently serves as the head of the Department of Educational Sciences. He earned his Ph.D. in Didactics of Mathematics from the Department of Science Education at the University of Copenhagen, Denmark. His research interests focus on digital technology in mathematics education, teachers' mathematical, didactic, and technological knowledge, the Anthropological Theory of the Didactic, Malay ethnomathematics, and research methodology in teaching and learning mathematics. He has been actively involved in international collaborations and has published widely in peer-reviewed journals, contributing to both theoretical and practical advancements in mathematics education.

Elif Saygı is an esteemed mathematics professor and educator at Hacettepe University, Faculty of Education. Her research primarily focuses on graph theory and mathematical sciences. She is an accomplished academic with numerous international publications and specialized studies on the contribution of mind games to mathematics education.

Şeyma Şengil Akar is a faculty member in the Department of Primary Education, with core academic expertise in Mathematics Education. Her research and academic studies primarily focus on Mathematical Modelling and Mathematical Creativity.

Berna Somuncu is currently working as a mathematics teacher while pursuing her Ph.D. in Mathematics Education at Hacettepe University. Both her professional practice and academic research focus on advancing mathematics education.

Index

A

C

D

E

G

H

I

K

M

N

O

P

R

S

T

U

V

W

www.ingramcontent.com/pod-product-compliance
Lightning Source LLC
LaVergne TN
LVHW080912300926
849162LV00003B/650